DIGITAL ELECTRONICS

By

PREETI SEHARAWAT

B.E., M.Tech. (ECE)

Associate Professor

Delhi Institute of Technology
and Management (DITM)
Gannaur (Sonepat)

S.K. KATARIA & SONS®

Publisher of Engineering Books

4885/109, Prakash Mahal, Dr. Subhash Bhargav Lane,
Opposite Delhi Medical Association, Daryaganj, New Delhi-110002 (India)
Phone: +91-11-43551243; Mobile:+91-9871775858, 7678425209
e-mail: katariabook@gmail.com; katariabooks@yahoo.com
Website : www.skkatariaandsons.com
(e-books, subscriptions and bulk buying also available on our website)

Published by :

S.K. Kataria & Sons®

4885/109, Prakash Mahal, Dr. Subhash Bhargav Lane,
Opposite Delhi Medical Association, Daryaganj, New Delhi-110002
Phone: +91-11-43551243; Mobile: +91-9871775858, 7678425209
e-mail: katariabook@gmail.com; katariabooks@yahoo.com

First Edition: August, 2012, Reprint: 2026

Price: ₹ 325/-

Laser Typesetting by: Aisa Enterprises | *Printed at : New A.S. Offset Printers*

Preface

This book is designed as a text book for the subject of Digital Electronics, Digital Logic Design as offered to second year students of various Universities for the Electronics and Communication Engineering, Computer Science and Engineering, Information Technology branches.

This book is divided in 8 chapters which are according to the current syllabus of Maharshi Dayanand University (Rohtak) and D.C.R.S.T. Murthal. Each chapter begins with introduction and ends with review questions and problems. All the topics are illustrated with simple language is used throughout the text to facilitate easy understanding of the concept. It includes numerical, theoretical as well as conceptutal questions. Spatiality of this book is that, this book is also convering experiments that covered in Digital Electronics Laboratory and model test paper, university paper, solved examples. Hence, it will be beneficial for the lab technician as well as associated teacher.

Any suggestions for the improvement of this book would be acknowledged any incoroprated in next edition.

PREETI SEHRAWAT

Acknowledgement

First and formost I bow to Almighty for giving me the opportunity to undertake the writing of this book and the strength and capacity to complete it.

I would like to express my gratitute to my father **Sh. Hawa Singh Sehrawat,** mother **Smt. Santra Devi Sehrawat,** elder sister **Jyoti,** younger brother **Rahul Sehrawat** for their exemplary patience, understanding and co-operation during the preparation of this book.

I deeply express my heartful thanks to the publishers **S.K. Kataria & Sons** for publishing this book in such beautiful getup and well in time.

Any constructive comments, suggestions, criticism from students and colleagues will be highly appreciated and gratefully acknowledged.

PREETI SEHRAWAT

Dedicated to

GOD and our parents

for giving us this priceless gift of

"Life"

Contents

UNIT 1: FUNDAMENTAL OF DIGITAL TECHNIQUES

1. FUNDAMENTAL OF DIGITAL TECHNIQUES **3**

1.1 Digital Signal 3
  1.1.1 Comparison between Analog and Digital Signal 3
  1.1.2 Digital Electronics 4
1.2 Logic Gates 4
  1.2.1 Basic Gate 5
  1.2.2 Universal Gate 7
  1.2.3 Exclusive Gate 10
1.3 Positive Logic and Negative Logic 14
1.4 Boolean Algebra 14
  1.4.1 Basic Rules of Boolean Algebra 15
  1.4.2 De Morgan's Theorem 16
1.5 Number System 17
  1.5.1 (1) Non-Positional Number System 17
  1.5.2 (2) Positional Number System 17
  1.5.3 Rules for Conversions of Number System 18
  1.5.4 Conversions with Example 19
  1.5.5 Fractional Part Conversion 21
  1.5.6 Some Different Conversion 22
1.6 Binary Arithmetic 23
  1.6.1 Fixed Point Arithmetic 23
1.7 Binary Codes 27
  1.7.1 Numeric Code 27
  1.7.2 Alphanumeric Code 31
  1.7.3 Error Detection Code 33
  1.7.4 Error Correction Method 35
1.8 BCD Addition 37
1.9 Subtraction Using 1's and 2's Complement 38
1.10 New IEEE Symbols 41
1.11 Solved Problems 41
1.12 *Unsolved Questions* 45

UNIT 2: COMBINATIONAL DESIGN USING GATES

2. COMBINATIONAL DESIGN USING GATES **51**

2.1 Logic Design 51

2.2 Design Using Gates 51
2.2.1 Simplification of Boolean Function 52
2.2.2 Representation of Boolean Function 52
2.3 Karnaugh-MAP (K-Map) 54
2.3.1 Pairing of 1's/0's 57
2.3.2 Do not-care Condition (*d*) 57
2.4 Solved Problems on *k*-map 58
2.5 Comparison between Minterm and Maxterm 64
2.6 Quine-Mc Clusky Method (Tabular Method) 65
2.6.1 Steps in Q-M Method 65
Unsolved Questions 74

UNIT 3: COMBINATIONAL DESIGN USING MSI DEVICES

3. COMBINATIONAL DESIGN USING MSI DEVICES 77
3.1 Combinational Circuits 77
3.1.1 Different Combinational Circuits 77
3.2 Adder 78
3.2.1 Half Adder 78
3.2.2 Full Adder 78
3.2.3 Parallel Binary Adder 79
3.3 Subtractor 80
3.3.1 Half Subtractor 81
3.3.2 Full Subtractor 81
3.4 Multiplexer (Data Selector) 82
3.4.1 Cascading of Multiplexer 84
3.4.2 Applications of Multiplexer 85
3.4.3 Important Example of Multiplexer 85
3.5 Demultiplexer: (Data Distributor) 88
3.5.1 Cascading of Demultiplexer 90
3.5.2 Demultiplexer Applications 91
3.5.3 Comparison between MUX and DEMUX 91
3.6 Magnitude Comparator 91
3.6.1 1-bit Comparator 92
3.6.2 2-Bit Comparator 92
3.7 Encoder 95
3.7.1 Types of Encoder 95
3.7.2 Octal to Binary Encoder 95
3.7.3 Decimal to BCD Encoder 96
3.7.4 Gray to Binary Encoder 97
3.7.5 BCD to Excess-3 Encoder 100
3.7.6 Excess-3 to BCD Encoder 102
3.8 Priority Encoder 103
3.8.1 Decimal to BCD Priority Encoder 104
3.8.2 Octal to Binary Priority Encoder 106
3.9 Decoder 106
3.9.1 Binary to Octal (3 to 8 Line Decoder) 107
3.9.2 BCD to Decimal Decoder: (4 to 10 line) 108
3.9.3 BCD to Gray Code Decoder 109
3.10 BCD to 7-segment Display Devices Decoder 110

3.11 BCD Arithmetic Element 114
3.11.1 BCD Adder 114
3.11.2 BCD Subtractor 115
3.12 Arithmetic Logic Unit (ALU) 116
3.13 *Some Unsolved Questions* 117

UNIT 4: SEQUENTIAL CIRCUITS

4. SEQUENTIAL CIRCUITS 121
4.1 Introduction 121
4.1.1 Comparison between Combinational and Sequential 122
4.1.2 Latch 122
4.2 Flip-Flop 123
4.2.1 Comparison between Latch and Flip-Flop 123
4.2.2 Types of Flip-Flop 124
4.2.3 S-R Flip-Flop 125
4.2.4 J-K Flip-Flop 126
4.2.5 Preset and Clear Signal 127
4.2.6 D-Type Flip-Flop 128
4.2.7 Race Around Condition 129
4.2.8 Edge-Triggered Flip-Flop 129
4.2.9 Master-Slave Flip-Flop 130
4.2.10 Application of Flip-Flop 131
4.3 Conversion of one Flip-Flop into Another Flip- Flop 131
4.3.1 Excitation Table of Flip-Flops 131
4.4 Register 135
4.4.1 Shift Register 136
4.4.2 Types of Shift Register 136
4.4.3 Bidirectional Shift Register 140
4.4.4 Universal Shift Register 142
4.4.5 Application of Shift Register 142
4.5 Counter 143
4.5.1 Classification of Counter 143
4.5.2 Basic Concept of Counter 144
4.5.3 Ring Counter 144
4.5.4 Johnson Counter 145
4.5.5 Divide By N Counter/Ripple Counter 147
4.6 Application of Counter 148
4.7 Sequence Generator 149
4.8 Designing of Sequential Circuits (Counters) 149
4.8.1 Design of Modulus Counter 149
Unsolved Questions 155

UNIT 5: DIGITAL LOGIC FAMILIES

5. DIGITAL LOGIC FAMILIES 159
5.1 Switching Mode Operation of P-N Junction 159
5.1.1 Switching Characteristics of P-N Junction 161

5.2 Bipolar and MOS Devices 163
5.2.1 Bipolar Devices 163
5.2.2 MOS Devices 165
5.3 Bipolar Logic Family 165
5.4 Characteristics of Digital ICs 166
5.5 RTL (Resistor-Transistor Logic) 168
5.6 DTL (Diode Transistor Logic) 169
5.7 DCTL (Direct Coupled Transistor Logic) 170
5.8 HTL (High Threshold Logic) 171
5.9 TTL (Transistor-Transistor Logic) 172
5.10 IIL (Integrator Injection Logic) 176
5.11 ECL (Emitter Coupled Logic) 176
5.12 MOS Family 177
5.12.1 NMOS Logic Family 177
5.12.2 PMOS Logic Family 178
5.13 CMOS Logic Family 179
5.14 Interfacing of CMOS and TTL Families 180
5.14.1 CMOS Driving TTL 180
5.14.2 TTL Driving CMOs 181
5.15 Comparison of Logic Families 182
Unsolved Questions 182

UNIT 6: A/D CONVERTER AND D/A CONVERTER

6. A/D CONVERTER AND D/A CONVERTER 187
6.1 Introduction 187
6.2 Types of Converter 188
6.3 Sample and Hold 188
6.3.1 Sample and Hold Circuit 189
6.4 D/A Converter (Digital to Analog Converter) 189
6.4.1 Weighted Resistor D/A Converter 190
6.4.2 R-2R Ladder Type DAC 192
6.5 Specification of D/A Converter 194
6.6 A/D Converter (Analog to Digital Converter) 195
6.6.1 Quantization 195
6.6.2 Types of A/D Converter 197
6.6.3 Parallel Comparator (Simultaneous Type) 197
6.6.4 Counter Type A/D Converter 198
6.6.5 Successive Approximation Type A/D Converter 200
6.6.6 Dual Lope A/D Converter 201
6.7 Specifications of A/D Converter 203
Unsolved Questions 204

UNIT 7: PROGRAMMABLE LOGIC DEVICES

7. PROGRAMMABLE LOGIC DEVICES 207
7.1 PLD Introduction 207
7.1.1 Types of PLD 208
7.1.2 Structure of PLD 208

7.2	ROM (Read Only Memory)	208
	7.2.1 Designing of Circuit Using ROM	210
7.3	PLA (Programmable Logic Array)	210
	7.3.1 Designing Using PLA	212
7.4	PAL (Programmable Array Logic)	213
7.5	Comparison between ROM, PLA, PAL	217
7.6	FPGA (Field Programmable Gate Array)	217
7.7	CPLD (Complex Programmable Logic Device)	219
	7.7.1 Block Diagram of CPLD	219
	Unsolved Questions	221

UNIT 8: SEQUENTIAL LOGIC DESIGN

8.	**SEQUENTIAL LOGIC DESIGN**	**225**
8.1	Sequential Logic Based Design Circuit	225
8.2	Design at RTL (Register Transfer Level)	225
8.3	ASM (Algorithmic State Machines)	227
8.4	Asynchronous Sequential Logic: (ASLC)	229
	8.4.1 Comparison of ASLC and SSLC	231
8.5	Analysis Procedure	231
8.6	Design Procedure	233
8.7	Reduction of State and Flow Table	233
8.8	Race Free State Assignment	238
8.9	Hazards	239
8.10	State Reduction and Assignment	241
	Unsolved Questions	242
	EXPERIMENTS	**243**
	PAPERS	**263**
	QUESTION BANK	**269**
	MODEL TEST PAPERS	**275**
	INDEX	**279**

UNIT 1

FUNDAMENTAL OF DIGITAL TECHNIQUES

Topics are:

- Digital Signal
- Logic Gates: AND, OR, NOT, NAND, NOR, Ex-OR, EX-NOR
- Positive Logic and Negative Logic
- Boolean Algebra
- Number System
- Binary Arithmetic
- Binary Codes: BCD, Excess-3, Gray, EBCDIC, ASCII
- BCD Addition
- Subtraction Using 1's and 2's Complement
- New IEEE Symbols

CHAPTER 1

Fundamental of Digital Techniques

1.1 DIGITAL SIGNAL

Digital means binary that is 0 and 1. Signal is nothing but transmission of data.

A digital signal uses discrete values (either 0 or 1) to represents, information in discrete or continuous form.

Discrete information *e.g.* numbers, letters, etc.

Continuous information *e.g.* sound, images, etc.

Digital signal uses only binary data that is (0 or 1). Representation of digital signal is shown in Fig. 1.1.

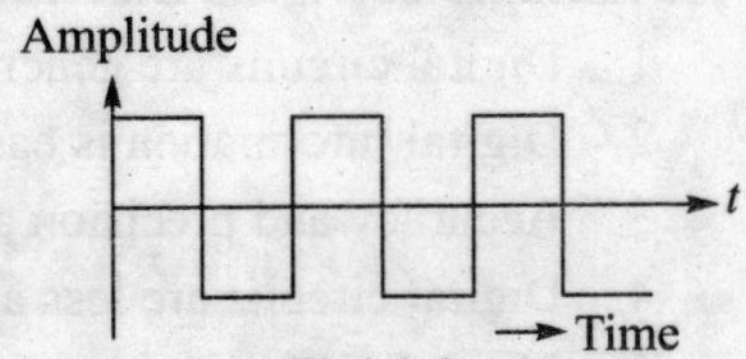

Fig. 1.1

But when we compare digital signal with analog signal then analog signal is varying continuous with time it means all the natural examples are analog signal. Analog signals are continuous in nature.

An analog signal uses some medium to convey the information. If we have information in form of some physical form like sound, light, etc. It can be converted from physical form to electrical signal by using transducer. As all the variation of analog signals are significant.

Hence, any disturbances is equivalent to a change in original signal and so appears noise. That is unwanted source of energy.

1.1.1 Comparison between Analog and Digital Signal

Analog Signal	Digital Signal
1. Analog signals are continuously varying. *e.g.*: Temperature, pressure.	1. Digital signals are discrete based on 0's and 1's *e.g.*: ON/OFF switch.
2. These signals are much more susceptible to noise. Means noise can cause corruption of information.	2. These are based on only two different values either 0 or 1, chances of noise are less in this. 0 = OFF, 1 = ON

3. Analog signals are difficult to transmit.	3. Digital signals are easier to transmit.
4. Chances of erros are much more because of continuous varying nature.	4. It offers less error to occur because of binary information.
5. It transmits less accurate data.	5. It transmits accurate data.
6. Transmission rate of data is not fast and productivity is less.	6. Transmission rate of data is faster and productivity is better.
7. It is less reliable.	7. It is more reliable.
8. It is represented as: 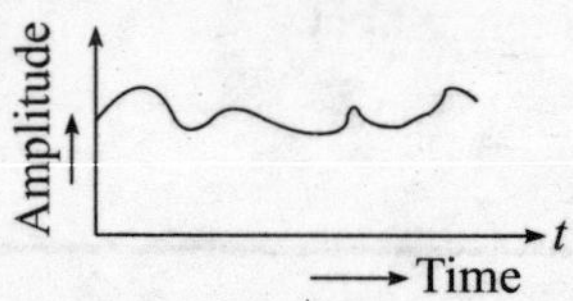	8. It is represented as: 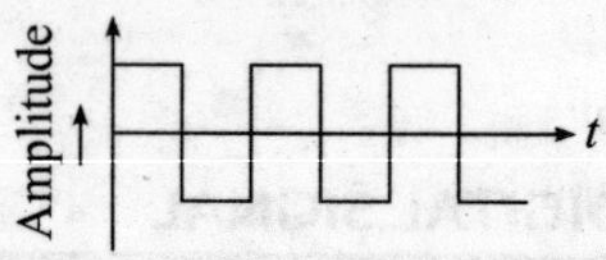

1.1.2 Digital Electronics

It is a system, which processes digital signals. It is a combination of devices designed to perform logical information (on physical quantities that are represented in digital form.)

Advantages of Digital Electronics

1. Digital circuits are generally easier to design.
2. Digital information is easy to store.
3. Accuracy and precision are greater.
4. Digital circuits are less affected by noise.
5. More digital circuits can be performed on single IC chips.

1.2 LOGIC GATES

Logic Gate is a digital circuit that is able to operate on a number of binary inputs in order to perform perticular logic function.

Logic Gates are solid state electronic circuit that are used to perform certain mathematical function, *i.e.* multiplication, addition, inversion by using binary digits (0 and 1). *E.g.*: AND gate for multiply, OR gate for addition.

Three common classification of Logic Gate are popularly used to designed the different circuits and to solve all the mathematical operation.

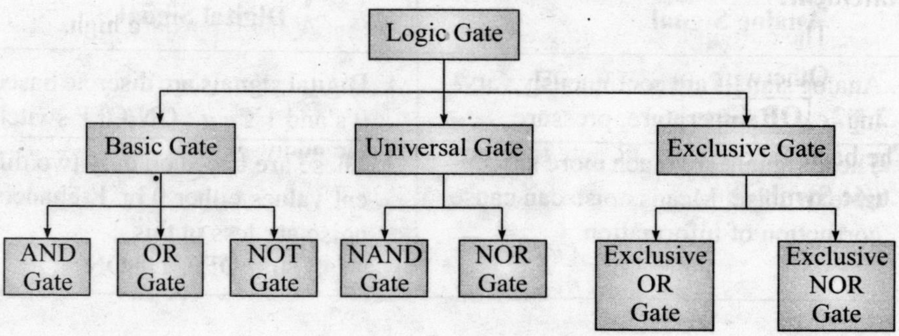

1.2.1 Basic Gate

These are common gate to solve the basic mathematical function like multiplication, addition and inversion operation.

1.2.1.1 AND Gate

The basic function of AND gate is to perform logic multiplication.

Logic Symbol

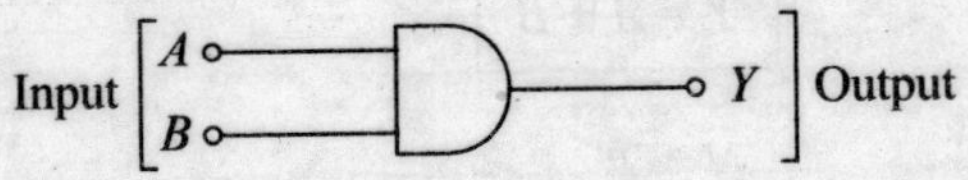

AND gate takes two or more input and gives a single output.

Inputs are on left hand side.

Output are on right hand side.

Logic Expression

Multiply A, B and equivalent to Y.

$$Y = A \cdot B$$

Truth Table (for 2 input AND gate)

$$N = 2^n$$

where N = No. of combinations.

n = No. of input variables.

when $n = 2$

$$N = 2^2$$

$$N = 4$$

No. of I/P variables = 2 [A, B]

No. of combinations = 4 [00, 01, 10, 11]

A	B	Y = A · B
0	0	0 (Low)
0	1	0 (Low)
1	0	0 (Low)
1	1	1 (High)

0 = Low

1 = High

Statement: According to the truth table.

The output of AND gate is high only when all the inputs are high.

Otherwise, output is Low.

1.2.1.2 OR Gate

The basic function of OR gate is to perform logic addition.

Logic Symbol:

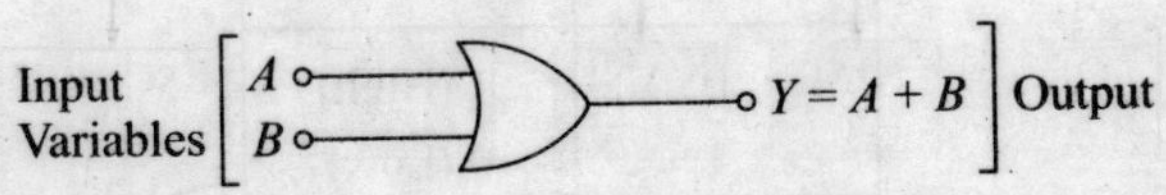

Similar to AND gate, OR gate takes two or more inputs and gives a single output.

Inputs are given on left hand side.

Outputs are given on right hand side.

Logic Expression:

Add A, B and equivalent to Y.

$$Y = A + B$$

Truth Table:

$N = 2^n$

$n = 2\ (A, B)$

$N = 4\ (00, 01, 10, 11)$

A	B	$Y = A + B$
0	0	0 (Low)
0	1	1 (High)
1	0	1 (High)
1	1	1 (High)

Statement: The output of OR gate is high when any of the inputs are high.

The output of OR gate is low only, when all of the inputs are low.

1.2.1.3 NOT Gate

It perform the function of inversion or complement.

Function: To change one logic level to opposite level.

$0 \rightarrow 1$

$1 \rightarrow 0$

Logic Symbol:

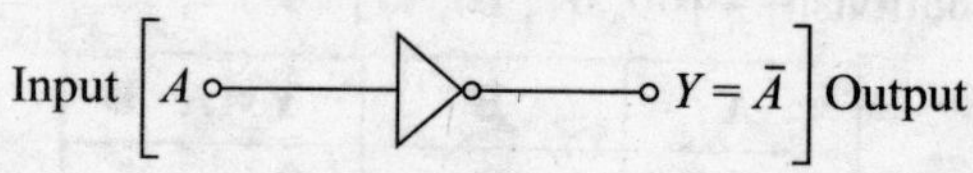

OR

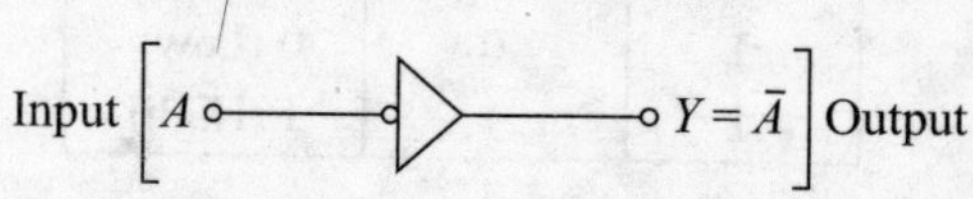

0 = Bubble will invert the level.

Logic Expression:

$$Y = \bar{A}$$

Truth Table:

A	$Y = \bar{A}$
0	1 (High)
1	0 (Low)

Statement: A high level to an inverter produces a low level and a low level to an inverter produces a high level.

1.2.2 Universal Gate

Universal Gate are those gate which can be used to construct an AND gate, an OR gate, an inverter or any combination of these function.

1.2.2.1 NAND Gate

It is the combination of AND and NOT gate.

NAND = AND + NOT

Logic Symbol:

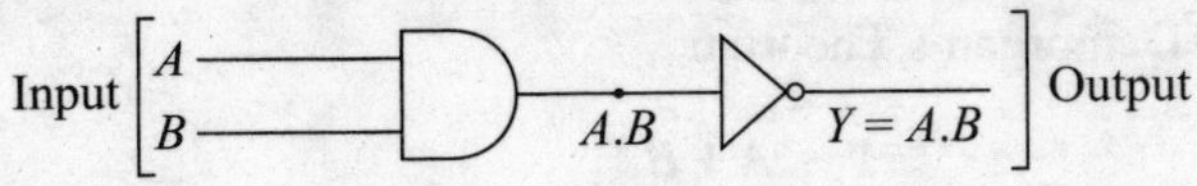

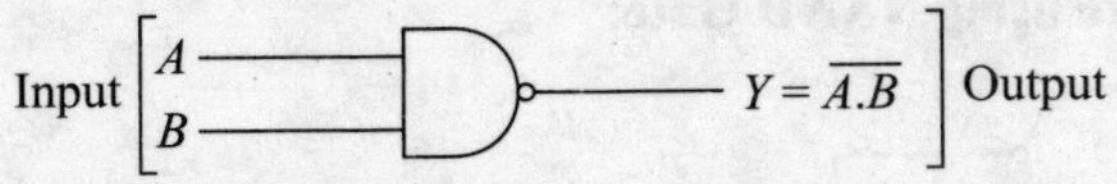

Logic Expression:

$$Y = \overline{A \cdot B}$$

Complement of AND gate output expression.

Truth Table:

A	B	$A \cdot B$	$Y = \overline{A \cdot B}$
0	0	0	1 (High)
0	1	0	1 (High)
1	0	0	1 (High)
1	1	1	0 (Low)

Statement: The output of NAND gate is low when all the inputs are high. Otherwise, output is high.

Universal Property of NAND Gate:

(*i*) **NAND Gate as an Inverter:**

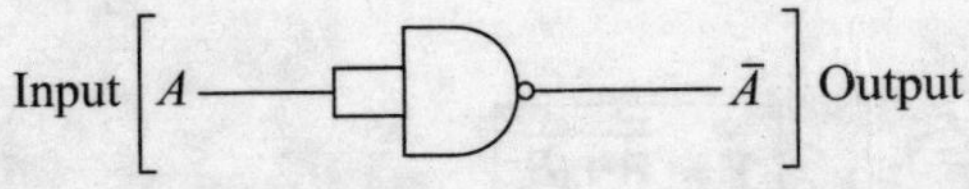

$$\overline{A \cdot A} = \bar{A}$$

(*ii*) **AND Gate using NAND Gate:**

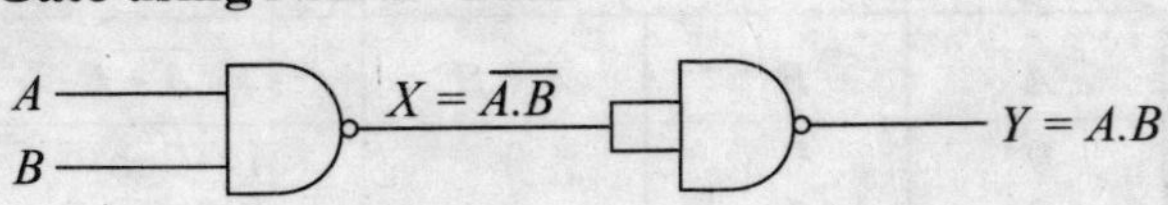

$$X = \overline{A \cdot B}$$

$$Y = \overline{\overline{A \cdot B}}$$

$$\boxed{Y = A \cdot B}$$

(*iii*) OR Gate using NAND Gate:

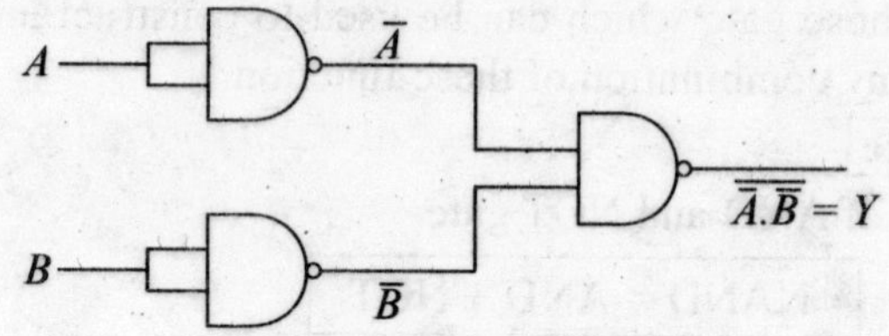

$$Y = \overline{\overline{A} \cdot \overline{B}}$$

Using Demorgan's Theorem:

$$Y = \overline{\overline{A}} + \overline{\overline{B}}$$

$$\boxed{Y = A + B}$$

(*iv*) NOR Gate using NAND Gate:

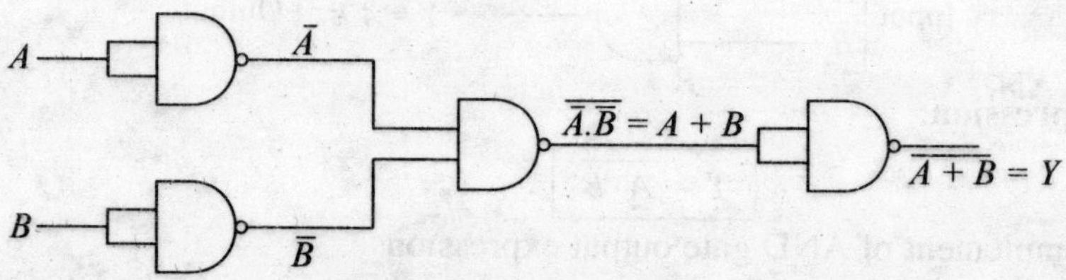

$$\boxed{Y = \overline{A + B}}$$

1.2.2.2 NOR Gate   OR + NOT

The basic function of NOR gate is to perform logical addition and then producing complemented output.

$$\boxed{\text{NOR} = \text{OR} + \text{NOT}}$$

Logic Symbol:

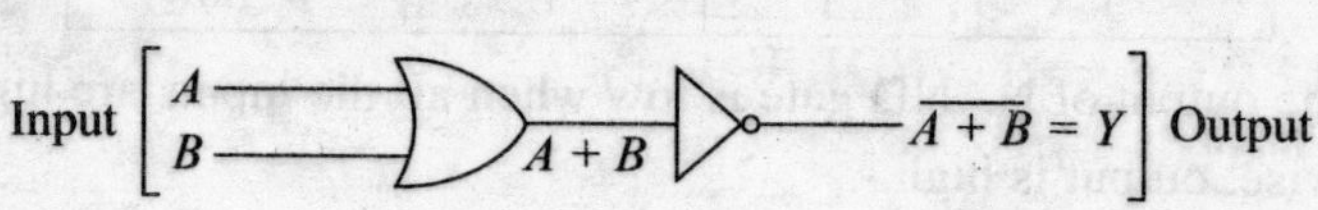

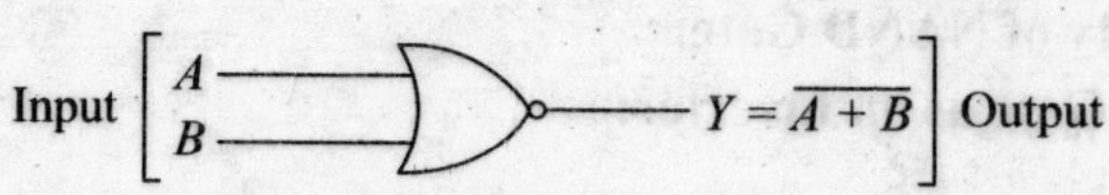

Logic Expression:

$$\boxed{Y = \overline{A + B}}$$

Complement of OR gate output expression.

Truth Table:

A	B	$A+B$	$Y = \overline{A+B}$
0	0	0	1 (High)
0	1	1	0 (Low)
1	0	1	0 (Low)
1	1	1	0 (Low)

Statement: The output of NOR gate is low, when any of its inputs are high.

and

The output is high, when all of its inputs are low.

Universal Property of NOR Gate:

(*i*) **NOR Gate used as Inverter:**

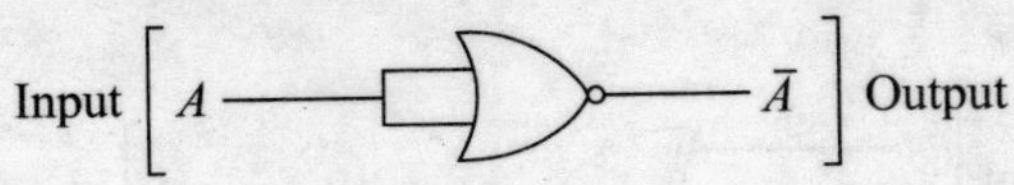

$$Y = \overline{A \cdot A}$$

$$\boxed{Y = \overline{A}}$$

(*ii*) **OR Gate using NOR Gate:**

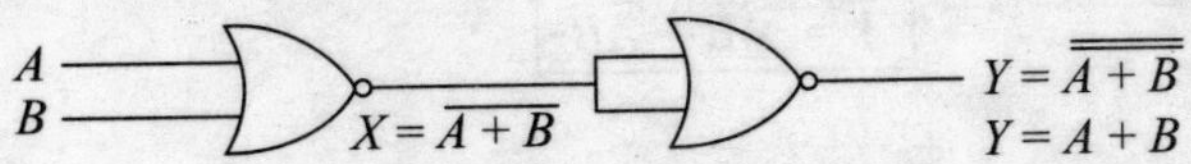

$$Y = \overline{\overline{A + B}} \qquad [\because \text{ Bar is cancel out to Bar}]$$

$$\boxed{Y = A + B}$$

(*iii*) **AND Gate using NOR Gate:**

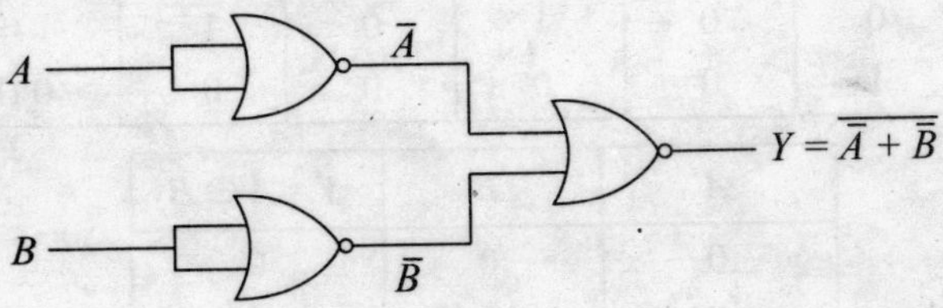

$$Y = \overline{\overline{A} + \overline{B}}$$

$$= \overline{\overline{A}} \cdot \overline{\overline{B}} \qquad [\because \text{ Bar } (-) \text{ can change sign}]$$

$$\boxed{Y = A \cdot B}$$

(*iv*) **NAND Gate using NOR Gate:**

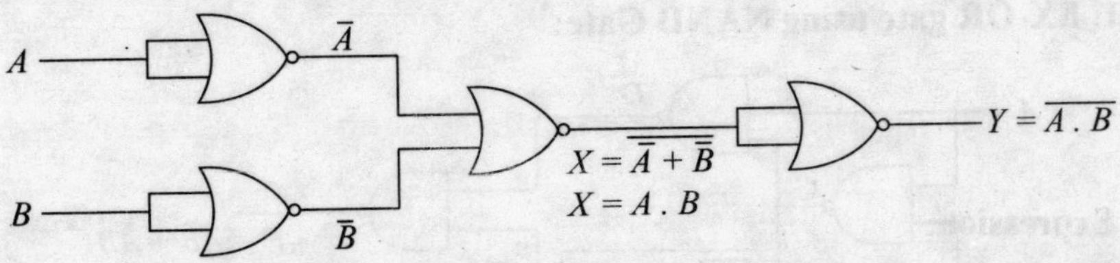

$$X = \overline{\overline{A} + \overline{B}}$$

$$= A \cdot B$$

$$\boxed{Y = \overline{A \cdot B}}$$

This is expression of NAND Gate.

1.2.3 Exclusive Gate

These gate are different from basic and universal gate but using the properties of basic gates only.

1.2.3.1 Exclusive–OR Gate

Exclusive–OR gate is abbreviated as Ex–OR gate, XOR gate.

Logical Symbol:

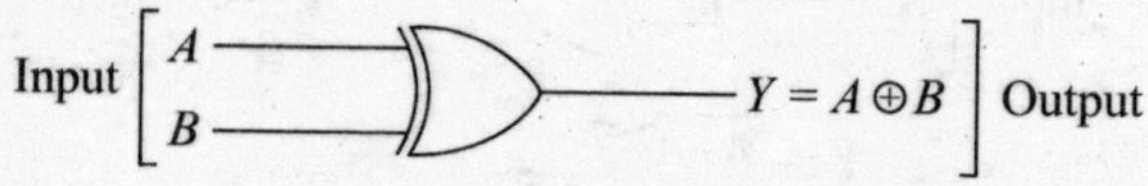

Logic Expression:

$$Y = A \oplus B \quad \Rightarrow \quad A \text{ XOR } B.$$

$$\boxed{Y = A\overline{B} + \overline{A}B}$$

Truth Table:

A	B	$\overline{A}$	$\overline{B}$	$\overline{A}B$	$A\overline{B}$	$Y = A\overline{B} + \overline{A}B$
0	0	1	1	0	0	0 (Low)
0	1	1	0	1	0	1 (High)
1	0	0	1	0	1	1 (High)
1	1	0	0	0	0	0 (Low)

A	B	$Y = A \oplus B$
0	0	0
0	1	1
1	0	1
1	1	0

Statement: The output is high only when the two inputs are at opposite level either 0, 1 and 1, 0.

Two Important designing of EX-OR gate:

1. EX-OR gate using NAND Gate:

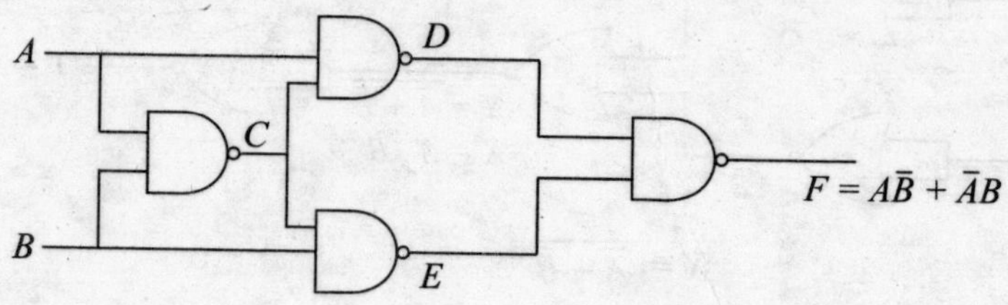

$$C = \overline{A \cdot B}$$

$$D = \overline{A \cdot C} \quad \Rightarrow \quad \overline{A \cdot \overline{A \cdot B}} \quad \Rightarrow \quad \overline{A} + A \cdot B$$

$$E = \overline{B \cdot C} \quad \Rightarrow \quad \overline{B \cdot \overline{A \cdot B}} \quad \Rightarrow \quad \overline{B} + A \cdot B$$

$$F = \overline{D \cdot E} \quad \Rightarrow \quad \overline{(\overline{A} + AB) \cdot (\overline{B} + AB)}$$

$$\Rightarrow \overline{(\overline{A}+A\cdot B)}+\overline{(\overline{B}+A\cdot B)}$$

$$\Rightarrow (\overline{\overline{A}}\cdot\overline{A\cdot B})+(\overline{\overline{B}}\cdot\overline{A\cdot B})$$

$$\Rightarrow [A\cdot(\overline{A\cdot B})]+[B\cdot(\overline{A\cdot B})]$$

$$\Rightarrow [A\cdot(\overline{A}+\overline{B})]+[B\cdot(\overline{A}+\overline{B})]$$

$$\Rightarrow A\cdot\overline{A}+A\cdot\overline{B}+B\cdot\overline{A}+B\cdot\overline{B} \qquad [A\cdot\overline{A}=0,\ B\cdot\overline{B}=0]$$

$$\Rightarrow A\cdot\overline{B}+\overline{A}\cdot B \quad\Rightarrow\quad A\oplus B$$

2. EX-OR gate using only NOR Gate:

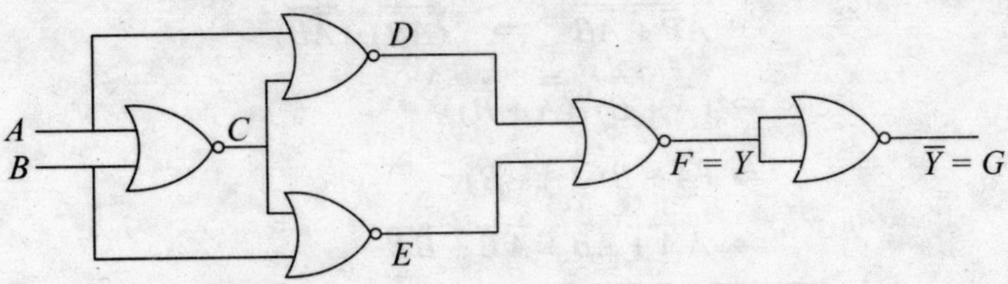

$$\boxed{C=\overline{A+B}}$$

$$D=\overline{A+C} \quad\Rightarrow\quad \overline{A+\overline{\overline{A+B}}} \quad\Rightarrow\quad \overline{A}\cdot(A+B)$$

$$\Rightarrow A\cdot\overline{A}+\overline{A}B$$

$$\boxed{D=\overline{A}B}$$

$$E=\overline{B+C} \quad\Rightarrow\quad \overline{B+\overline{\overline{A+B}}} \quad\Rightarrow\quad \overline{B}\cdot(A+B)$$

$$\Rightarrow A\cdot\overline{B}+B\cdot\overline{B}$$

$$\boxed{E=A\overline{B}}$$

$$F=Y=\overline{D+E} \quad\Rightarrow\quad \overline{\overline{A}B+A\overline{B}} \quad\Rightarrow\quad \overline{\overline{A}B}\cdot\overline{A\overline{B}}$$

$$\Rightarrow (\overline{\overline{A}}+\overline{B})\cdot(\overline{A}+\overline{\overline{B}}) \quad\Rightarrow\quad (A+\overline{B})\cdot(\overline{A}+B)$$

$$\Rightarrow A\cdot\overline{A}+\overline{A}\,\overline{B}+B\overline{B}+AB$$

$$\boxed{F=Y \quad\Rightarrow\quad AB+\overline{A}\,\overline{B}}$$

$$G=\overline{Y} \quad\Rightarrow\quad \overline{AB+\overline{A}\cdot\overline{B}}$$

$$\Rightarrow \overline{AB}\cdot\overline{\overline{A}\cdot\overline{B}}$$

$$\Rightarrow (\overline{A}+\overline{B})\cdot(\overline{\overline{A}}+\overline{\overline{B}})$$

$$\Rightarrow (\overline{A}+\overline{B})\cdot(A+B)$$

$$\Rightarrow A\overline{A}+A\overline{B}+\overline{A}B+B\cdot\overline{B} \qquad [A\cdot\overline{A}=0,\ B\cdot\overline{B}=0]$$

$$G\Rightarrow A\cdot\overline{B}+\overline{A}B$$

$$\boxed{G\Rightarrow A\oplus B}$$

1.2.3.2 EX-NOR Gate

The complement of EX-OR function is Exclusive-NOR gate.

It is abbreviated as X NOR.

Logic Symbol:

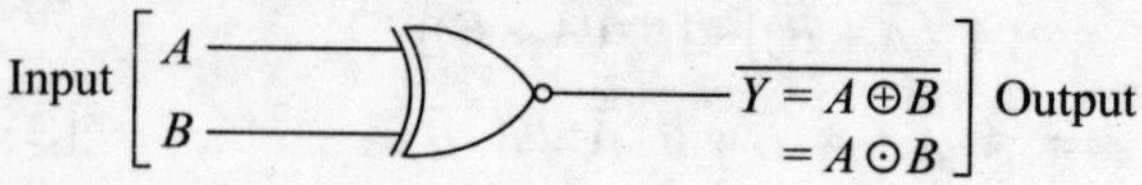

Logic Expression:

$$Y = \overline{A \oplus B}$$

$$= \overline{A\bar{B} + \bar{A}B} \quad \Rightarrow \quad \overline{(A\bar{B})} \cdot \overline{(\bar{A}B)}$$

$$\Rightarrow (\bar{A} + \bar{\bar{B}}) \cdot (\bar{\bar{A}} + \bar{B})$$

$$\Rightarrow (\bar{A} + B) \cdot (A + \bar{B})$$

$$\Rightarrow A\bar{A} + AB + \bar{A}\bar{B} + B\bar{B}$$

$$\boxed{Y \Rightarrow AB + \bar{A}\bar{B}}$$

$$\boxed{Y = A \odot B}$$

Truth Table:

A	B	$A \cdot \bar{B} + \bar{A} \cdot B$	$Y = AB + \bar{A}\bar{B}$
0	0	0	1 (High)
0	1	1	0 (Low)
1	0	1	0 (Low)
1	1	0	1 (High)

Statement: The output is high only, when the two inputs are alike (00, 11).

Two Important designing of EX-NOR gate:

1. X NOR Gate Using NAND Gate:

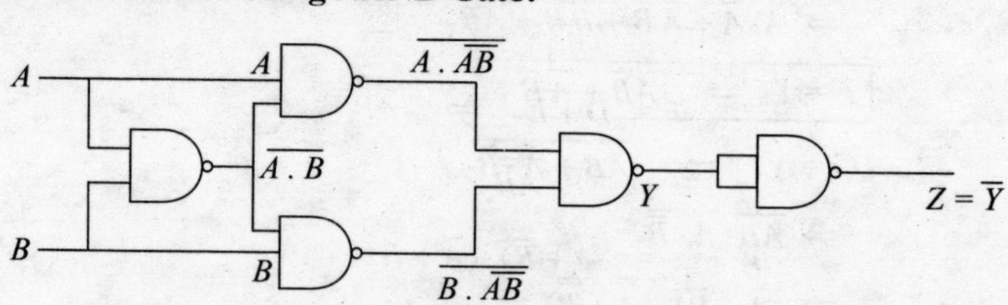

$$Y = \overline{\overline{A \cdot \overline{AB}} \cdot \overline{B \cdot \overline{AB}}}$$

$$= \overline{\overline{A \cdot \overline{AB}}} + \overline{\overline{B \cdot \overline{AB}}}$$

$$= A \cdot \overline{AB} + B \cdot \overline{AB}$$

$$= A \cdot (\bar{A} + \bar{B}) + B \cdot (\bar{A} + \bar{B})$$

$$= A\cdot\bar{A} + A\bar{B} + B\bar{A} + B\bar{B}$$

$$= A\bar{B} + \bar{A}B$$

$$Z = \overline{Y}$$

$$= \overline{A\bar{B} + \bar{A}B}$$

$$= \overline{A\cdot\bar{B}}\cdot\overline{\bar{A}\cdot B}$$

$$= (\bar{A} + \bar{\bar{B}})\cdot(\bar{\bar{A}} + \bar{B})$$

$$= (\bar{A} + B)(A + \bar{B}) \quad \Rightarrow \quad A\bar{A} + AB + B\bar{B} + \bar{A}\bar{B}$$

$$\boxed{Z \Rightarrow AB + \bar{A}\bar{B}}$$

$$\boxed{Z = A \odot B}$$

2. X NOR Gate using NOR Gate:

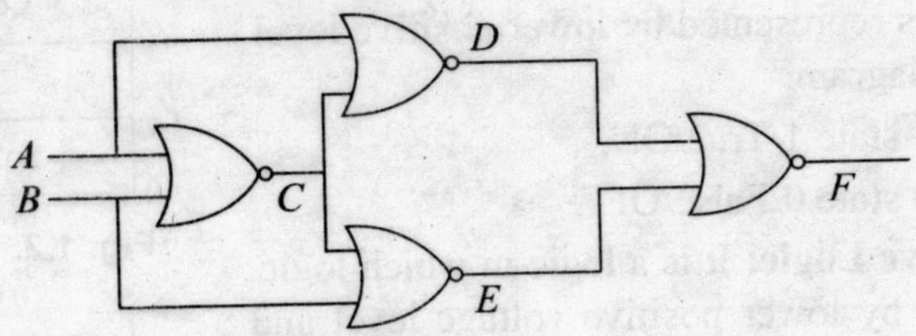

$$\boxed{C = \overline{A+B}}$$

$$D = \overline{A+C} \quad \Rightarrow \quad \overline{A + \overline{A+B}} \quad \Rightarrow \quad \bar{A}\cdot(A+B)$$

$$\Rightarrow A\cdot\bar{A} + \bar{A}B \quad \Rightarrow \quad \bar{A}B$$

$$\boxed{D = \bar{A}B}$$

$$E = \overline{B+C} \quad \Rightarrow \quad \overline{B + \overline{A+B}} \quad \Rightarrow \quad \bar{B}\cdot(A+B)$$

$$\Rightarrow A\bar{B} + B\bar{B} \quad \Rightarrow \quad A\bar{B}$$

$$\boxed{E = A\bar{B}}$$

$$F = \overline{D+E} \quad \Rightarrow \quad \overline{\bar{A}B + A\bar{B}}$$

$$\Rightarrow \overline{\bar{A}B}\cdot\overline{A\bar{B}}$$

$$\Rightarrow (\bar{\bar{A}} + \bar{B})\cdot(\bar{A} + \bar{\bar{B}})$$

$$\Rightarrow (A + \bar{B})\cdot(\bar{A} + B)$$

$$\Rightarrow A\bar{A} + \bar{A}\bar{B} + B\bar{B} + AB$$

$$\boxed{F \Rightarrow AB + \bar{A}\bar{B}}$$

$$\boxed{F = A \odot B}$$

1.3 POSITIVE LOGIC AND NEGATIVE LOGIC

Digital electronics is based on binary '0' and '1' and practically the binary values are represented by voltage levels. Under normal operating conditions, the voltage applied to any input terminal of a logic circuit is restricted to have one of the two nominal values.

Need of Logics: For identification of associativity of logic variables with electrical signals.

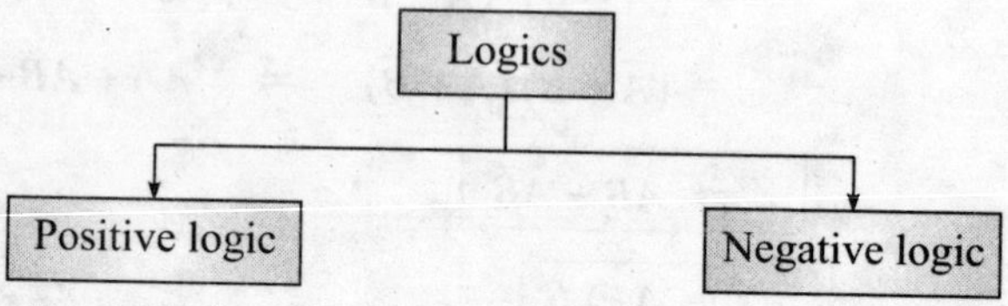

(*i*) **Positive Logic:** It is a logic in which logic '1' is represented by higher positive voltage level.

Logic '0' is represented by lower positive level that is shown in diagram.

Logic '1' = state 1/True/ON.

Logic '0' = state 0/False/OFF.

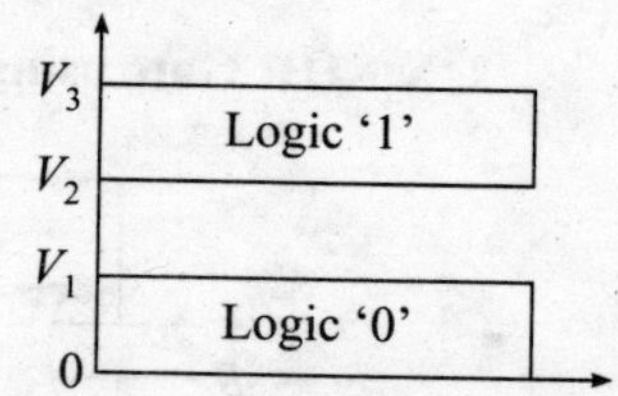

Fig. 1.2. Positive Logic

(*ii*) **Negative Logic:** It is a logic in which logic '1' is represented by lower positive voltage level and Logic '0' is represented by high positive voltage level.

State 1/True/ON = Logic '0'

State 0/False/OFF = Logic '1'

An electronic logic gate can perform different operations depends upon logic.

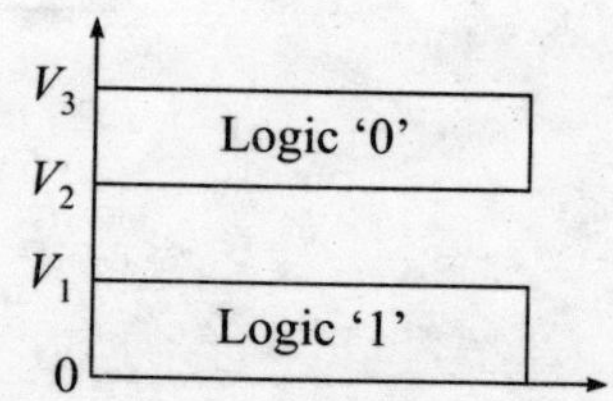

Fig. 1.3. Negative Logic

For example:

A gate can work as AND gate – for Positive Logic

OR gate – For Negative Logic.

1.4 BOOLEAN ALGEBRA

It is one tool to reduce logical expressions and mathematics of logical expression. The reduced expression can then be implemented with a smaller, simpler circuit, which in turn saves the prices of unnecessary gates, reduces the number of gates needed and reduces the power and the amount of space required by those gates.

In Boolean algebra there are two levels:

Binary 1 – High level.

Binary 0 – Low level.

Basic Laws of Boolean Algebra:

(*i*) Commutative Law

(*ii*) Associative Law

(*iii*) Distributive Law.

Explanation:

Commutative Law

$A + B = B + A$ → Commutative law of addition (OR Logic)

$A . B = B . A$ → Commutative law of multiplication (AND Logic)

Associative Law

Associative law of addition → $A + (B + C) = (A + B) + C$

Associative law of multiplication → $A (BC) = (AB) C$

Distributive Law

$A (B + C) = AB + AC$ → OR function

$(A + B)C = AC + BC$ → OR function

1.4.1 Basic Rules of Boolean Algebra

1. $A + 0 = A$

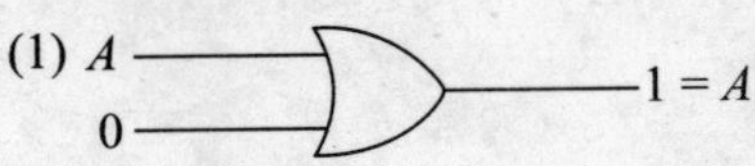

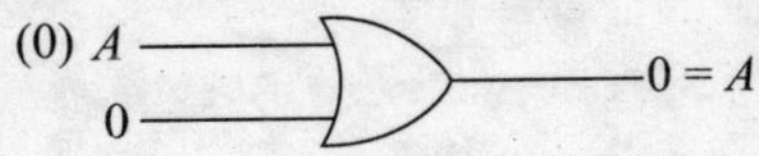

2. $A + 1 = 1$

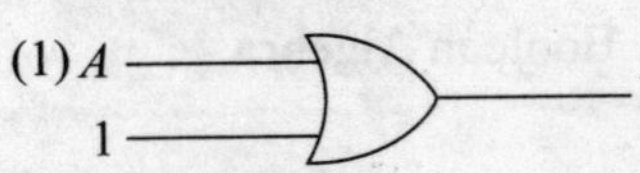

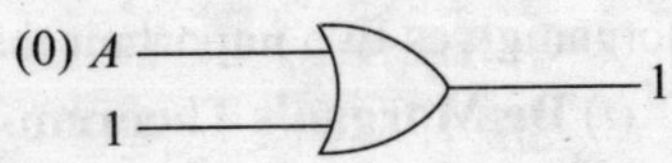

3. $A \cdot 0 = 0$

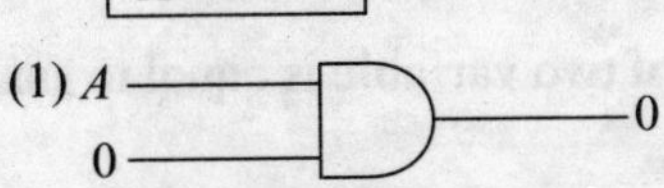

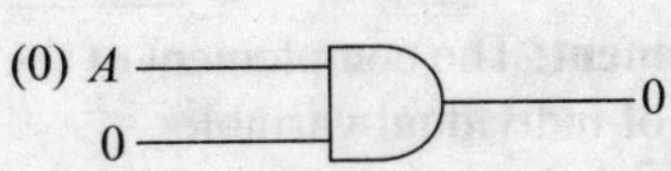

4. $A \cdot 1 = A$

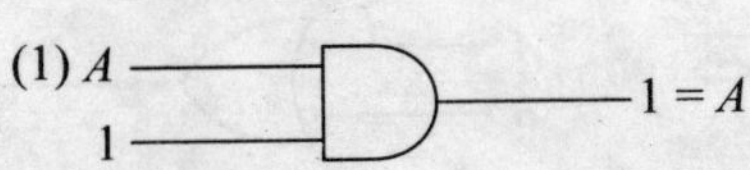

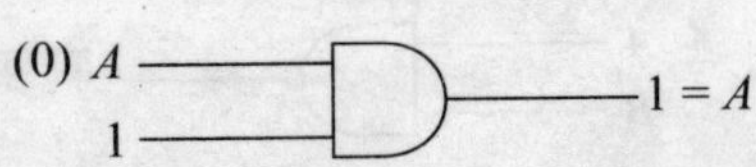

5. $A + A = A$

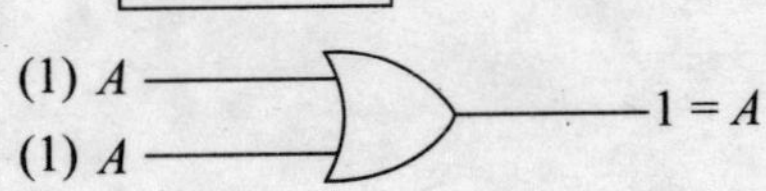

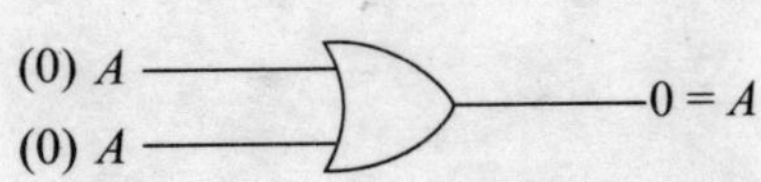

6. $\boxed{A + \bar{A} = 1}$

7. $\boxed{A \cdot \bar{A} = 0}$

8. $\boxed{A \cdot A = A}$

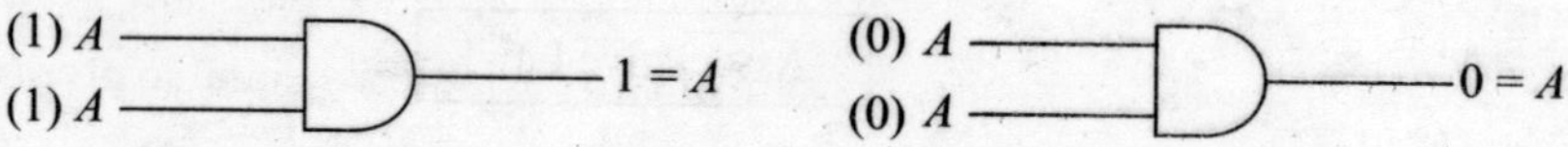

9. $\boxed{\bar{\bar{A}} = A}$

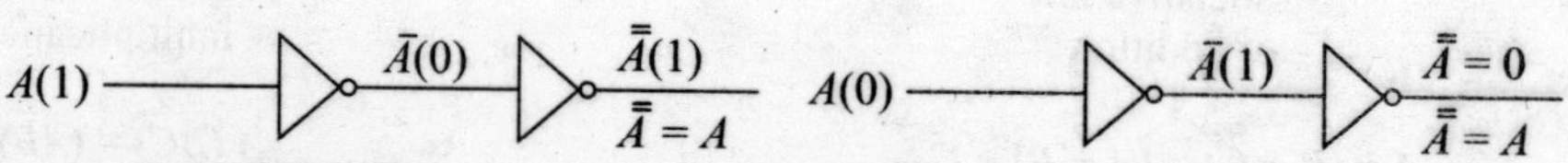

10. (*i*) $\boxed{A + AB = A}$

Proof:

$$\begin{aligned} & A + AB \\ &= A(1+B) && \text{As } 1 + B = 1 \\ &= A(1) && A \cdot 1 = A \\ &= A \end{aligned}$$

(*ii*) $\boxed{A + \bar{A}B = A + B.}$

Proof:

$$\begin{aligned} & A + \bar{A}B \\ &= A + AB + \bar{A}B && \text{As } A + AB = A \\ &= A + B(A + \bar{A}) && A + \bar{A} = 1 \\ &= A + B. \end{aligned}$$

1.4.2 De Morgan's Theorem

De Morgan gives two important theorems for Boolean Algebra.

(*i*) **De Morgan's Theorem-I:**

$$\boxed{\overline{AB} = \bar{A} + \bar{B}}$$

Statement: The complement of the product of two variable is equal to sum of complement of individual variable.

Gate Implementation:

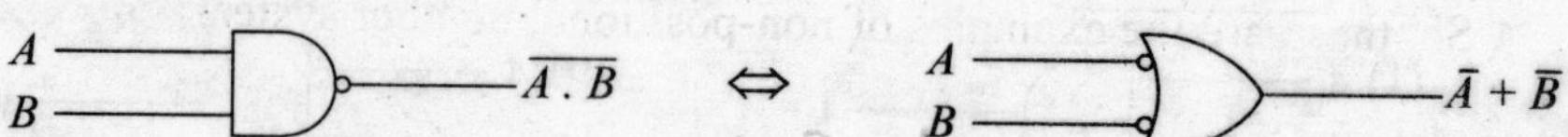

Truth Table:

A	B	$\overline{AB}$	$\overline{A}+\overline{B}$
0	0	1	1
0	1	1	1
1	0	1	1
1	1	0	0

(*ii*) **De Morgan's Theorem-II:**

$$\overline{A+B} = \overline{A}\cdot\overline{B}$$

Statement: The complement of the sum of two variables is equal to product of complement of individual variables.

Gate Implementation:

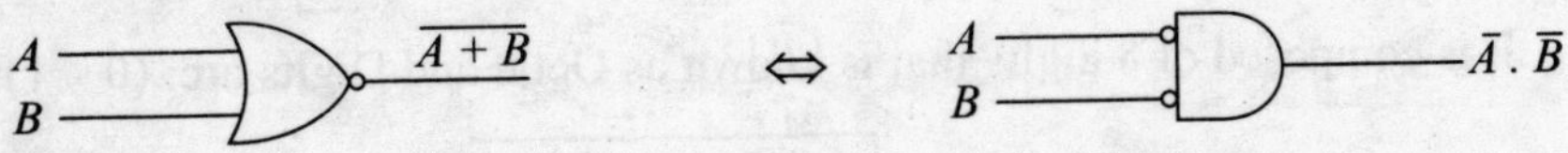

Truth Table:

A	B	$\overline{A+B}$	$\overline{A}\cdot\overline{B}$
0	0	1	1
0	1	0	0
1	0	0	0
1	1	0	0

1.5 NUMBER SYSTEM

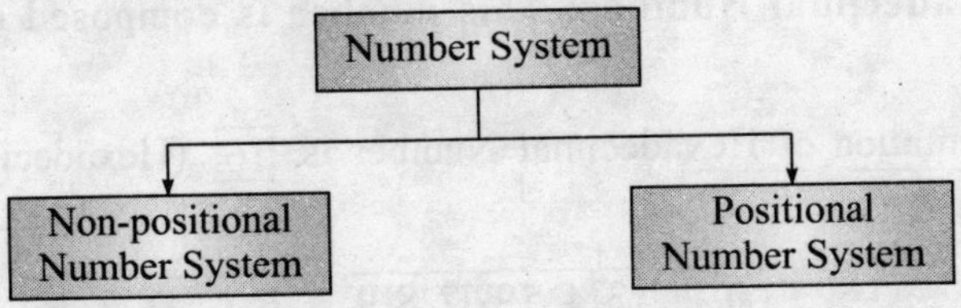

1.5.1 (1) Non-Positional Number System

It is the Roman numerals number system, where same digits are assigned a unique alphabetic symbol (*I*, *V*, *L*, *C* etc.) and the value of the number is given by the "sum or difference" of digits.

Example: *I*(1), *II*(1 + 1), *IV*(5 – 1 = 4), *XX*(10 + 10).

So, these are the examples of non-positional number system.

1.5.2 (2) Positional Number System

In positional number system each digit is assigned a unique symbol and the number of symbols used is called the radix (or base) of the number system and each position in a number is assigned a weight as a power of radix.

An the basis of power of Radix there are different types of number system are:

(*i*) Binary Number

(*ii*) Octal Number

(*iii*) Decimal Number

(*iv*) Hexadecimal Number.

(*i*) **Binary Number:** It is composed of only two digits either '0' or '1' and these are known as Bits.

Representation of Binary Number is [2]

Values under the Binary Number are [0, 1]

Binary weights $2^x = 2^{-2}\ 2^{-1}\ 2^0\ 2^1\ 2^2 \ldots$

(*ii*) **Octal Number:** Representation of octal number is [8]

It is composed of 8 digits that is known as Octal and Digits are: (0 – 1)

[0, 1, 2, 3, 4, 5, 6, 7]

It can represent 3 bit binary Number.

Octal weights $8^x = 8^{-2}\ 8^{-1}\ 8^0\ 8^1\ 8^2 \ldots$

(*iii*) **Decimal Number:** It is composed of ten digits. Representation of Decimal number is [10]

Due to these ten digits it is known as Decimal and Digits are: (0 – 9)

[0, 1, 2, 3, 4, 5, 6, 7, 8, 9]

Decimal weights $10^x = 10^{-2}\ 10^{-1}\ 10^0\ 10^1\ 10^2 \ldots$

(*iv*) **Hexadecimal Number:** This number is composed of 16 digits and characters.

Representation of Hexadecimal Number is [16] (Hexadecimal Number are: 0–9, *A* – *F* (10 – 15)

[0, 1, 2, 3, 4, 5, 6, 7, 8, 9, *A*, *B*, *C*, *D*, *E*, *F*]

Hexadecimal number system is convenient in digital because each hexadecimal represents, a 4-bit binary Number.

1.5.3 Rules for Conversions of Number System

1. **To Convert:**

Binary (2)
Octal (8)
Hexadecimal (16)
→ Decimal (10)

For this conversion of number system **multiplication with weight method** is used or **sum of weight method**.

2. **To Convert:**

Decimal (10) ⟶ Binary (2) / Octal (8) / Hexadecimal (16)

We have to use the **division method by radix** for conversion of number system.

Explanations of Conversion:

1. Binary to Decimal 2. Octal to Decimal 3. Hexadecimal to Decimal	Sum of Weight Method
4. Decimal to Binary 5. Decimal to Octal 6. Decimal to Hexadecimal	Division Method by Radix

1.5.4 Conversions with Example

1. **Binary to Decimal:** (2 to 10)

It means values given in the binary number have to convert in the decimal values.

Example:

$$(0\ 1\ 0\ 1\ 1)_2 = (?)_{10}.$$

Solution:

$$(0\ 1\ 0\ 1\ 1)_2 = (\ )_{10}.$$

↑ ↑ ↑ ↑ ↑

4 3 2 1 0

$$\text{Sum of Weight Method} = 2^0 \times 1 + 2^1 \times 1 + 2^2 \times 0 + 2^3 \times 1 + 2^4 \times 0$$
$$= 1 \times 1 + 2 \times 1 + 4 \times 0 + 8 \times 1 + 16 \times 0$$
$$= 1 + 2 + 8$$
$$= 11$$

$$\boxed{(01011)_2 = (11)_{10}}$$

2. **Octal to Decimal**: (8 to 10)

It means values given in the octal number have to convert in the decimal values.

Example:

$$(4\ 5\ 7)_8 = (\ )_{10}.$$

Solution:

$$(4\ 5\ 7)_8 = (\ )_{10}.$$

↑ ↑ ↑

2 1 0

$$= 8^0 \times 7 + 8^1 \times 5 + 8^2 \times 4$$
$$= 7 \times 1 + 8 \times 5 + 64 \times 4$$
$$= 7 + 40 + 256$$
$$= 303$$

(Sum of weight Method)

$$\text{Answer} = (457)_8 = (303)_{10}$$

3. **Hexadecimal to Decimal:** (16 to 10)

It means values given in the hexadecimal number have to convert in the decimal values.

Example: $(A\,7\,B)_{16} = (\,)_{10}$

Solution:

$$(A\ 7\ B)_{16} = (\,)_{10}.$$
$$\uparrow\ \uparrow\ \uparrow$$
$$2\ 1\ 0$$
$$= (10\ 7\ 11)_{16} = (\,)_{10}.$$
$$\uparrow\ \uparrow\ \uparrow$$
$$2\ 1\ 0$$
$$= 16^0 \times 11 + 16^1 \times 7 + 16^2 \times 10$$
$$= 11 + 112 + 256 \times 10$$
$$= 123 + 2560$$
$$= 2683$$

Sum of weight Method

Answer $(A\,7\,B)_{16} = (2683)_{10}$

4. **Decimal to Binary:** (10 to 2)

Now, values are given in the decimal no. and have to convert in the binary values. Using division by radix method.

Example: $(25)_{10} = (?)_2$

Solution: Now divide 25 by 2.

2	25
2	12 – 1
2	6 – 0
2	3 – 0
2	1 – 1
	0 – 1

Divide by Radix Method

For result with values from bottom to top.

$$= 11001$$

Answer: $(25)_{10} = (11001)_2$

5. **Decimal to Octal:** (10 to 8)

Now, values given in the decimal number have to convert in the octal values by using the division by radix method.

Example: $(89)_{10} = (?)_8$

Solution: Divide 89 by 8.

8	89
8	11 – 1
8	1 – 3
	0 – 1

= 131

Answer: $(89)_{10} = (131)_8$

6. **Decimal to Hexadecimal:** (10 to 16)

In this values given in the decimal values have to convert in the hexadecimal values.

Example: $(289)_{10} = (?)_{16}$

Solution: Divide 289 by 16

16	289
16	18 – 1
16	1 – 2
	0 – 1

= 121

Answere: $(289)_{10} = (121)_{16}$

Example: $(678)_{10} = (?)_{16}$

Solution: Divide 678 by 16.

16	678
16	42 – 2
16	2 – 10
	0 – 2

= 2 10 2] Hexadecimal value.
↑ ↑ ↑
2 A 2

Answer: $(678)_{10} = (2A2)_{16}$

1.5.5 Fractional Part Conversion

(*i*) Binary to Decimal Number
(*ii*) Octal to Decimal Number
(*iii*) Hexadecimal to Decimal Number

] All are same as previous method.

(*i*) **Binary to Decimal Number (2 to 10):** The method used for conversion for fractional part is same that is sum of weight method.

Example: $(101.101)_2 = (?)_{10}$

Solution:

$$(1\ 0\ 1\ .\ 1\ 0\ 1)_2 = (?)_{10}.$$

$$\uparrow\ \uparrow\ \uparrow\ \uparrow\ \uparrow\ \uparrow$$

$$2\ \ 1\ \ 0\ -1\ -2\ -3$$

$$= 2^0 \times 1 + 2^1 \times 0 + 2^2 \times 1 + 2^{-1} \times 1 + 2^{-2} \times 0 + 2^{-3} \times 1$$

$$= 1\times1 + 2\times0 + 4\times1 + \frac{1}{2}\times1 + \frac{1}{4}\times0 + \frac{1}{8}\times1$$

$$= 1 + 0 + 4 + \frac{1}{2} + 0 + \frac{1}{8}$$

$$= 5 + \frac{1}{2} + \frac{1}{8}$$

$$= 5 + \frac{01}{2} + \frac{1}{8}$$

$$= \frac{40+4+1}{8} \Rightarrow \frac{45}{8} \Rightarrow 5.625$$

] Sum of weight Method.

Answer: $(101.101)_2 = (5.625)_{10}$

Some Unsolved Example:

(*i*) $(47.65)_8 = (?)_{10}$

(*ii*) $(A2.C7)_{16} = (?)_{10}$

(*iii*) Decimal to Binary Number

(*iv*) Decimal to Octal Number

(*v*) Decimal to Hexadecimal Number

(*ii*) **Decimal to Binary Number:** The method used for the conversion of fractional part is same in these case that is division by Radix.

Examples: $(18.8125)_{10} = (?)_2$.

Solution:

18 . 8125

18 → Division Method

8125 → By Multiply method.

2	18
2	9 – 0
2	4 – 1
2	2 – 0
2	1 – 0
	0 – 1

$$\begin{array}{r} .8125 \\ \times\ 2 \\ \hline 1 \leftarrow 1).6250 \\ \times\ 2 \\ 1 \leftarrow 1).2500 \\ \times\ 2 \\ 1 \leftarrow 0).5000 \\ \times\ 2 \\ 1 \leftarrow 1).0000 \end{array}$$

Multiply will 2 till the result will not give zero after point and if the value is repeating then write complement (–) one the repeated values.

$(.1101)_2$

Answer: $(18.8125)_{10} = (10010.1101)_2$

Some Unsolved Example:

(*i*) $(14.\ 765)_{10} = ()_8$

(*ii*) $(27.\ 745)_{10} = ()_{16}$

1.5.6 Some Different Conversion

(*i*) Binary to Octal Number

(*ii*) Binary to Hexadecimal Number.

For these two method we will use the same methods used previously.

(*i*) **Binary to Octal Number (2 to 8):** For this first convert (2 to 10) binary to decimal number and then convert the (10 to 8) decimal values in octal number.

For 2 to 10] use sum of weight method.

10 to 8] use division by Radix method.

And this is same for the second conversion.

Some unsolved Examples are:

(*i*) $(10110)_2 = (?)_8$

(*ii*) $(476)_8 = (?)_2$

(*iii*) $(10111)_2 = (?)_{16}$

(*iv*) $(ABC)_{16} = (?)_{16}$

(*v*) $(452)_8 = (?)_{16}$

1.6 BINARY ARITHMETIC

It is fundamental to all digital computers and most other digital system.

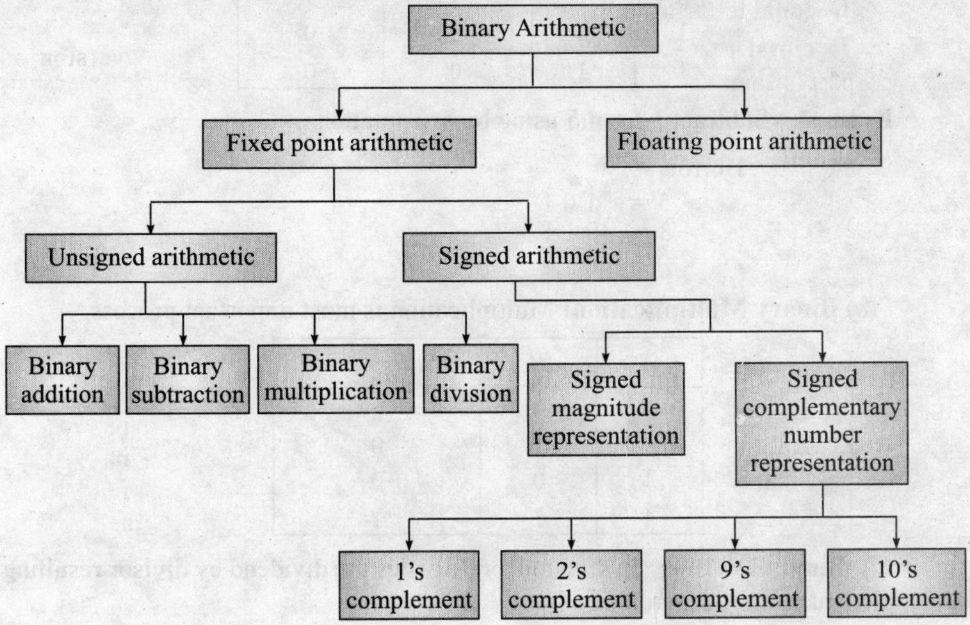

1.6.1 Fixed Point Arithmetic

1. Unsigned Arithmetic

(*a*) **Binary Addition:** Addition is the most important binary arithmetic process because it can be used to perform all other arithmetic operations such as subtraction, multiplication and division.

Table for Binary addition:

A	*B*	*Sum*	*Carry*
0	0	0	0
0	1	1	0
1	0	1	0
1	1	0	1

Rules of Binary Addition

Example: Add 5 and 3 using binary notation.

$$\begin{array}{rcr} & & 1\,1 \text{ (Carry)} \\ 5 & = & 1\,0\,1 \\ 3 & = & 0\,1\,1 \\ \hline & & 1\,0\,0\,0 \end{array}$$

(*b*) **Binary Subtraction:** Subtraction is also most important process.

A	*B*	*Difference*	*Borrow*
0	0	0	0
0	1	1	1
1	0	1	0
1	1	0	0

Example: Subtract 3 from 5 using binary notation.

Borrow =

$$\begin{array}{rr} 5 = & 1\,0\,1 \\ 3 = & -0\,1\,1 \\ \hline & 0\,1\,0 \end{array}$$

(*c*) **Binary Multiplication:** Multiplication is most important process.

A	*B*	*Multiplication*
0	0	0
0	1	0
1	0	0
1	1	1

(*d*) **Binary Division:** In division, one divides the dividend by divisor resulting in the quotient and the remainder.

Example: Divide 1111 by 110.

$$\begin{array}{r} 110\overline{)1111.0}(10.1 \\ 110 \\ \hline 00110 \\ 110 \\ \hline 000 \end{array}$$

2. **Signed Arithmetic:**

(*a*) **Signed Magnitude Representation:** Since computers work on binary system and there are only 2 signs (+, –1) and (0, 1) are used to represent the sign of numbers.

0 → +ve Number

1 → –ve Number.

The MSB or the left most bit is used to represent or hold the sign bit and rest of the bits represent the magnitude of number.

Thus in a 4 bit signed representation the numbers in computer will be:

+4 = 0100, –4 = 1100

while 8 bit representation will be:

+4 = 00000100, –4 = 10000100.

The comma has been used to differentiate between the sign bit and magnitude bit.

(*b*) **Signed Complementary Number Representation:**

Complement method is used to represent negative number only.

→ 1's complement
→ 2's complement
→ 9's complement
→ 10's complement

(*i*) **1's Complement:** For a binary number 1's complement 1's obtained by changing all

1's to 0's

and

0's to 1's

Binary value = 1 0 0 1 0
↓ ↓ ↓ ↓ ↓
1's complement = 0 1 1 0 1

(*ii*) **2's Complement:** 2's complement of a binary number is obtained by finding out 1's complement and then by adding 1 to it at LSB (Least significant bit).

MSB LBS
Binary value = 1 0 0 1 0
↓ ↓ ↓ ↓ ↓
1's complement = 0 1 1 0 1
2's complement = + 1
0 1 1 1 0

MSB = Most significant bit
LSB = Least significant bit.

(*iii*) **9's Complement:** It is obtained by subtracting each digit in number from 9.

Example:

Number = 652
9's complement = 999
of Number = –652
347

347 is the 9's complement of 652.

(*iv*) **10's Complement:** 10's complement of decimal number

= 9's complement + 1

Example: Find out 10's complement of 52.

(*i*) Find 9's complement of 52:

99
– 52
47

(*ii*) Add 1 to it:

47
+ 1
48

So, 48 is the 10's complement of 52.

Some Unsolved Example:

(*i*) 1's complement of 11101.

(*ii*) 2's complement of 00110.

(*iii*) 9's complement of 478.

(*iv*) 10's complement of 676.

***Mathematical formula to find Complement of Number**

Formula is: For ODD Value's Complement.

$$\boxed{(r^n - r^{-m})_{10} - (N)_r}$$

where:

n = no. of digits in integer part

m = no. of digits in fractional part

r = Radix.

N = no. of that Radix.

Example: *Find 1's complement of* $(110.101)_2$.

Solution:

$r = 2, n = 3, m = 3, N = 110.101$

$= (r^n - r^{-m})_{10} - (N)_r$

$= (2^3 - 2^{-3})_{10} - (110.101)_2$

$= (8 - .125)_{10} - (110.101)_2$

$= (7.875)_{10} - (110.101)_2$

$= (111.111)_2 - (110.101)_2$

[∵ Convert decimal to binary value]

$= \boxed{001 \cdot 010}$

Example: *Find 7's complement of* $(162)_8$.

Solution:

$r = 8, n = 3, m = 0, N = 162$

$= (r^n - r - m)_{10} - (N)_r$

$= (8^3 - s^{-0})_{10} (162)_8$

$= (512 - 1)_{10} - (162)_8$

[∵ Converts decimal to octal value]

$= (511)_{10} - (162)_8$

$= (777)_8 - (162)_8$

$\Rightarrow \boxed{(615)_8}$

Formula for Even Number Complement:

$$\text{Formula} = \boxed{(r^n)_{10} - (N)_r}$$

where n = Number of digits in integer part.

r = is radix.

N = Number in that radix.

Example: *Find 2's complement of $(110.101)_2$.*

Solution:

$$r = 2, n = 3, N = 110.101$$

$$(r^n)_{10} - (N)_r = (2^3)_{10} - (110.101)_2$$
$$= (8)_{10} - (110.101)_2$$
$$= (1000)_2 - (110.101)_2$$

($\because$ Convert decimal to Binary Number)

$$= \boxed{001 \cdot 011}$$

Example: *Find 8's complement of $(162)_8$*

Solution:

$$r = 8, n = 3, N = 162$$

$$[(r^n)_{10} - (N)_r] = (.8^3)_{10} - (162)_8$$
$$= (.512)_{10} - (162)_8$$
$$= (1000)_8 - (162)_8 \quad [\because \text{ Convert decimal into octal}]$$
$$= \boxed{(616)_8}$$

1.7 BINARY CODES

Coding is done because digital circuits and computers process data inform of 0 and 1 only and coding is easy to detect and correct errors.

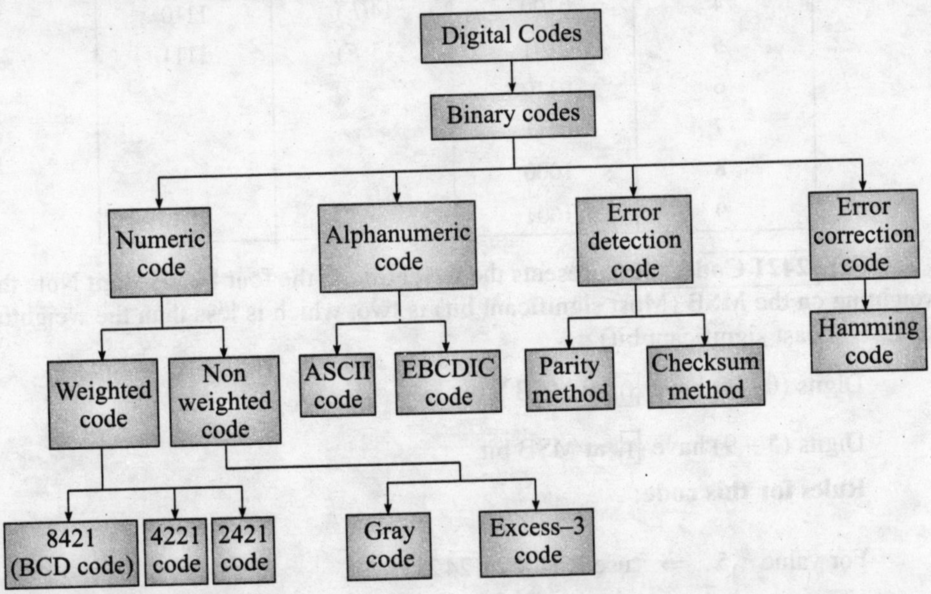

1.7.1 Numeric Code

These codes are used extensively in digital logic applications (*i.e.*, arithmetic circuits, code converter) and in computer communication application to provide error detection and correction.

1.7.1.1 Weighted Code

In these types of codes each position is assigned a weight and the total number is the sum of digits multiplied with their weight.

(*i*) **8421 Code:** In this code structures, each of decimal digits (0 – 9) is converted to its 4-bit pure binary equivalent.

The 8421-binary code is most common form used for decimal representation of numbers, where 8, 4, 2, 1 represent the weights assigned to each bits with 4 bit the number of unique representation is

$$2^4 = 16$$

(0-9) 10 values are valid 8421 code.

(A-F) 6 values are invalid 8421 code.

8421 code is sequential, non-reflective code means where the code for each successive digit is in binary sequences. These are also known as 'unit distance code.'

The valid 8421 code are known as BCD (Binary coded decimal) value.

Valid codes		***Invalid codeds***	
Decimal	***8421 Code***	***Decimal***	***8421 Code***
0	0000	10(*A*)	1010
1	0001	11(B)	1011
2	0010	12(*C*)	1100
3	0011	13(*D*)	1101
4	0100	14(*E*)	1110
5	0101	15(*F*)	1111
6	0110		
7	0111		
8	1000		
9	1001		

(*ii*) **2421 Code:** It represents the weighting of the four binary digit Note that weighting on the MSB (Most significant bit) is two, which is less than the weighting for LSB (Least significant bit)

Digits (0 – 4) have 0 at MSB bit.

Digits (5 – 9) have 1 at MSB bit.

Rules for this code:

For value < 5 ⇒ use R.H.S 2 (24$\overset{\downarrow}{2}$1)

value > 5 ⇒ use L.H.S. 2 ($\overset{\downarrow}{2}$421)

value = 5 ⇒ use 2, 2, 1 but not 4, 1.

Decimal Digit	*2421 Code*
0	0000
1	0001
2	0010
3	0011
4	0100
5	1011
6	1100
7	1101
8	1110
9	1111

(*iii*) **4221 Code:** This code is another binary code, where each bit is weighted by 4, 2, 2, 1. These code also shows values from (0 – 9) and do not have invalid code.

Rules for this code is similar to 2421 code.

Decimal Digit	*4221 Code*
0	0000
1	0001
2	0010
3	0011
4	1000
5	0111
6	1100
7	1101
8	1110
9	1111

1.7.1.2 Non-Weighted Code

For this code, the digits are not assigned any weight. Each code was formed for a specific purpose and the advantage it offered for certain applications.

(*i*) **Gray Code:** This is an important code and is used for its speed. It is free from errors.

These codes are designed in such a way that only one bit changes from any one count to the next. It is a minimum change code.

These codes are suitable and very fast for application where a count sequence is needed. It is not used for arithmetic operations. It is example of unit distance code.

Conversion for Gray Code:

(*a*) Binary number to Gray code conversion.

(*b*) Gray code to binary number conversion.

(*c*) **Binary Value to Gray Code Conversion:**

Procedure:

(*i*) The binary MSB (Most significant bit) and Gray code MSB will always be the same. So, note that MSB of binary is same as MSB of Gray code.

(*ii*) To get the next gray bit, add the binary MSB and bit next to MSB. Record the sum but ignore all the carry.

(*iii*) Continue in this manner right through to the end.

Example: $(100011)_2 = (?)_{\text{Gray code}}$

$$\begin{array}{rcccccc} & \text{MSB} & & & & & \text{LBS} \\ \text{Binary} = & 1 + & 0 + & 0 + & 0 + & 1 + & 1 \\ & \downarrow & \downarrow & \downarrow & \downarrow & \downarrow & \downarrow \\ \text{Gray code} = & 1 & 1 & 0 & 0 & 1 & 0 \end{array}$$

$$\boxed{(100011)_2 = (110010)_{\text{Gray code}}}$$

Example: $(28)_{10}$ $(?)_{\text{Gray code}}$

Firstly convert $(28)_{10} \rightarrow (0)_2$ by the number system division method then into Gray.

$$(28)_{10} = (?)_2$$

2	28
2	14 – 0
2	7 – 0
2	3 – 1
2	1 – 1
	0 – 1

$$= (1 + 1 + 1 + 0 + 0) \rightarrow (?)_{\text{Gray}}$$

$$\begin{array}{ccccc} \downarrow & \downarrow & \downarrow & \downarrow & \downarrow \\ 1 & 0 & 0 & 1 & 0 \end{array} \quad (28)_{10} = (10010)_{\text{Gray code}}$$

(*b*) **Gray Code to Binary value Conversion:**

Procedure:

(*i*) The MSB is same in both case.

(*ii*) The next bit to MSB in binary value comes from adding the MSB is binary number to next-to-MSB bit in Gray code. Record the sum and ignore the carry means diagonally addition.

(*iii*) Generally moving from Left to Right, repeat the above process.

Example: $(10010)_{\text{Gray code}} = (?)_{\text{Binary value}}$

$$\begin{array}{rccccc} & \text{MSB} & & & & \text{LSB} \\ \text{Gray code} = & 1 & 0 & 0 & 1 & 0 \\ \text{Binary value} = & 1 & 1 & 1 & 0 & 0 \end{array}$$

or we can use EX – OR gate directly

$$\begin{array}{rccccc} & \text{MSB} & & & & \text{LSB} \\ \text{Gray code} = & 1 & 0 & 0 & 1 & 0 \\ & \downarrow & \oplus & \oplus & \oplus & \oplus \\ \text{Binary value} = & 1 & 1 & 1 & 0 & 0 \end{array}$$

(*ii*) **Exces-3 code:** In this code a digit is represented by adding 3 to each number and then converting it to a 4-bit binary number.

Also Represent by XS3 BCD code. This is used to perform subtraction operation in digital system.

Example: *Convert* $(359)_{10} = (?)_{XS\text{-}3\ code}$

Solution: Decimal = 359

$$
\begin{array}{r}
\text{Decimal} = 3\ 5\ 9 \\
\underline{3\ 3\ 3} \\
6\ 8\ 1\ 2 \\
\downarrow\ \downarrow\quad \downarrow \\
0110\ 1000\ 1100
\end{array}
$$

Decimal	*Excess-3 code*
0	0011
1	0100
2	0101
3	0110
4	0111
5	1000
6	1001
7	1010
8	1011
9	1100

1.7.2 Alphanumeric Code

This is the combination of $\begin{bmatrix} \text{Alphabet values } (A - Z, a - z) \\ + \\ \text{Numeric value } (0 - 9) \end{bmatrix}$

Need for the communication and keyboards, teletypewriters, paper tape reader etc. are using this code.

Types:

(*i*) ASCII Code.

(*ii*) EBCDIC Code.

1.7.2.1 ASCII Code

Pronounced as "as-kCl".

It stands for "**American standard code for information Interchange**".

This code is standard alphanumeric code for keyboards and a host of other data interchange tasks. Letters, numbers and single keystroke commands are represented by seven-bit word.

It is a 7-bit code and according to this, it has $2^7 = 128$ possible code groups. There is separate code for every value.

Example: Alphabets are $=(A - Z), (a - z)$

Numbers are = $(0 - a)$

Some specific values = Blank, ., C, +, $, *,), /, , , = , 0, Return.

Character	7-bit ASCII	Hexadecimal Code
A	1000001	41
B	1000010	42
C	1000011	43
:	:	:
I	1001001	49
J	1001010	4A
:	:	:
:	:	:
Z	1011010	5A

Character	7-bit ASCII	Hexadecimal Code
a	1100001	61
b	1100010	62
c	1100011	63
:	:	:
:	:	:
:	:	:
:	:	:
:	:	:
z	1111010	7A

Numeric Values	7 bit ASCII	Hexadecimal Code
0	0110000	30
1	0110001	31
2	0110010	32
3	0110011	33
4	0110100	34
5	0110101	35
6	0110110	36
7	0110111	37
8	0111000	38
9	0111001	39

1.7.2.2 EBCDIC Code

Pronounced as "eb-si-dik"

It stands for **"Extended binary coded decimal Interchange code"**

It is an 8-bit alphanumeric code which has been extensively used by IBM in its mainframe application.

It is used by large computer and simple hardware requirement.

Example: Alphabets = $(A-Z)$

Numbers = (0 – 9)

Alphabet	*Hexadecimal Code*	*EBCDIC Code*
$(A-I)$	(C_1-C_9)	(11000001 – 11001001)
$(J-R)$	(D_1-D_9)	(11010001 – 11011001)
$(S-Z)$	(E_2-E_9)	(11100010 – 11101000)

Numbers	*Hexadecimal Code*	*EBCDIC Code*
0	F_0	11110000
1	F_1	:
2	F_2	:
3	F_3	:
4	F_4	:
5	F_5	:
6	F_6	:
7	F_7	:
8	F_8	:
9	F_9	11111001

1.7.3 Error Detection Code

During transmission of data it is possible that same bits can get altered due to noise or other defects, thus data received will not be same as data sent.

To prevent problems due to this some codes or methods were developed to detect the errors.

Due to noise problem, the data received will not be same as data transmitted. So, for this problem error detection is required so that error is tranceable and further can be corrected at reception end.

Two methods of error detection are:

(*i*) Parity Method.

(*ii*) Checksum Method.

1.7.3.1 Parity Method

An extra bit(1/0) known as parity bit that is added with the code, to make the total number of 1s in the combined code odd or even. Extra bit can be added in starting or ending of code.

Two Method:

(*a*) Even Parity

(*b*) ODD Parity

(*a*) **Even Parity:** It means addition of parity bit to code produce an even number of 1's.

Example: *Code: 0110111 Detect error with the help of even parity.*

Solution. Count number of 1's in this code

= 5.

It is an odd number of 1's but we have to use even parity. So, add 1 at end or starting of code to make even number of 1's.

Code after extra bit = 01101111 / 10110111

Now, Number of 1's = 6

(*ii*) **ODD Parity:**

It means addition of parity bit to code produce an ODD number of 1's.

Parity can be added at starting or end of code.

Parity (0/1) according to requirement can be added.

Example: *Code = 1001, detect error with even and ODD parity.*

Solution. For **even Parity:**

No. of 1's = 2

Add 0 at end of code = 10010

For **ODD Parity:**

No of 1's = 2

Add 1 at end of code = 10011.

Disadvantages of Parity Method: We can detect error only if there is a change of 1 bit, when more than two error occurs, then this method fails.

1.7.3.2 Checksum Method

This method is based upon adding redundant bits (means adding extra bits which is transferred along the data bits).

It is divided into two parts:

(*i*) Checksum Generator.

(*ii*) Checksum checker.

(*i*) **Checksum Generator:**

- Checksum code is generated at transmitter side.
- It divides data into equal segments.
- Then all the data bits of equal segments are added together.
- Then take the complement of sum values.
- The complemented sum will be acting as the checksum.
- Then this checksum is sent with data bits.

That is: D_1 and D_2 are two data bits.

$$\begin{array}{r} D_1 \\ + D_2 \\ \hline \text{Sum} = \; D_1 + D_2 \end{array}$$

$$\text{Sum} \Rightarrow \text{complemented } i.e., \; \overline{D_1 + D_2} \Rightarrow \text{Checksum.}$$

(*ii*) **Checksum checker:**

- It works at receiver side.
- Here all the data bits along with checksum is added to produce sum.
- The sum is complemented.
- If result is zero after complemented, then there is no error, otherwise error is present that is: D_1 and D_2 produces the sum and which is complemented to form checksum.

$$\begin{array}{r} D_1 \\ +\,D_2 \\ \hline \text{Sum} \rightarrow \text{Checksum} \end{array}$$

Complemented of result → If result is zero, no error
→ If result have one (1) then error.

Example 1. *Find the checksum of following 16 bits.*

10101001 00111001

Solution: Distribute in two equal parts

$$\begin{array}{rr} \text{Data bits} \rightarrow & 00111001 \\ & 10101001 \\ \hline \text{Sum} \rightarrow & 11100010 \end{array}$$

Checksum = $\overline{\text{Sum}}$ = 00011101

So, data is sent as:

10101001 00111001 00011101

(10101001 00111001 → Data bits; 00011101 → Data bits)

Example 2. *At receiver side:*

10101001, 00111001, 00011101, check error is there or not

Solution: Add all data bits:

$$\begin{array}{rrl} \text{Data bits} \rightarrow & 10101001 & \\ & 00111001 & \\ & 00011101 & \rightarrow \text{Checksum} \\ \hline \text{Sum} \rightarrow & 11111111 & \\ \overline{\text{SUM}} & 00000000 & \end{array}$$

↓

It means pattern is OK and has no error.

1.7.4 Error Correction Method

Parity detection method just tells that there must be some error while transmission. It fails to explain that which bit is incorrect. Hence it is not possible to correct the corrupted data.

So, to solve above problem, we can use error correction method or hamming code.

1.7.4.1 Hamming Code

This code detects an error and also indicates which bit is having error. Hence, bits can be corrected to get desired data.

So, that is why this code also called "**self correcting code.**"

It uses 3 parity bits for transmitting 4 bit data. So, it is also called "**Distributed Parity Check Code.**"

But only single data bit error can be detected using this code.

Formula to Calculate the number of Parity bits:

$$2^P \geq m + P + 1$$

where m = number of message or data bits.

P = Parity bits.

We can calculate values of parity bits by hit and trial method when number of data bits are given.

If

$$m = 4, P = ?$$
$$2^P \geq 4 + P + 1$$
$$2^P \geq 5 + P$$

if put $P = 3$, then formula is satisfied.

$$2^3 \geq 5 + 3$$
$$\boxed{8 = 8}$$

So, number of parity bits are 3

when $m = 4\ (D_0, D_1, D_2, D_3)$

$P = 3\ (P_0, P_1, P_2)$

3 parity bits are : P_0, P_1, P_2.

Position for Parity $P_0 = 2^0 \Rightarrow$ 1st position.

Parity $P_1 = 2^1 \Rightarrow$ 2nd position.

Parity $P_2 = 2^2 \Rightarrow$ 4th position.

4 Data bits are:	D_0,	D_1,	D_2,	D_3
	↓	↓	↓	↓
Positions are:	3	5	6	7

Total number of parity bits are depends upon number of data bits.

Total values = 7 on transmitter.

001	010	011	100	101	110	111
1	2	3	4	5	6	7
P_0	P_1	D_0	P_2	D_1	D_2	D_3

Parity values are given by

For P_0 positions are those where all the values from (1–7) have 1 at LSB:

$$\boxed{1,\ 3,\ 5,\ 7}$$

001, 001, 101, 111

For P_1 positions are those where all the values from (1–7) have 1 at center value.

$$\boxed{2,\ 3,\ 6,\ 7}$$

010, 011, 110, 111

For P_2 positions are those where all the values from (1-7) have 1 at MSB.

4, 5, 6, 7

100, 101, 110, 111

Example 1: *Suppose we want to transmit data 1010 with even parity. The data will be positioned according to hamming code.*

Solution:

1	2	3	4	5	6	7
P_0	P_1	1	P_2	0	1	0

$$P_0 = 1, 3, 5, 7 \Rightarrow P_0\,100 \Rightarrow 1100$$

so, according to even parity P_0 must be 1

$$P_1 = 2, 3, 6, 7 \Rightarrow P_1\,110 \Rightarrow 0110$$

So, according to even parity P_1 must be 0

$$P_2 = 4, 5, 6, 7 \Rightarrow P_2\,010 \Rightarrow 1010$$

So, according to even parity P_2 must be 1

So, full hamming code will be:

1011010

Example 2: *If hamming code is received as 0010001 and even parity is used to detect and correct the code.*

Solution:

1	2	3	4	5	6	7
0	0	1	0	0	0	1
P_0	P_1	D_0	P_2 ↑	D_1	D_2	D_3

$P_0 = 1, 3, 5, 7 \Rightarrow$ 0101 [It is even] 0

$P_1 = 2, 3, 6, 7 \Rightarrow$ 0101 [It is even] 0

$P_2 = 4, 5, 6, 7 \Rightarrow$ 0101 [It is odd then to make even put 1] ↑

100

↓ It is 4th position.

So, error is there on 4th position.

So, to correct the error, change the bit from $0 \rightarrow 1$ on 4th position.

So, correct hamming code:

0011001

1.8 BCD ADDITION

This is binary coded decimal have value from 0 to 9 (valid values). This is a 4 bit code.

After addition if the result have more than 9 then add code of 6 (0110) in that value which is greater than 9.

Example: *38 + 24 by BCD addition.*

Solution:

$$\begin{array}{r} 38 \\ +24 \\ \hline 62 \end{array}$$

BCD Code for 38:

0011 1000

BCD Code for 24:

00100100

$$\begin{array}{r} 00111000 \\ 00100100 \\ \hline 01011100 \end{array}$$

$\downarrow$ 5 $\quad\downarrow$ 12 This is more than 9, so add 6 (0110) code in 12.

$$\begin{array}{r} 01011100 \\ +0100 \\ \hline 01100010 \end{array}$$

$\downarrow$ 6 $\quad\downarrow$ 2 This is BCD addition.

1.9 SUBTRACTION USING 1's AND 2's COMPLEMENT

To make subtraction firstly write binary value or BCD value for given number and then take 1's and 2's complement of that value which have to subtract, then after complement add both values.

Example: (27) – (13) using 1's complement.

$$\begin{array}{r} 27 \\ -13 \\ \hline +14 \end{array}$$

Then write binary value of 27 and 13 using division method.

2	27
2	13 – 1
2	6 – 1
2	3 – 0
2	1 – 1
	0 – 1

$(27) = (11011)_2$

2	13
2	6 – 1
2	3 – 0
2	1 – 1
	0 – 1

$(13) = (1101)_2$

(27) = 11011
(13) = 01101 $\longrightarrow$ add extra 0 at MSB to make equal values

Then take 1's complement of 13.

= 10010

then add both values.

```
    27 =  1 1 0 1 1
    13 =  1 0 0 1 0
         ----------
EOC  (1) 0 1 1 0 1
     ↳         + 1
         ----------
          0 1 1 1 0
         ----------
```

If carry generate the End around the carry means add that carry at LSB.

Now, take the decimal code of result.

$(0\ 1\ 1\ 10)_2 = (\ )_{10}$

↑ ↑ ↑ ↑ ↑

4 3 2 1 0

$= 2^0 \times 0 + 2^1 \times 1 + 2^2 \times 1 + 2^3 \times 1 + 2^4 \times 0$

$= 0 + 2 + 4 + 8 + 0$

$= 14$

(27) – (13) ⇒ (14)

Example: $(13 - 27)_{BCD}$ by 1's complement

```
   13    →    0 0 0 1 0 0 1 1
 – 27         – 0 0 1 0 0 1 1 1
 ----         -----------------
 – 14                 ?
```

1's complement of 27 → 11011000 then add this to 13

```
   0 0 0 1 0 0 1 1
 + 1 1 0 1 1 0 0 0
 -----------------
   1 1 1 0 1 0 1 1 ]
 -----------------
```

If no carry generate it means it is –ve value so, again take 1's complement of result.

```
 0 0 0 1 0 1 0 0 ]  →  [–14]
 └──┬──┘ └──┬──┘
    1       4
```

Examples of 2's complement:

(*i*) $(27 - 13)_{BCD}$ by 2's complement.

```
   27  →  0 0 1 0 0 1 1 1
 – 13  →  0 0 0 1 0 0 1 1
 ----     └──────┬──────┘
 + 14            ↓ 2's complement

          1 1 1 0 1 1 0 0
                      + 1
          ---------------
          1 1 1 0 1 1 0 1
          ---------------
```

Add 2's complement value of 13 in 27.

```
    0 0 1 0 0 1 1 1
  + 1 1 1 0 1 1 0 1
 (1) 0 0 0 1 0 1 0 0
     └──┬──┘ └──┬──┘
        ↓       ↓
        1       4
```

If carry generate in case of 2's complement it means result is +ve.

Example: *(18 – 27) by 2's complement.*

Solution:

```
   18
 – 27
 ----
 – 09
```

2	18
2	9 – 0
2	4 – 1
2	2 – 0
2	1 – 0
	0 – 1

(18) = 10010

2	27
2	13 – 1
2	6 – 1
2	3 – 0
2	1 – 1
	0 – 1

(27) = 11011

Now take 2's complement of 27.

$$\begin{array}{lrl} 11011 \rightarrow & 00100 & \rightarrow \text{1's complement of 27} \\ & +1 & \\ \hline & 00101 & \rightarrow \text{2's complement of 27} \end{array}$$

Add this value to 18:

$$\begin{array}{r} 10010 \\ +00101 \\ \hline 10111 \end{array}$$

If no carry generate then again take 2's complement of result.

$$\begin{array}{rl} 10010 & \rightarrow \text{1's complement of 10111} \\ +00101 & \\ \hline 10111 & \rightarrow \text{2's complement of 10111} \end{array}$$

Again take decimal value.

0 1 0 0 1
↑ ↑ ↑ ↑ ↑
4 3 2 1 0

$= 2^0 \times 1 + 2^1 \times 0 + 2^2 \times 0 + 2^3 \times 1 + 2^4 \times 0$

$= 1 + 8$

$= ⑨$

So, 9 is final result.

Example: $(18 - 27)_{BCD}$ using 2's complement.

Solution:

$$\begin{array}{rl} 18 & \xrightarrow{BCD} 00011000 \\ -27 & \xrightarrow{BCD} 00100111 \\ \hline -09 & \end{array}$$

Take 2's complement of 27.

$$\begin{array}{r} 11011000 \\ +1 \\ \hline 11011001 \end{array}$$

then add this value to 18.

$$\begin{array}{lcr} 00011000 & \longrightarrow & 00011000 \\ 00100111 & \xrightarrow[\text{complement}]{\text{2's}} & +11011001 \\ \hline & & 11110001 \end{array}$$

if no carry then take 2's complement.

```
00001110
      +1
--------
00001111
   ↓
```

15 that is greater than 9.

So, again add the 2's complement of 6(0 1 1 0) $\xrightarrow[\text{complement}]{\text{2's}}$

```
 1001
   +1
-----
 1010
```

So, add this value to the final result.

```
 00001111
     1010
---------
 000(1)1001
```

Discard carry ← (1); 9

Note: When in subtraction we have to take borrow (22–17) or so on, then we have to add 2's complement of 6 that is (1010) to value which is greater than 9.

1.10 NEW IEEE SYMBOLS

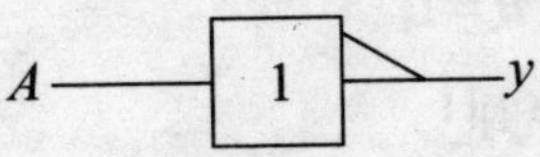

NOT Gate

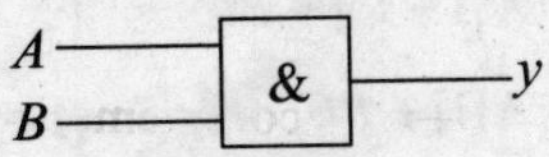

AND Gate

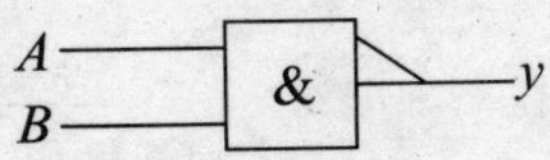

NAND Gate

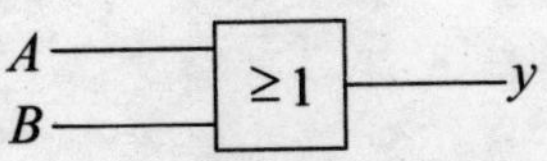

OR Gate

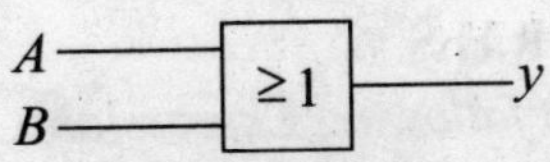

NOR Gate

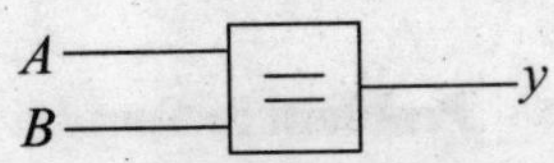

XOR Gate

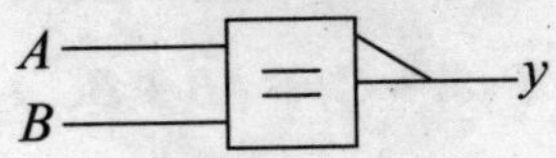

XNOR Gate

1.11 SOLVED PROBLEMS

Problems on Boolean Algebra

Problem 1: *Prove that:*

$A + \bar{A}B = A + B.$

Solution: L.H.S. $= A + \bar{A}B$

$= A.1 + \bar{A}B$

$= A(1 + B) + \bar{A}B$

$= A + AB + \bar{A}B$

$= A + B(A + \bar{A})$

$= A + B.1$ $[\because \; A + \bar{A} = 1]$

$= A + B.$

= R.H.S.

Problem 2: *Prove that:*

$$(A+\bar{B}+\bar{C})(A+\bar{B}C) = A+\bar{B}C.$$

Solution:

$$\begin{aligned}
\text{L.H.S.} &= (A+\bar{B}+\bar{C})(A+\bar{B}C) \\
&= AA+A\bar{B}C+\bar{B}A+\bar{B}\bar{B}C+\bar{C}A+\bar{C}\bar{B}C \\
&= A+A\bar{B}C+\bar{B}A+\bar{B}C+\bar{C}A \quad [\because \ \bar{C}.C\bar{B}=0] \\
&= A[1+\bar{B}C+\bar{B}]+\bar{B}C+\bar{C}A \\
&= A[1+\bar{B}C+\bar{B}+\bar{C}]+\bar{B}C \\
&= A[(1+\bar{B})(1+C)+\bar{B}+\bar{C}]+\bar{B}C \\
& \qquad [\because \ 1+\bar{B}C=(1+\bar{B})(1+C)] \\
&= A[(1)(1)+\bar{B}+\bar{C}]+\bar{B}C \quad \left[\because \begin{matrix} 1+\bar{B}=1 \\ 1+C=1 \end{matrix}\right] \\
&= A[1+\bar{B}+\bar{C}]+\bar{B}C \\
&= A[1+\bar{C}]+\bar{B}C \quad [\because \ 1+\bar{B}=1] \\
&= A[1]+\bar{B}C \quad [1+\bar{C}=1] \\
&= A+\bar{B}C \\
&= \text{R.H.S.}
\end{aligned}$$

Problem 3: *Simplify the following Expression:*

$$B(A+C)+A\bar{B}+B\bar{C}+C.$$

Solution:

$$\begin{aligned}
& B(A+C)+A\bar{B}+B\bar{C}+C \\
&= BA+BC+A\bar{B}+B\bar{C}+C \\
&= AB+A\bar{B}+BC+B\bar{C}+C \\
&= A(B+\bar{B})+B(C+\bar{C})+C \quad \left[\because \begin{matrix} (B+\bar{B})=1 \\ (C+\bar{C})=1 \end{matrix}\right] \\
&= A+B+C.
\end{aligned}$$

Problem 4: *Simplify the following expression:*

$$ABC+\bar{A}BC+A\bar{B}C+\bar{A}\bar{B}C+AB\bar{C}.$$

Solution:

$$\begin{aligned}
& ABC+\bar{A}BC+A\bar{B}C+\bar{A}\bar{B}C+AB\bar{C} \\
&= BC(A+\bar{A})+\bar{B}C(A+\bar{A})+AB\bar{C} \\
&= BC+\bar{B}C+AB\bar{C} \quad [\because \ A+\bar{A}=1] \\
&= C(B+\bar{B})+AB\bar{C} \quad [\because \ B+\bar{B}=1] \\
&= C+AB\bar{C} \\
&= (C+\bar{C})(C+AB) \\
&= 1(C+AB) \\
&= C+AB.
\end{aligned}$$

Problem 5: *Simplify following expression:*

$$\overline{(\bar{A}+C).(B+\bar{\bar{D}})}.$$

Solution:

$$\overline{(\bar{A}+C).(B+\bar{\bar{D}})}$$

$$= \overline{(\bar{A}+C)+(B+\bar{\bar{D}})} \qquad (\because \ \overline{A.B} = \bar{A}+\bar{B})$$

$$= (\bar{\bar{A}}.\bar{C})+(\bar{B}-\bar{\bar{D}})$$

$$= (A.\bar{C})+(\bar{B}.D)$$

$$= A\bar{C}+\bar{B}D.$$

Problems on Number System:

Problem 1: *Find 9's and 10's complement of following Number*

(i) 25 *(ii) 155* *(iii) 333*

Solution: (*i*) Number = 25

9's complement = 99 − 25 = 74

$$\begin{array}{r} 99 \\ -25 \\ \hline 74 \\ \hline \end{array}$$

10's complement = 9's complement + 1

= 74 + 1 ⇒ **75**.

(*ii*) Number = 155

9's complement =

$$\begin{array}{r} 999 \\ -155 \\ \hline 844 \\ \hline \end{array}$$

10's complement = 9's complement + 1

= 844 + 1 ⇒ **845**.

(*iii*) Number = 333

9's complement =

$$\begin{array}{r} 999 \\ -333 \\ \hline 666 \\ \hline \end{array}$$

10's complement = 9's complement + 1

= 666 + 1 ⇒ **667**

Problem 2: *(–39 + 25) using 9's complement and 10's complement.*

Solution. Subtraction using 9's complement:

$$\begin{array}{rl} 25 & \text{Minuend} \\ -39 & \text{Subtrahend.} \\ \hline \end{array}$$

Take 9's complement of subtrahend:

i.e. 39 ⇒

$$\begin{array}{r} 99 \\ -39 \\ \hline 60 \\ \hline \end{array}$$

Then add the 9's complement of subtrahend to minuend and is given:

$$\begin{array}{r} 25 \\ +60 \\ \hline 85 \\ \hline \end{array}$$

No end around carry

It means answer is negative and in 9's complement from which is given as:

$$\begin{array}{r} 99 \\ -85 \\ \hline 14 \\ \hline \end{array}$$

Hence the result is $\boxed{-14}$

Subtraction using 10's Complement:

$$\begin{array}{rl} 25 & \text{Minuend} \\ -38 & \text{Subtrahend.} \\ \hline \end{array}$$

Take 10's complement of subtrahend:

9's complement ⇒ $\begin{array}{r} 99 \\ -39 \\ \hline 60 \\ \hline \end{array}$

10's complement = $\begin{array}{r} 60 \\ +1 \\ \hline 61 \\ \hline \end{array}$

Add 10's complement of subtrahend to minuend

$$\begin{array}{r} 25 \\ +61 \\ \hline 86 \\ \downarrow \end{array}$$

No end around carry, the answer is negative and 10's complement form.

9's complement of 86 = $\begin{array}{r} 99 \\ -86 \\ \hline 13 \\ \hline \end{array}$

10's complement of 86 ⇒ 13 + 1 ⇒ 14.

The answer is $\boxed{-14}$.

Problem 3: *Subtract 15 from 25 using 1's and 2's complement.*

Solution: Subtract using 1's complement:

$$\begin{array}{rl} 25 & \rightarrow \text{Minuend} \\ -15 & \rightarrow \text{Subtrahend} \\ \hline \end{array}$$

Represent both value in binary representation

25 → 11001

15 → 01111

Take 1's complement of subtrahend: 10000. Then add minuend with 1's complement of subtrahend:

```
                          1 1 0 0 1 → Minuend
                          1 0 0 0 0 → is complement of Subtrahend
End around carry →      ① 0 1 0 0 1
Add this carry to result:   0 1 0 0 1
                                 + 1
                            0 1 0 0 1 = ( 10)
```

Subtract using 2's Complement:

25 → 11001

15 → 01111

Take 2's complement of 15 and 1's given

15 → 1's complement = 1000

2's complement = +1

10001

So, add 2's complement of 15 *i.e.*, 10001 to minuend:

```
   1 1 0 0 1
   1 0 0 0 1
 ① 0 1 0 1 0
 ↓
Drop this
carry
```

Hence result = (01010) = 10

1.12 UNSOLVED QUESTIONS

(Boolean Algebra, Number system, Digital codes)

BOOLEAN ALGEBRA

Q.1. Prove the following identities using Boolean Algebra

(*a*) $(A + B\overline{C} + C)\,\overline{C} = ABC + A\overline{B}\,.\,\overline{C} + \overline{A}B\overline{C}$

(*b*) $A(\overline{A} + C)\,(\overline{A}B + \overline{C}) = 0$

(*c*) $AB + \overline{AC} + A\overline{B}C(AB + C) = 1$

(*d*) $A[B + \overline{C}\,(AB + A\overline{C})]$

(*e*) $(A + \overline{B}\overline{C})\,(A\overline{B} + ABC)$

(*f*) $A\overline{B}\,(A + C) + \overline{A}B\,(A + \overline{B} + C)$

(*g*) $A + \overline{A}B\overline{C} + (\overline{B} + \overline{C})$

(*h*) $AB + ABC + ABCD + ABCDE + ABCDEF$

(*i*) $Z = \overline{A}\overline{B}\overline{C} + \overline{A}B + AB\overline{C} + AC$

(*j*) $Z = \overline{A}B + B\overline{C} + BC + A\overline{B}\overline{C}$

(*k*) $Y = (B + CA)\,(C + \overline{A}B)$

(*l*) $Y = \overline{AB.(\overline{B}C + AC)}$

(*m*) $Y = AB + AC + BD + CD$

(*n*) $Y = \overline{A}\overline{B}\overline{C}\overline{D} + \overline{A}\overline{B}\overline{C}D + A\overline{B}\overline{C}\overline{D} + AB\overline{C}D$.

Q.2. Prove that function $y = xy + yz + zx$ is self dual.

Q.3. Implement XOR gate using minimum number of NAND gates and show that circuit drawn realizes the XOR gate.

Q.4. Show how a XOR gate can be used as an inverter.

Q.5. Show how a XNOR gate can be used as an inverter.

Q.6. Show that dual of XOR gate is equivalent to its complement.

Q.7. Draw the circuit for following function using only NOR gates:

$$A.\overline{A}\overline{B} + AB.$$

NUMBER SYSTEM

Q.1. Convert the following decimal numbers to binary, octal and Hexadecimal Number.

(*a*) 36 (*b*) 79

(*c*) 147 (*d*) 888

(*e*) 746 (*b*) 1086.

Q.2. Convert the following number to decimal Number.

(*a*) $(A_3)_{16}$ (*b*) $(275)_8$

(*c*) $(1101)_2$ (*d*) $(54.4)_8$.

Q.3. Convert following octal number to binary and hexadecimal decimal number.

(*a*) 36.25 (*b*) 27.125

(*c*) 376.4.

Q.4. Convert following hexadecimal number to binary, octal and decimal number.

(*a*) $A_{32}.B$ (*b*) $9C.A$

(*c*) $1FA.4$.

Q.5. Convert following binary number to octal, decimal and Hexadecimal format.

(*a*) 10101 (*b*) 110011

(*c*) 100011.1101 (*d*) 1011.001

(*e*) 101.11 (*f*) 11011.101

Q.6. Convert decimal number $1\frac{81}{512}$ into octal and hexadecimal format.

Q.7. Find the value of radix *r* in each of following:

(*a*) $(132)_r = (30)_r$ (*b*) $(35)_r + (21)_r = (100)_r$

(*c*) $(1000)_r = (40)_r$.

DIGITAL CODES

Q.1. Give BCD, Excess–3 and gray code for the following number:

(*a*) 236 (*b*) 841

(*c*) 902 (*d*) 57.23

(*e*) 34.

Q.2. Find the following decimal number in BCD using 2421, 8421, 4241 code.

(*a*) 123 (*b*) 89

(*c*) 44.75 (*d*) 40.07.

Q.3. Explain the significance of ASCII code.

RADIX ARITHMETIC

Q.1. Perform the following addition:

(*a*) $(107)_8 + (73)_8$ (*b*) $(2C3)_{16} + (DA9)_{16}$.

Q.2. Perform the following subtraction:

(*a*) $(137)_8 - (45)_8$ (*b*) $(B_3)_{16} - (A9)_{16}$

(*c*) $(151)_8 - (54)_8$.

Q.3. Add the following binary number:

(*a*) 11001 + 1001 (*b*) 101100 + 110011

(*c*) 10011 + 111 (*d*) 10010 + 1101.

Q.4. Subtract the following binary number:

(*a*) 10000 – 1011 (*b*) 101100 – 10011

(*c*) 10001 – 111 (*d*) 10110 – 101.

Q.5. Solve in binary using is complement.

(*a*) $(11 - 7)_{10}$ (*b*) $(86 - 39)_{10}$

(*c*) $(17 - 30)_{10}$.

Q.6. Solve in binary unsing 2's complement.

(*a*) $(62 - 29)_{10}$ (*b*) $(57 - 37)_{10}$

(*c*) $(21 - 55)_{10}$.

Q.7. Solve using 9's complement.

(*a*) $(135 - 89)_{10}$ (*b*) $(2179 - 364)_{10}$

(*c*) $(75 - 863)_{10}$.

Q.8. Evaluate using BCD arithmetic.

(*a*) 139 + 973 (*b*) 48 + 964.

Q.9. Solve the 1101110 with the help of parity method (Even and ODD)

Q.10. Solve the code 011001 100111 with checksum method.

Q.11. Suppose we want to transmit data 1100 with odd parity. The data will be positioned according to hamming code.

UNIT 2

COMBINATIONAL DESIGN USING GATES

Topics are:

- Logic Design
- Design Using Gates
- Karnaugh-MAP for Simplification
- Solved Problems on *k*-map
- Comparison between Minterm and Maxterm
- Quine-Mc Clusky Method (Tabular Method)

CHAPTER 2

Combinational Design Using Gates

By combining the laws of Boolean algebra and knowledge of logic gates we can simplify Boolean algebra expressions. Minimising terms and expressions is also important because electrical circuit consist of individual components that are implemented for each term or literal for a given expression. This allows designers to make use of fewer components, thus reducing the cost of particular system.

2.1 LOGIC DESIGN

It is a process in which we design a system which work under certain logic. For designing we need certain thing such as:

(*i*) Truth Table

or

(*ii*) Boolean Expression.

After simplification of Boolean expression or function we implement it using basic gates or universal gates.

2.2 DESIGN USING GATES

We can design a Boolean function with the help of Basic and universal gates. If the Boolean function is large then we can minimize the boolean function with the help of boolean algebra and then designing can be done with the help of logic gates (Basic or universal gate.)

Example 1: $y = AB + \bar{A}B$ Design using Logic gate.

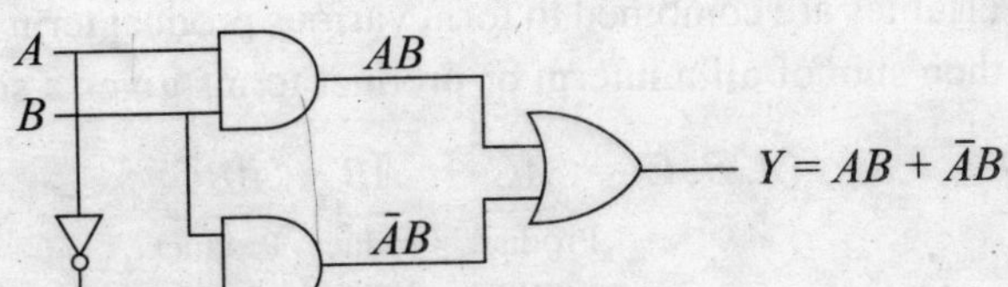

Example 2: $y = (A+B)(\bar{A}+B)$ Design using Logic gate.

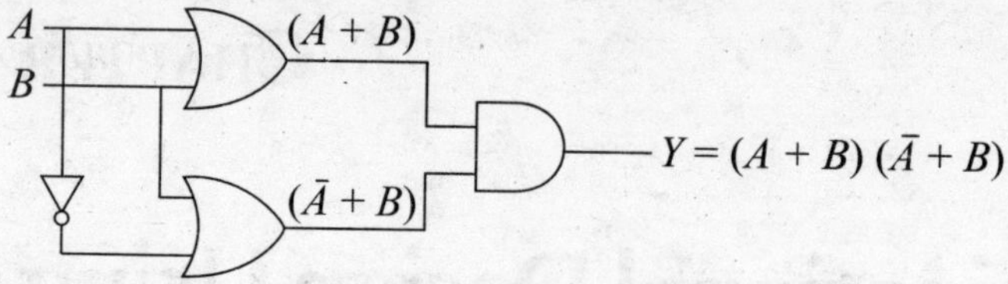

Example 3: $y = AB + BC$ Design using universal gate.

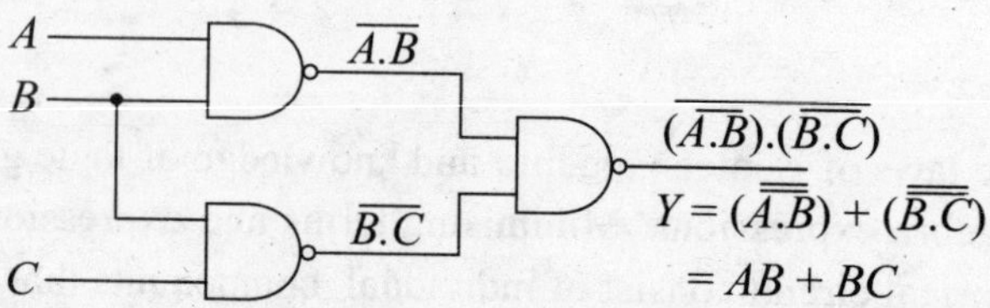

2.2.1 Simplification of Boolean Function

For the simplification or minimization of Boolean function following method are given:

(*i*) Boolean Algebra method.

(*ii*) Karnaugh-map (*k*-Map) Method.

(*iii*) Quine-Mccluskey (Q-M) Method.

Boolean algebra is based on the boolean algebra different properties.

In the above method, the *k*-map is the simplest one and widely used.

2.2.2 Representation of Boolean Function

Boolean function consist of Boolean variables (*i.e.*, *A*, *B*, *C*, *D*, ...). A **variable** is a quantity, which changes by taking on the value of any constant in algebraic system. Each variable have only two value either it is zero (0) or it is one (1).

Boolean function can be expressed into two standard forms as:

(*i*) Sum of Product (SOP) form.

(*ii*) Product of sum (POS) form.

2.2.2.1 Sum of Product Form (SOP)

In SOP form all variables are combined to form various product term which are known as minterms and then sum of all minterm or product terms gives a sop form.

Example: $f(A, B, C) = \underbrace{AC}_{\text{Product term}} + \underbrace{\bar{A}B}_{\text{Product term}} + \underbrace{\bar{A}B\bar{C}}_{\text{Product term}}$

SOP form is Implemented using Basic gates.

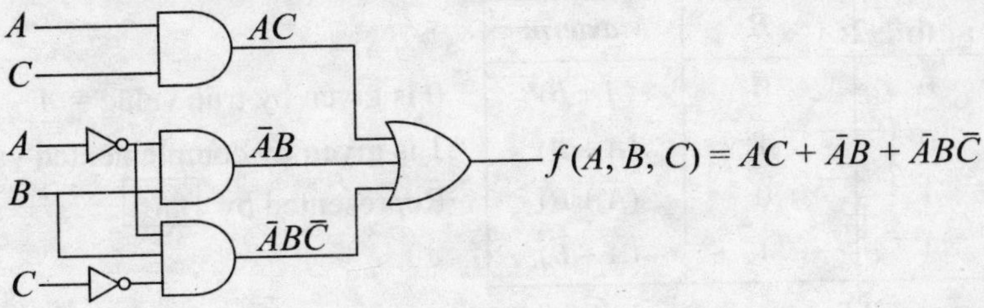

SOP form can be easily designed using NAND gate.

Minterm–(*m*)/Canonical Product form: The product term or minterm consist of product of all variable in complemented form to true form.

0 is given by complemented value = $\bar{A}$
1 is given by true value = *A*.

A	*B*	*Minterm*
0	0	$\bar{A}\bar{B}$
0	1	$\bar{A}B$
1	0	$A\bar{B}$
1	1	AB

Represented by = $\boxed{\Sigma m}$

Numerical representation of SOP form:

$$f(A, B, C) = \Sigma m\ (2, 3, 4, 5)$$

2.2.2.2 Product of Sum Form (POS)

In POS form all the variables are summed into their true form or in complemented form and this summed term are known as Maxterm (M) and then these maxterm are produced to form POS form.

Example: $f(A, B, C) = (A + C)\ (\bar{A} + B)\ (\bar{A} + B + \bar{C})$

It is designed using Basic gates.

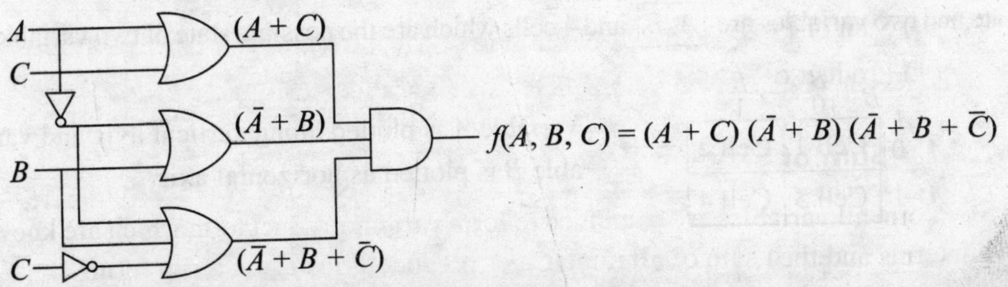

- POS form can be easily designed using NOR gate.

Maxterm–(M)/Canonical Sumform: It is opposite to minterm, here summed term or Maxterm consist of sum of all variables in their true form or in complemented form.

A	B	*Maxterm*
0	0	$(A+B)$
0	1	$(A+\bar{B})$
1	0	$(\bar{A}+B)$
1	1	$(\bar{A}+\bar{B})$

0 is given by true value = A

1 is given by complemented value = $\bar{A}$

Represented by $\boxed{\pi M}$

Numerical form:

$$y(A, B, C) = \pi M(0, 1, 2, 6)$$

2.3 KARNAUGH-MAP (K-MAP)

K-Map provides a systematic method for simplifying a Boolean expression or truth table function when used properly, the *k*-map will produce the simplest SOP or POS expression possible. It is based on gray code.

(*a*) K-map is a graphical representation of truth table.

(*b*) Rules of Boolean algebra is not required.

(*c*) Simplification is done using mapping technique graphically.

(*d*) The *k*-map is a table consisting of $\boxed{N = 2^n}$ cells, where n = Number of input variable.

(*e*) The table format is such that there is single variable changes between any adjacent cells.

(*f*) For variable > 6, the *k*-map technique becomes Impractical.

According to the number of variables *k*-map is divided as:

(*a*) 2-variable *k*-map

(*b*) 3-variable *k*-map

(*c*) 4-variable *k*-map

(*d*) 5-variable *k*-map

(*e*) 6-variable *k*-map

(*a*) **2 -variable *k*-map:** (A, B) 2-variable have maximum $2^n \Rightarrow 2^2 = 4$ possible state and two variables are $\boxed{A, B}$ and 4 cells which are the possible state of two variables.

$A \backslash B$	$\bar{B}$ 0	B 1
$\bar{A}$ 0	Cell 1	Cell 2
A 1	Cell 3	Cell 4

$\longrightarrow$ Variable A is plotted along vertical axis and variable B is plotted as horizontal axis.

$B \backslash A$	$\bar{A}$ 0	A 1
$\bar{B}$ 0	Cell 1	Cell 3
B 1	Cell 2	Cell 4

$\longrightarrow$ Variable A is plotted along horizontal axis and variable B is plotted along vertical axis.

A is MSB bit

$$00 = 0,\ 01 = 1,\ 10 = 2,\ 11 = 3$$

A \ B	0	1
0	0	1
1	2	3

B \ A	0	1
0	0	2
1	1	3

(*ii*) **3-variable *k*-map:** (A, B, C) It have possible number of state $2^3 = 8$ cells. We will use gray code in the *k*-map.

A = M.S.B. bit, C = L.S.B. bit.

MSB LSB

A	B	C	Value
0	0	0	0
0	0	1	1
0	1	0	2
0	1	1	3
1	0	0	4
1	0	1	5
1	1	0	6
1	1	1	7

when BC at horizontal or vertical:

00 → 01 → 10 → 11] Not possible because k-map is based on 1-bit change.

1 bit change 2 bit change 1 bit change

So, interchange the position:

00 → 01 → 11 → 10] Possible

1 bit change

A \ BC	00	01	11	10
0	Cell 1 0	Cell 2 1	Cell 3 3	Cell 4 2
1	Cell 5 4	Cell 6 5	Cell 7 7	Cell 8 6

A = Vertical axis
BC = Horizontal axis

BC \ A	0	1
00	Cell 1 0	Cell 5 4
01	Cell 2 1	Cell 6 5
11	Cell 3 3	Cell 7 7
10	Cell 4 2	Cell 8 6

A = Horizontal axis.
BC = Vertical axis.

(*iii*) **4-variable *k*-map:** (A, B, C, D) It have possible number of states $2^4 = 16$ cells, A is MSB, D = *LSB*, the 4-variable K-map is:

AB \ CD	00	01	11	10
00	Cell 1 0	Cell 2 1	Cell 3 3	Cell 4 2
01	Cell 5 4	Cell 6 5	Cell 7 7	Cell 8 6
11	Cell 9 12	Cell 10 13	Cell 11 15	Cell 12 14
10	Cell 13 8	Cell 14 9	Cell 15 11	Cell 10 10

AB = Vertical axis
CD = Horizontal axis

CD \ AB	00	01	11	10
00	Cell 1 0	Cell 5 4	Cell 9 12	Cell 13 8
10	Cell 2 1	Cell 6 5	Cell 10 13	Cell 14 9
11	Cell 3 3	Cell 7 7	Cell 11 15	Cell 15 11
10	Cell 4 2	Cell 8 6	Cell 12 14	Cell 16 10

AB = Horizontal axis
CD = Vertical axis

MSB			*LSB*		
A	***B***	***C***	***D***	***Values***	***Gray code***
0	0	0	0	0	0
0	0	0	1	1	1
0	0	1	0	2	3
0	0	1	1	3	2
0	1	0	0	4	6
0	1	0	1	5	7
0	1	1	0	6	5
0	1	1	1	7	4
1	0	0	0	8	12
1	0	0	1	9	13
1	0	1	0	10	15
1	0	1	1	11	14
1	1	0	0	12	10
1	1	0	1	13	11
1	1	1	0	14	9
1	1	1	1	15	8

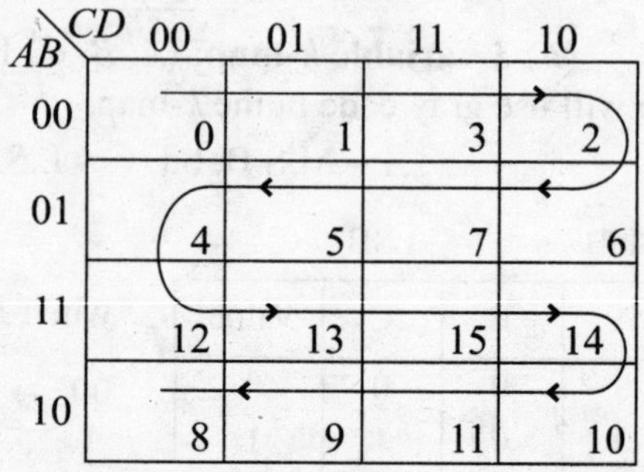

We filled gray code of binary value in all the 16 cells.

(*iv*) **5-variable *k*-map:** (*A*, *B*, *C*, *D*, *E*) As the number of variable five then number of states are $2^5 = 32$ cells.

A

$\bar{A} = 0$

BC \ *DE*	00	01	11	10
00	0	1	3	2
01	4	5	7	6
11	12	13	15	14
10	8	9	11	10

$A = 1$

BC \ *DE*	00	01	11	10
00	16	17	19	18
01	20	21	23	22
11	28	29	31	30
10	24	25	27	26

A is MSB in this, we can combine two 4-variable *k*-map then total 32 states are there keep $\bar{A}$ on *A* means face to face values are kept.

(*v*) **6-variable *k*-map:** (*A*, *B*, *C*, *D*, *E*, *F*) Here possible states are $2^6 = 64$ cells. Here we can combine 4-four variable *k*-map then total 64 states are possible.

$A = MSB, \quad F = LSB$

A \ B	CD \ EF (B = 0)	00	01	11	10	CD \ EF (B = 1)	00	01	11	10
A = 0	00	0	1	3	2	00	16	17	19	18
	01	4	5	7	6	01	20	21	23	22
	11	12	13	15	14	11	28	29	31	30
	10	8	9	11	10	10	24	25	27	26
A = 1	00	32	33	35	34	00	48	49	51	50
	01	36	37	39	38	01	52	53	55	54
	11	44	45	47	46	11	60	61	63	62
	10	40	41	43	42	10	56	57	59	58

2.3.1 PAIRING OF 1's/0's

Minterms are filled by 1's and maxterms are filled by 0's. Pairing should start from the largest value as in 4-variable *k*-map states are $2^4 = 16$, so, highest pairing will be of 16, then 8, then 4, then 2, then individual value is left.

Pairing will be of adjacent values when naming of grouping is written then always their pair have one less term than two originals terms.

2.3.2 DO NOT-CARE CONDITION (*d*)

In digital we have two values of a variables *i.e.*, 0 or 1.

If any variable does not have value 0 hence it have some value, that is not always true because there are certain combination of input variable that never occur and we cannot fill that value by 0/1. They are known as Don't-care condition and filled in *k*-map by $\boxed{X}$ mark.

They are used if they are helpful in minimizing the equation otherwise they are not used in pair or in grouping.

The mark (*X*) denotes that is never 1 nor 0. It never come alone it always come with minterm or with maxterm.

2.4 SOLVED PROBLEMS ON K-MAP

2-variable *k*-map:

Q.1. Solve the following Boolean function using *k*-map:

$$f(A, B) = \Sigma m\ (0, 3)$$

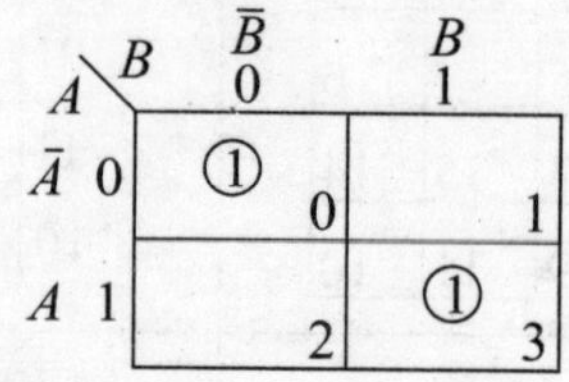

$$f(A, B) = \overline{A}\overline{B} + AB.$$

Q.2. Solve problem using *k*-map:

$$f(A, B) = \pi M\ (0, 1, 2)$$

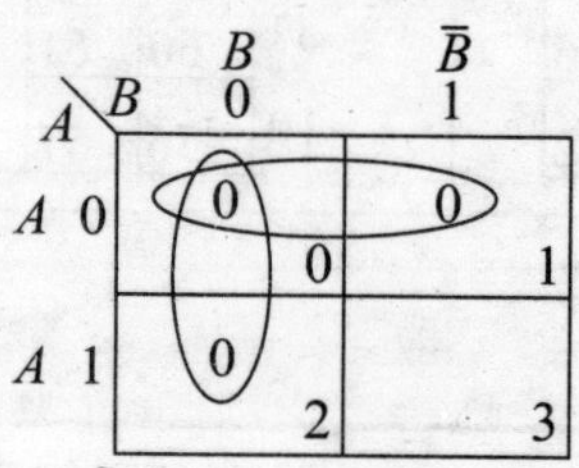

$$f(A, B) = A\ .\ B$$

3-variable *k*-Map

Q.3. Solve the problem using *k*-map:

$$f(A, B, C) = \Sigma m(0, 2, 3, 4, 6, 7)$$

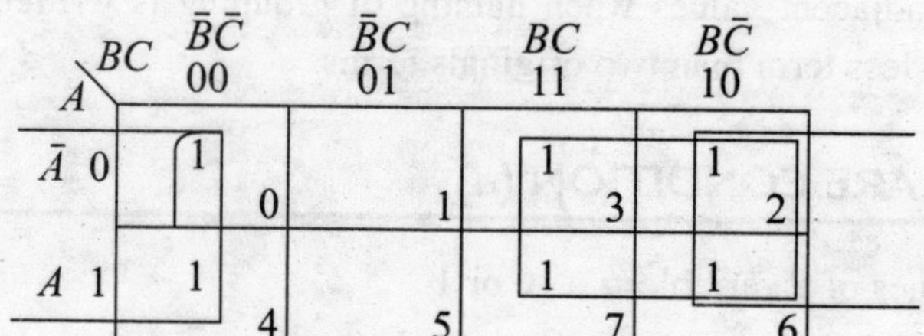

$$f(A, B, C) = \bar{C} + B$$

Q.4. Solve the problem using *k*-map:

$$f(A, B, C) = \pi M(0, 1, 6, 7) + d(2)$$

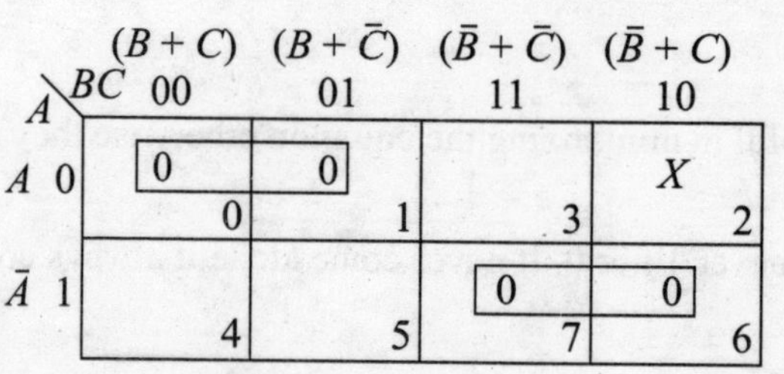

$$f(A, B, C) = (\bar{A} + \bar{B})\ (A + B)$$

Q.5. Solve the problem using *k*-map:

$$f(A, B, C) = \Sigma m(0, 1, 2, 5) + d\ (3, 6, 7)$$

and Realize using gates.

A \ BC	00 $(B+C)$	01 $(B+\bar{C})$	11 $(\bar{B}+\bar{C})$	10 $(\bar{B}+C)$
A 0	0 (0)	1 (1)	X (3)	1 (2)
$\bar{A}$ 1	(4)	1 (5)	X (7)	X (6)

$$f(A, B, C) = (A)\cdot(B+\bar{C})$$

Realization:

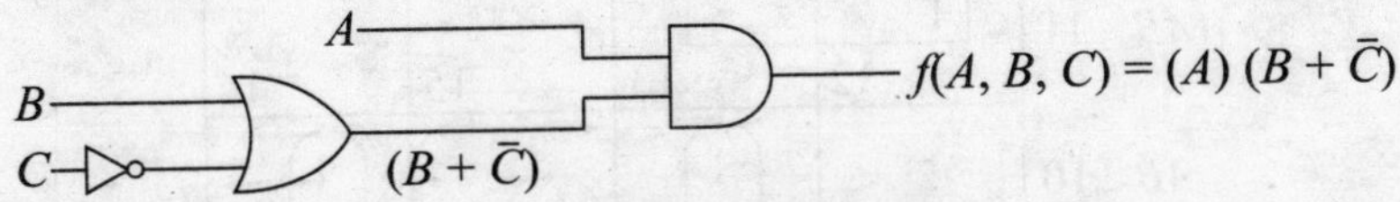

4-variable *k*-map

Q.1. Solve the function using *k*-map:

$$f(A, B, C, D) = \Sigma m(1, 3, 5, 7, 9, 11, 13, 15)$$

AB \ CD	$\bar{C}\bar{D}$ 00	$\bar{C}D$ 01	CD 11	$C\bar{D}$ 10
$\bar{A}\bar{B}$ 00	(0)	1 (1)	1 (3)	(2)
$\bar{A}B$ 01	(4)	1 (5)	1 (7)	(6)
AB 11	(12)	1 (13)	1 (15)	(14)
$A\bar{B}$ 10	(8)	1 (9)	1 (11)	(10)

$$f(A, B, C, D) = D.$$

Q.2. Solve the function using *k*-map:

$$f(A, B, C, D) = \Sigma m(0, 1, 3, 4, 6, 7, 9, 10, 12, 13)$$

AB \ CD	$\bar{C}\bar{D}$ 00	$\bar{C}D$ 01	CD 11	$C\bar{D}$ 10
$\bar{A}\bar{B}$ 00	1 (0)	1 (1)	1 (3)	(2)
$\bar{A}B$ 01	1 (4)	(5)	1 (7)	1 (6)
AB 11	1 (12)	1 (13)	(15)	(14)
$A\bar{B}$ 10	(8)	1 (9)	(11)	1 (10)

$$f(A, B, C, D) = \overline{A}\overline{B}\overline{C} + \overline{A}CD + \overline{A}B\overline{D} + A\overline{C}D + A\overline{B}C\overline{D}$$

Q.3. Simplify the following function using *k*-map:

$$f(A, B, C, D) = \Sigma m\,(0, 1, 6, 9, 10, 12) + d\,(2, 3, 7, 13, 15)$$

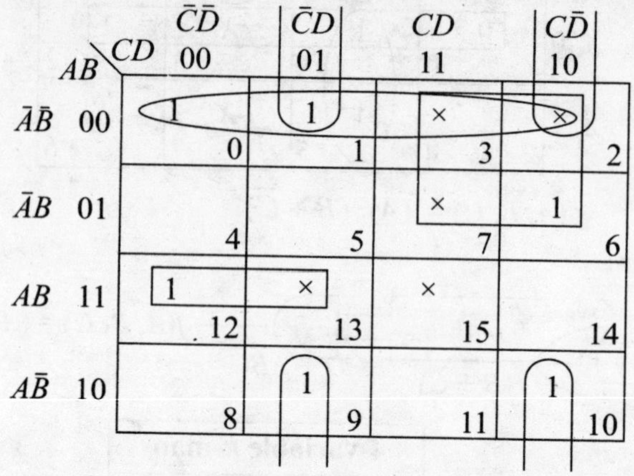

$$f(A, B, C, D) = \bar A\bar B + \bar A C + \bar B\bar C D + AB\bar C + \bar B C\bar D$$

Q.4. Solve the function using *k*-map:

$$f(A, B, C, D) = \pi M\,(0, 1, 4, 5, 6, 7, 12, 14)$$

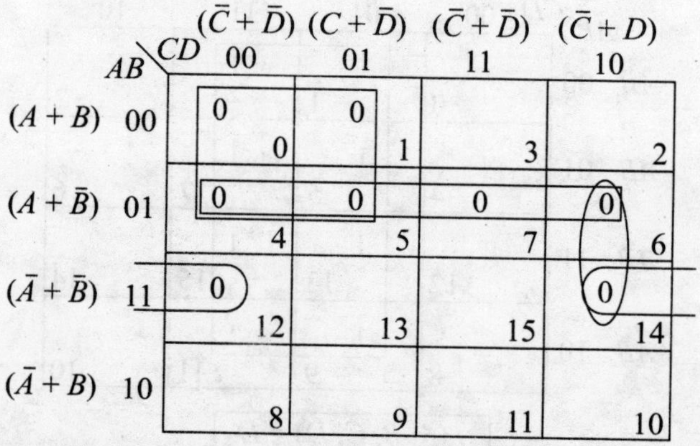

$$f(A, B, C, D) = (A + C)\,(A + \bar B)\,(\bar A + \bar B + D)\,(\bar C + D + \bar B)$$

Q.5. Solve the function using *k*-map and realize using Gates:

$$f(A, B, C, D) = \pi M(0, 1, 3, 6, 7, 12, 14) + d\,(4, 5, 8, 10\ 15)$$

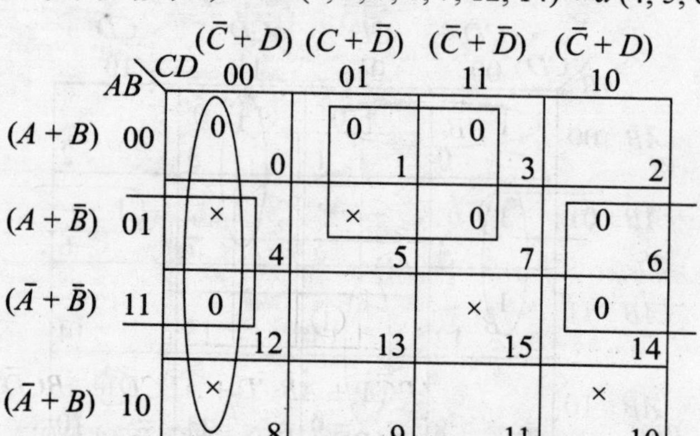

$$f(A, B, C, D) = (A + \bar D)\,(C + D)\,(\bar B + D)$$

$$f(A, B, C, D) = (A + \overline{D})(\overline{B} + D)(C + D)$$

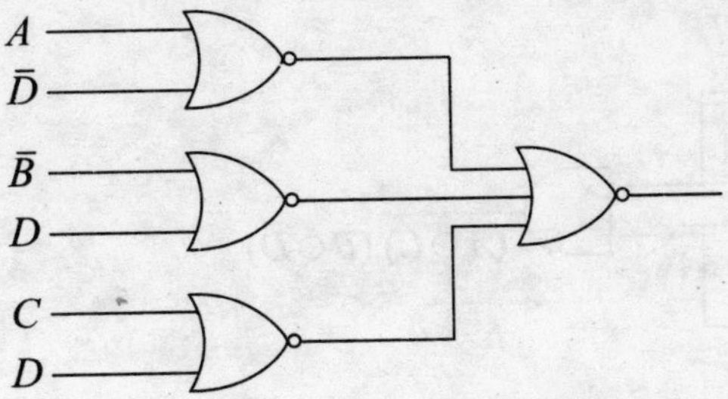

$f(A, B, C, D) = (A + \overline{D})(\overline{B} + D)(C + D)$

Q.6. Realize the expression using gate.

$$f = AB\overline{C}\overline{D} + A\overline{B}\overline{C}\overline{D} + \overline{A}\overline{B}CD + \overline{A}BC\overline{D}$$

AB \ CD	$\overline{C}\overline{D}$ 00	$\overline{C}D$ 01	CD 11	$C\overline{D}$ 10
$\overline{A}\overline{B}$ 00			1	
$\overline{A}B$ 01				1
AB 11	1			
$A\overline{B}$ 10	1			

$$f = A\overline{C}\overline{D} + \overline{A}\overline{B}CD + \overline{A}BC\overline{D}$$

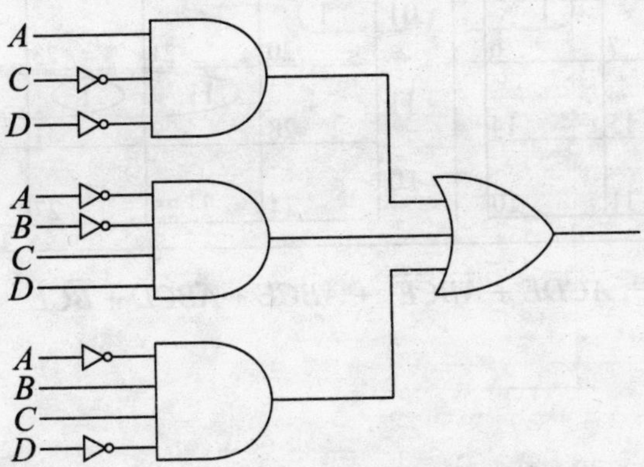

$f = A\overline{C}\overline{D} + \overline{A}\overline{B}CD + \overline{A}BC\overline{D}$

Q.7. Realize the expression using Ex-or gate:

$$f = A \cdot B\overline{C}\overline{D} + A\overline{B}\overline{C}D + \overline{A}\overline{B}CD + \overline{A}BC\overline{D}$$

AB \ CD	$\overline{C}\overline{D}$ 00	$\overline{C}D$ 01	CD 11	$C\overline{D}$ 10
$\overline{A}\overline{B}$ 00			1	
$\overline{A}B$ 01				1
AB 11	1			
$A\overline{B}$ 10		1		

$$\begin{aligned} f &= AB\overline{C}\overline{D} + A\overline{B}\overline{C}D + \overline{A}\overline{B}CD + \overline{A}BC\overline{D} \\ &= A\overline{C}\,(B\overline{D} + \overline{B}D) + \overline{A}C(\overline{B}D + B\overline{D}) \\ &= A\overline{C}\,(B \oplus D) + \overline{A}C\,(B \oplus D) \end{aligned}$$

$$= B \oplus D\,[A\overline{C} + \overline{A}C]$$
$$= (B \oplus D)\,(A \oplus C)$$

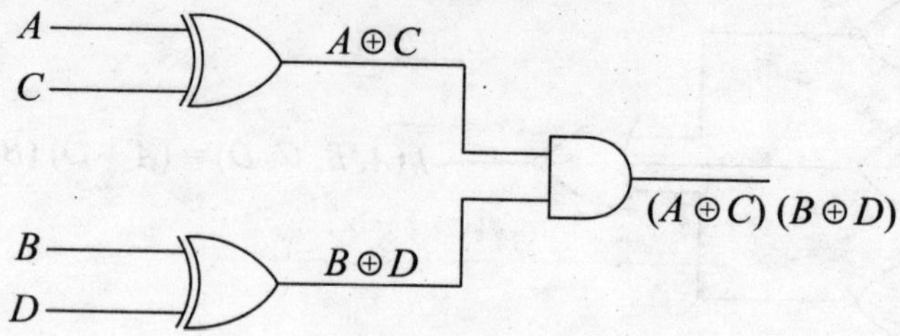

5-Variable *k*-map

Q.1. Reduced the following equation into SOP form and implement it using *N* and gates.

$$f(A, B, C, D, E) = \Sigma m(0, 1, 2, 6, 9, 15, 19, 20, 29, 30, 31)$$

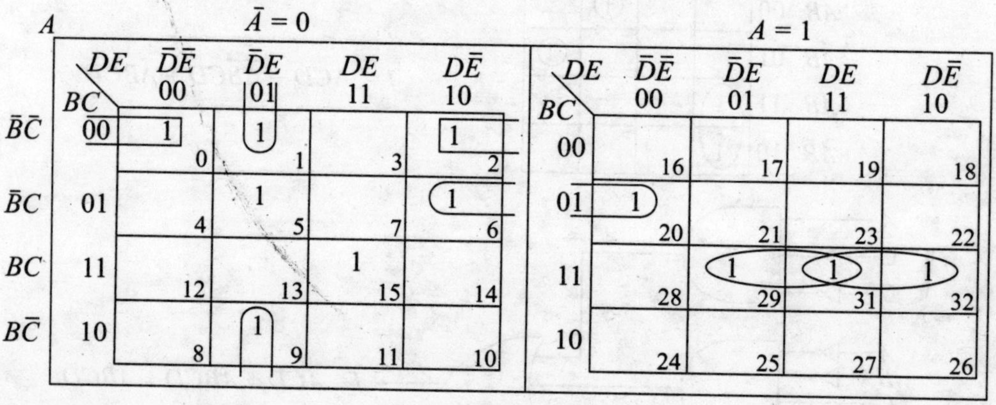

$$f(A, B, C, D, E) = \overline{A}\overline{C}\overline{D}E + \overline{A}\overline{B}\overline{C}\overline{E} + ABCE + ABCD + B\overline{C}\overline{E}$$

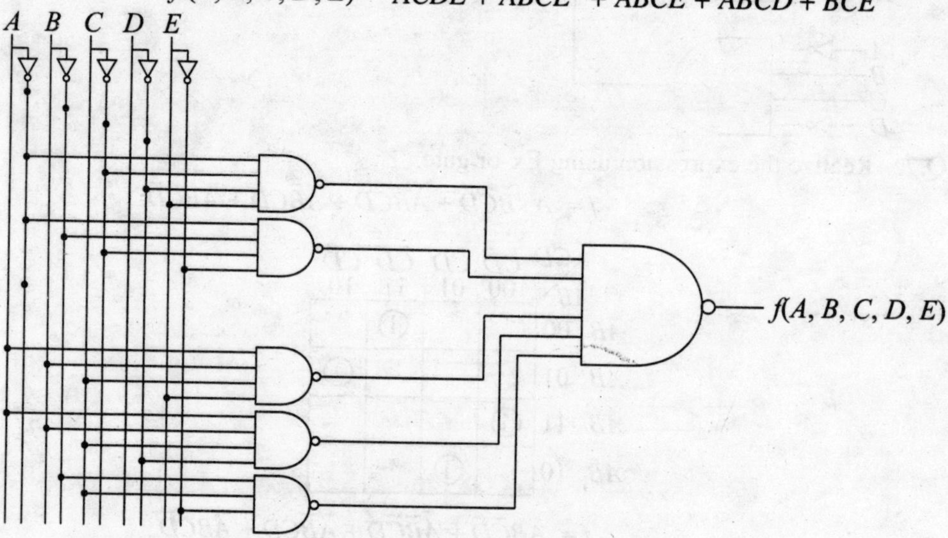

Realization of above function using NAND gate only.

Q2. Solve the given function using *k*-map.

$$y(A, B, C, D, E) = \Sigma m(0, 1, 2, 3, 5, 6, 7, 9, 10, 12, 13, 15, 16, 17, 18, 19, 24, 25, 26, 27, 28, 29, 30, 31)$$

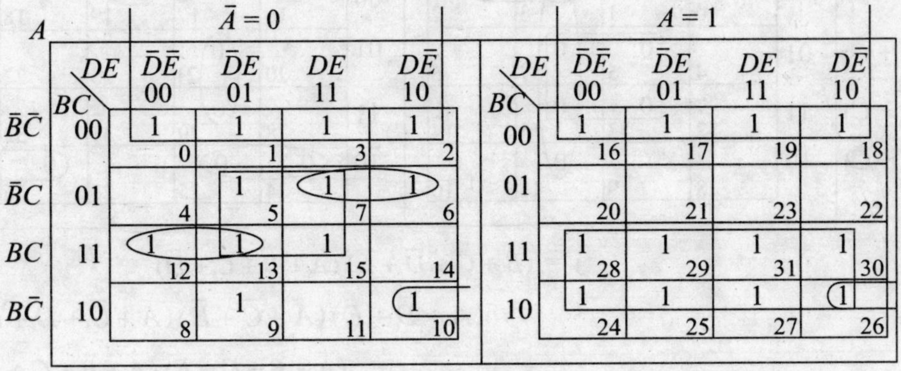

$$y(A, B, C, D, E) = \bar{B}\bar{C} + AB + B\bar{C}D + \bar{A}BC\bar{D} + \bar{A}\bar{B}CD + \bar{A}CE$$

Q3. Reduce the function into SOP and implement it using gate only.

$$y = \Sigma m(0, 2, 6, 8, 10, 12, 14, 17, 18, 21, 29) + d(1, 3, 4, 22, 25, 30$$

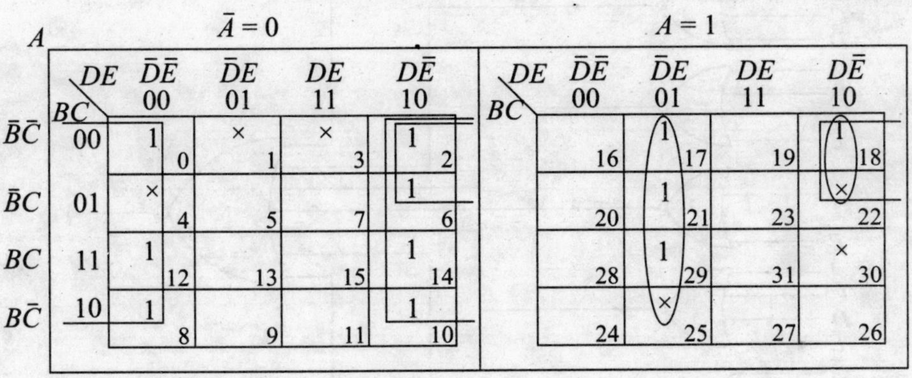

$$y = \bar{A}\bar{E} + A\bar{D}E + \bar{B}D\bar{E}$$

$$= A\bar{D}E + \bar{E}(\bar{A} + \bar{B}D)$$

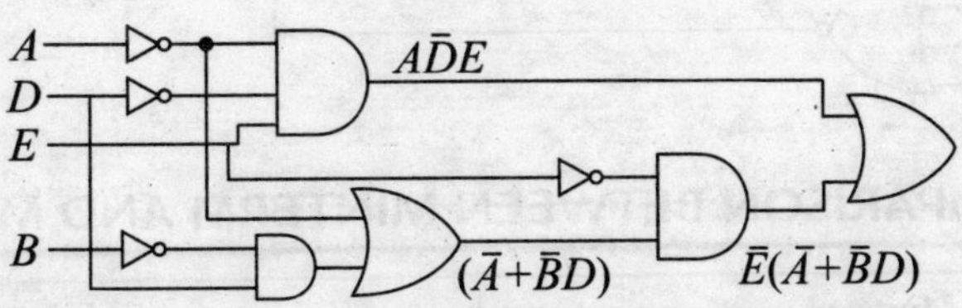

Q4. Reduced the given equation into POS form and implement using NOR gate.

$$y(A, B, C, D, E) = \pi M(2, 3, 5, 7, 11, 13, 15, 19, 21, 24, 25, 26, 29, 30, 31).$$

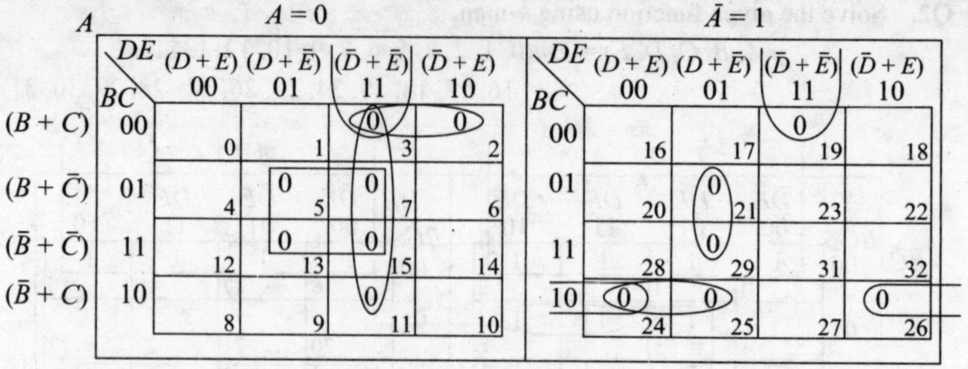

$$y = (B + C + \bar{D} + \bar{E})(A + B + C + \bar{D})$$
$$(A + \bar{D} + \bar{E})(A + \bar{C} + \bar{E})(\bar{A} + \bar{C} + D + \bar{E})$$
$$(\bar{A} + \bar{B} + C + E)(\bar{A} + \bar{B} + C + D)$$

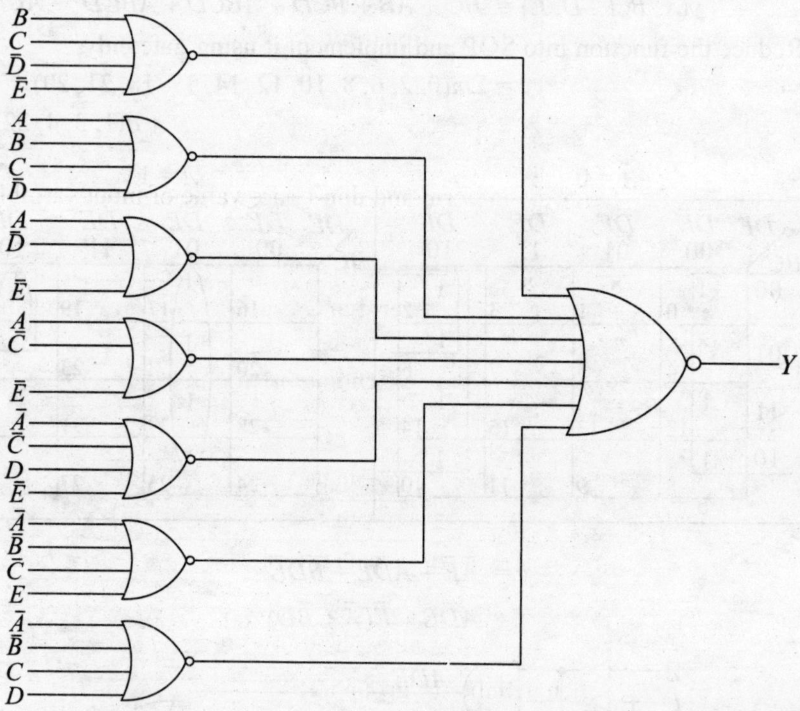

2.5 COMPARISON BETWEEN MINTERM AND MAXTERM

S.No.	Minterm	Maxterm
1.	It is denoted by Σm.	It is denoted by πM.
2.	Output is represented by sum of product (SOP) form.	Output is represented by product of sum (POS) form.
3.	'1' is used for representing the minterm.	'0' is used for representing the maxterm.

4.	It can be easily realized by NAND gate.	It can be easily realized by NOR gate.
5.	Expression is given as: *e.g.*, $f(A, B, C) = \Sigma m(0, 2, 5, 6, 7)$	Expression is given by: *e.g.*, $f(A, B, C) = \pi M(0, 2, 5, 6, 7)$
6.	Output expression can be: $f\overline{A}\overline{B} + A\overline{B}C$	Output expression can be: $f = (A+B)(\overline{A}+B+\overline{C})$

2.6 QUINE- Mc CLUSKY METHOD (TABULAR METHOD)

It is an algorithm method that finds prime implicant, essential prime implicant and minimum SOP expression for digital system with any number of variables.

- **It can be implemented as a computer program.**
- **It makes repeated use of law $\boxed{A+\overline{A}=1.}$ Thus two combining term must differ from each other in only 1 bit position.**
- **It is useful for minimizing functions with more than 6 variables means solving the problem of *k*-map.**

2.6.1 Steps in Q-M Method

Step 1: Represent the minterm/maxterm and don't care value of input variable in binary form.

$$1 = 0001, 2 = 0010$$

Step 2: Arrange the minterm/maxterm and don't care in groups according to the number of 1's in their binary equivalent form in the form of a table.

Important: They must be listed in ascending order.

Index	***Group***
0	Group with no 1's means all 0's.
1	Group with single '1'
2	Group with two "1"
3	Group with three "1"
4	Group with all "1"

Step 3: Compare each member of group to each member of higher succeeding index group. Check that there should be change in only 1 bit like 0000 and 0001, in this LSB are changed and other bits remain same.

Then mark (✓) to both minterm.

Place (–) where bit position are changed.

Index	***Minterm***	***Variable ABCD***	***Group***	***ABCD***	***Check***
0	0	0000	0,1	000–	✓
1	1	0001			

Step 4: Again compare the term generated in step (3) in same style *i.e.*, compare the two terms those have 2 dashes (–) are at same position and differ by single 1 only.

Step 5: Continue the above process till no further combinations are possible, the rules given in step (4).

Step 6: Name the items in columns, which have been left unchecked. Write 0 = complement, 1 = True term. So, this is prime implicant term.

Step 7: Prepare the Q-M reduction chart.

Important: Don't care minterm should not be included in the chart.

Step 8: Examine Q-M Reduction chart to find essential prime implicant:

Look at column of each minterm. If it is covered in only one group, then that group constitute an EPI (essential Prime implicant) means that is final result.

And take other minterms to cover all minterms.

Example 1: *Simplify the Boolean function using Q-M method.*

$$f(A, B, C, D) = \Sigma m(0, 2, 3, 4, 5, 7, 14, 15)$$

Binary Representation = (0000, 0010, 0011, 0100, 0101, 0111, 1110, 1111)

Solution: Convert minterm into binary representation form and arrange according to ascending order.

Table 2.1

Index	*Minterm*	*Variable ABCD*	*Check*
0	0	0 0 0 0	✓
1	2	0 0 1 0	✓
	4	0 1 0 0	✓
2	3	0 0 1 1	✓
	5	0 1 0 1	✓
3	7	0 1 1 1	✓
	14	1 1 1 0	✓
4	15	1 1 1 1	✓

Table 2.2

Now compare each term with lower index.

Index	*Group*	*Variable ABCD*	*Check*
0	(0, 2)	0 0 – 0	✓
	(0, 4)	0 – 0 0	✓
1	(2, 3)	0 0 1 –	✓
	(4, 5)	0 1 0 –	✓
2	(3, 7)	0 – 1 1	✓
	(5, 7)	0 1 – 1	✓
3	(7, 15)	– 1 1 1	✓
	(14, 15)	1 1 1 –	✓

Ignore those terms which have more than one bit change.

Now for table 2.2 compare each term to next lower index.

Now repeat again Table 2.3 step. Keeping mind that dash in two adjacent groups are on same position and there will be change in only one bit.

Table 2.3

Index	*Group*	*Variable* *A B C D*
	No More Grouping	

write all prime implicant as:

$$f = \bar{A}\bar{B}\bar{D} + \bar{A}\bar{C}\bar{D} + \bar{A}\bar{B}C + \bar{A}B\bar{C} + \bar{A}CD + \bar{A}BD + BCD + ABC$$

Now Q-M reduction chart to find essntial prime Implicant.

Minterm → / *Prime Implicant* ↓	*0*	*2*	*3*	*4*	*5*	*7*	*14*	*15*
$\bar{A}\bar{B}\bar{D}$	×	×						
$\bar{A}\bar{C}\bar{D}$	×			×				
$\bar{A}\bar{B}C$		×	×					
$\bar{A}B\bar{C}$				×	×			
$\bar{A}CD$			×			×		
$\bar{A}BD$					×	×		
BCD						×		×
ABC							⊗	×

Result: $y = ABC$

This is essential prime implicant but not complete solution, so, we take more values to cover all minterms.

$$y = ABC + \bar{A}\bar{B}\bar{D} + \bar{A}B\bar{C} + \bar{A}CD$$

Example 2: *Simplified the given function using Q-M method and realize using gate.*

$$y = \Sigma m(0, 2, 3, 4, 7, 9, 14) + d(1, 10, 11)$$

Solution:

Table 2.4

Index	*Minterm*	*Variable* *A B C D*	*Check*
0	0	0 0 0 0	✓
1	1*	0 0 0 1	✓
	2	0 0 1 0	✓
	4	0 1 0 0	✓
2	3	0 0 1 1	✓
	9	1 0 0 1	✓
	10*	1 0 1 0	✓
3	7	0 1 1 1	✓
	11*	1 0 1 1	✓
	14	1 1 1 0	✓

These are binary representation of all the minterms

Table 2.5

Index	*Group*	*Variable* *A B C D*	*Check*
0	(0, 1*)	0 0 0 –	✓
	(0, 2)	0 0 – 0	
	(0, 4)	0 – 0 0	✓
1	(1*3)	0 0 – 1	✓
	(1*, 9)	– 0 0 1	✓
	(2, 3)	0 0 1 –	✓
	(2, 10*)	– 0 1 0	✓
2	(3, 7)	0 – 1 1	✓
	(3, 11*)	– 0 1 1	✓
	(9, 11*)	1 0 – 1	✓
	(10*, 11*)	1 0 1 –	✓
	(10*, 14)	1 – 1 0	✓

Table 2.6: Now again Repeat the Table (2.5) steps

Index	*Group*	*Variable* *A B C D*	*Check*
0	(0, 1*, 2, 3)	0 0 – –	✓
	(0, 2, 1*, 3)	0 0 – –	✓
1	(1*,3, 9, 11*)	– 0 – 1	✓
	(1*, 9, 3, 11*)	– 0 – 1	✓
	(2, 3, 10*, 11*)	– 0 1 –	✓
	(2, 10*, 3, 11*)	– 0 1 –	✓

Prime Implicant:

$$y = \bar{A}\bar{B} + \bar{B}D + \bar{B}C$$

These are common terms in groop. Now make *Q-M* reduction.

Minterm → / *Prime Implicant* ↓	*0*	*2*	*3*	*4*	*7*	*9*	*14*
$\bar{A}\bar{B}$	⊗	×	×				
$\bar{B}D$			×			⊗	
$\bar{B}C$		×	×				

We do not write do not care value in *Q-M* Reduction chart.

Essential Prime Implicant is:

$$y = \bar{A}\bar{B} + \bar{B}D$$

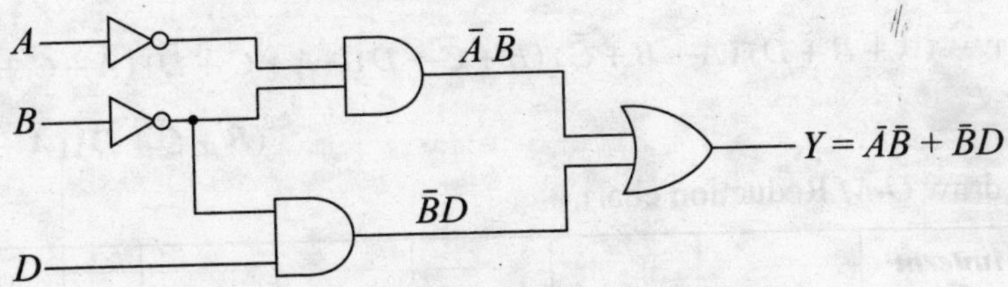

Realization of output expression is using Basic Gates.

Example 3: *Solve the given function using Q-M*

$$y = \pi M(0, 2, 3, 7, 10, 14, 15)$$

Solution:

Table 2.7

Index	*Minterm*	*Variable A B C D*	*Check*
0	0	0 0 0 0	✓
1	2	0 0 1 0	✓
2	3	0 0 1 1	✓
	10	1 0 1 0	✓
3	7	0 1 1 1	✓
	14	1 1 1 0	✓
4	15	1 1 1 1	

Now compare all term to next index:

Table 2.8

Index	*Group*	*Variable A B C D*	*Check*
0	(0, 2)	0 0 – 0	✓
1	(2, 3)	0 0 1 –	✓
	(2, 10)	– 0 1 0	✓

2	(3, 7)	0 – 1 1	✓
	(10, 14)	1 – 1 0	✓
3	(7, 15)	– 1 1 1	✓
	(14, 15)	1 1 1 –	✓

Table 2.9

Index	***Group***	***Variable*** *A B C D*
	No More Grouping	

Prime implicant are:

$$y = (A + B + D)\,(A + B + \overline{C})\,(B + \overline{C} + D)\;(A + \overline{C} + \overline{D})\,(\overline{A} + \overline{C} + D)$$

$$(\overline{B} + \overline{C} + \overline{D})\,(\overline{A} + \overline{B} + \overline{C})$$

Now draw *Q-M* Reduction chart.

***Prime Implicant* ↓ / *Minterm* →**	***0***	***2***	***3***	***7***	***10***	***14***	***15***
$(A + B + D)$	⊗	×					
$(A + B + \overline{C})$		×	×				
$B + \overline{C} + D$		×			×		
$(A + \overline{C} + \overline{D})$			×	×			
$(\overline{A} + \overline{C} + D)$					×	×	
$(\overline{B} + \overline{C} + \overline{D})$				×			×
$(\overline{A} + \overline{B} + \overline{C})$						×	×

Essential Prime Implicant:

$$y = (A + B + D)$$

Not final result, so, some more value to cover max term:

$$X = (A + B + D)\,(A + \overline{C} + \overline{D})\;(\overline{A} + \overline{C} + D)\,(\overline{A} + \overline{B} + \overline{C})$$

Example 4: *Design a parity generator to generate an ODD parity bit for a 4-bit word, use Ex-or and Ex-NOR gate.*

Solution: Truth table for 4-bit parity generator is:

4 bit word	*ODD Parity Bit Bit*	*Even Parity Bit Bit*
A B C D	P_0	P_E
0 0 0 0	1	0
0 0 0 1	0	1
0 0 1 0	0	1
0 0 1 1	1	0
0 1 0 0	0	1
0 1 0 1	1	0
0 1 1 0	1	0
0 1 1 1	0	1
1 0 0 0	0	1
1 0 0 1	1	0
1 0 1 0	1	0
1 0 1 1	0	1
1 1 0 0	1	0
1 1 0 1	0	1
1 1 1 0	0	1
1 1 1 1	1	0

K-map for P_0 and P_E is:

AB \ CD	00	11	11	10
00	(1) 0	1	(1) 3	2
01	4	(1) 5	7	(1) 6
11	(1) 12	13	(1) 15	14
10	8	(1) 9	11	(1) 10

P_O

AB \ CD	00	11	11	10
00	0	(1) 1	3	(1) 2
01	(1) 4	5	(1) 7	6
11	12	(1) 13	15	(1) 14
10	(1) 8	9	(1) 11	10

P_E

$$P_o = (A \oplus C) \odot (B \oplus D)$$

$$P_E = A \oplus B \oplus C \oplus D$$

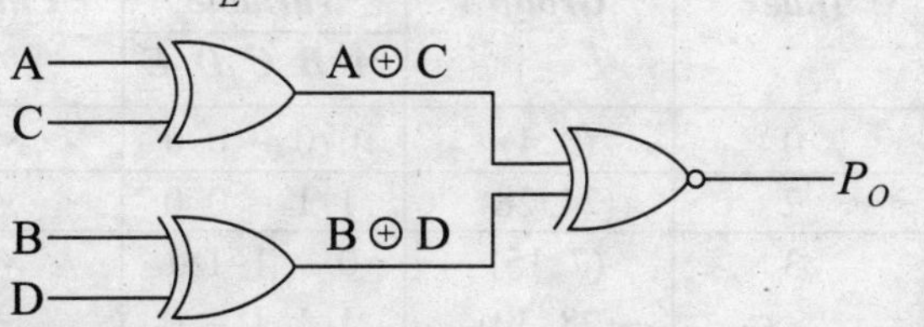

Circuit for ODD Pariy Generator

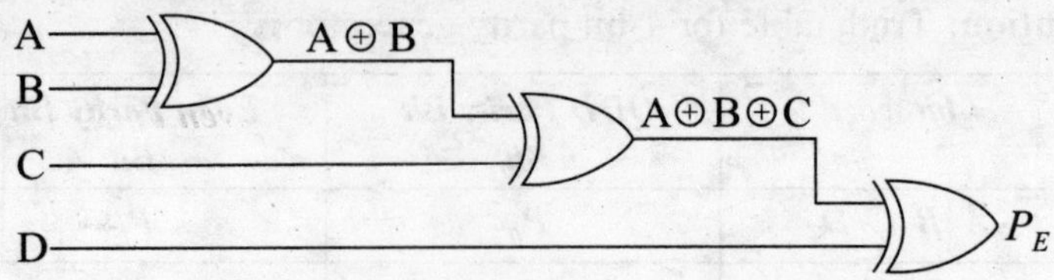

Circuit for Even Parity Generator

Example 5: *Simplify the given function. Using Tabular method and realize it using gate that is NAND.*

$$y = \Sigma m(0, 4, 7, 9, 15, 24, 28, 30, 31)$$

Solution: Binary Representation:

Minterm	*A B C D E*
0	0 0 0 0 0
4	0 0 1 0 0
7	0 0 1 1 1
9	0 1 0 0 1
15	0 1 1 1 1
24	1 1 0 0 0
28	1 1 1 0 0
30	1 1 1 1 0
31	1 1 1 1 1

Now Table 2.10

Index	*Minterm*	*Variable*	*check*
		A B C D E	
0	0	0 0 0 0 0	✓
1	4	0 0 1 0 0	✓
2	9	0 1 0 0 1	✓
	24	1 1 0 0 0	✓
3	7	0 0 1 1 1	✓
	28	1 1 1 0 0	✓
4	15	0 1 1 1 1	✓
	30	1 1 1 1 0	✓
5	31	1 1 1 1 1	✓

Table 2.11

Index	*Group*	*Variable*	*Check*
		A B C D E	
0	(0, 4)	0 0 – 0 0	✓
2	(24, 28)	1 1 – 0 0	✓
3	(7, 15)	0 – 1 1 1	✓
	(28, 30)	1 1 1 – 0	✓
4	(15, 31)	– 1 1 1 1	✓
	(30, 31)	1 1 1 1 –	✓

Table 2.12

Index	*Group*	*Variable*	*Check*
	No More Grouping		

So, prime implicant form Table 2.12

$$y = \overline{A}\overline{B}\overline{D}\overline{E} + AB\overline{D}\overline{E} + \overline{A}CDE + ABC\overline{E} + BCDE + ABCD$$

QM Reduction chart is have to draw for bring out the essential prime implicant for this *y* result.

Q-M Reduction Chart

Minterm → / *Prime Implicant* ↓	*0*	*4*	*7*	*9*	*15*	*24*	*28*	*30*	*31*
$\overline{A}\overline{B}\overline{D}\overline{E}$	⊗	⊗							
$AB\overline{D}\overline{E}$						⊗	×		
$\overline{A}CDE$			⊗		×				
$ABC\overline{E}$							×	×	
$BCDE$					×				×
$ABCD$								×	×

So, essential prime implicant is:

$$y = \overline{A}\overline{B}\overline{D}\overline{E} + AB\overline{D}\overline{E} + \overline{A}CDE.$$

But not final solution so, to cover all minterms the result is:

$$\boxed{y = \overline{A}\overline{B}\overline{D}\overline{E} + AB\overline{D}\overline{E} + \overline{A}CDE + ABCD}$$

So, this is final solution.

Now, we will do realization or designing of output expression using NAND gate only. As we know SOP form are easily designed by NAND gate.

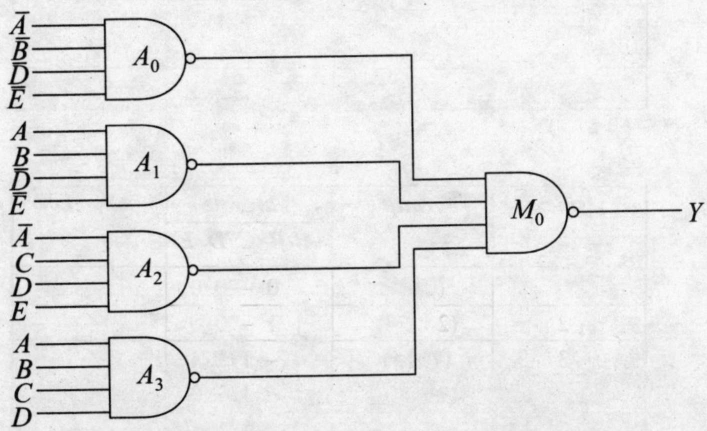

This is designing of output expression using universal gate NAND only.

UNSOLVED QUESTIONS

Q.1. Give truth table for XOR gate and hence derive the function and realize the circuit diagram in SOP and POS form.

Q.2. Simplify these function using *k*-map and Q-M

(*a*) $y = \Sigma m(0, 2, 3, 5, 7)$

(*b*) $y(A, B) = \Sigma m(0, 2, 3)$

(*c*) $y(A, B) = \pi m(0, 3)$

(*d*) $y(A, B, C) = \pi m(0, 1, 4, 6, 7) + d(2, 3)$

(*e*) $y(A, B, C) = \Sigma m(0, 4, 5) + d(2, 6)$

(*f*) $y(A, B, C, D) = \Sigma m(0, 2, 4, 7, 9, 12, 14)$

(*g*) $y(A, B, C, D) = \pi M(1, 3, 5, 6, 11, 13)$

(*h*) $y(A, B, C, D) = \Sigma m(0, 3, 6, 10, 11) + d(2, 7, 9, 15)$

(*i*) $y(A, B, C, D) = \pi M(0, 2, 5, 7, 12) + d(4, 9, 11)$

(*j*) $y(A, B, C, D, E) = \Sigma m(0, 1, 3, 19, 24, 27) + d(18, 25)$

Q.3. Simplify using *k*-map method:

$$y(A, B, C, D) = A\bar{B} + A\bar{B}C + A\bar{B}C\bar{D} + \bar{A}BCD + \bar{A}B\bar{C}$$

Q.4. Simplify using *k*-map method:

$$y(A, B, C, D, E) = \bar{B}C + B\bar{D}E + A\bar{B}CD + AC\bar{D} + BCDE$$

Q.5. Solve it using *k*-map and *Q-M* method and realize output expression using gates:

(*i*) $y = \Sigma m(0, 4, 6, 7, 13, 16, 20, 22, 23, 29)$

(*ii*) $y = \pi M(1, 3, 5, 7, 8, 9, 19, 23, 24, 25)$

Q.6. Solve 6 variable function by Tabular method:

(*a*) $y = \Sigma m(2, 9, 30, 31, 41, 45, 53, 62, 63)$

(*b*) $y = \pi M(0, 4, 22, 23, 30, 32, 36, 54, 63)$

UNIT 3

COMBINATIONAL DESIGN USING MSI DEVICES

Topics are:

- Combinational Circuits
- Adder
- Subtractor
- Multiplexer (Data Selector)
- Demultiplexer (Data Distributor)
- Magnitude Comparator
- Encoder
- Priority Encoder
- Decoder
- BCD to Gray Code Decoder
- BCD to 7-segment Display Devices Decoder
- BCD Arithmetic Element
- Arithmetic Logic Unit (ALU)

CHAPTER 3

Combinational Design Using Msi Devices

Digital system consists of two types of circuits:

1. Combinational circuits.
2. Sequential circuits.

3.1 COMBINATIONAL CIRCUITS

It is a set of logic gates that transform a set of binary inputs to a set of binary outputs where output depends only on the values of inputs not on the state of circuit.

Model of Combinational Circuit:

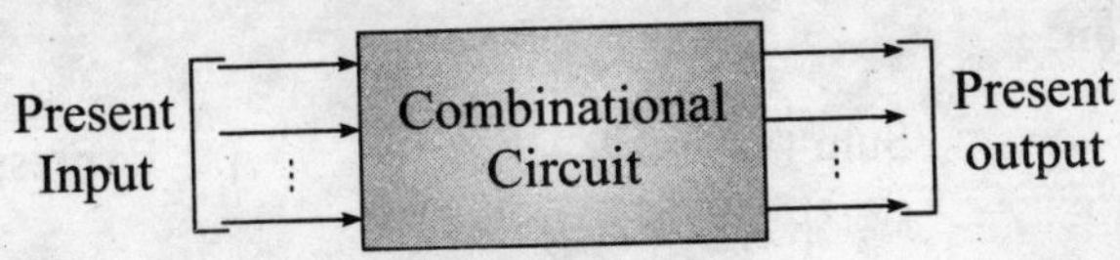

Characteristics of Combinational Circuit:

(*a*) Every output will be function of inputs.

(*b*) For the given same input, the output will always be same.

(*c*) The output is a combination of inputs using basic circuit components such as AND gate, OR gate, Ex-OR gate, etc.

(*d*) The characteristics of combinational circuit is completely defined by truth table that define all possible input combination.

There is no memory element to store the data for future use.

3.1.1 Different Combinational Circuits

1. Adder
2. Subtractor
3. Multiplexer
4. Demultiplexer
5. Comparator
6. Encoder
7. Decoder
8. BCD Arithmetic element

3.2 ADDER

It is basically to add the binary digits two or more than two.

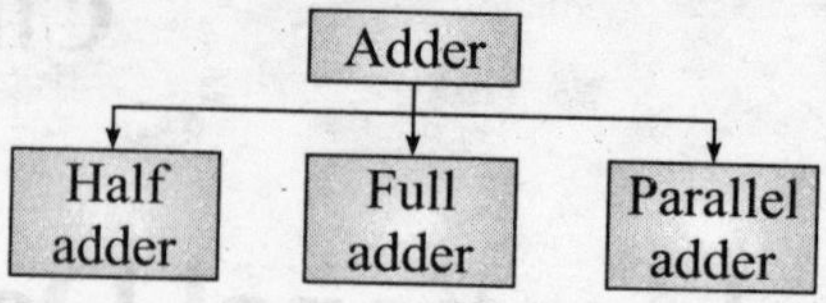

3.2.1 Half Adder

It is one which adds two binary digits simultaneously.

Two Inputs = A, B

Two outputs = Sum (s), Carry (c).

Truth table to add two binary digits.

Truth Table

Inputs		*Outputs*	
A	***B***	***Sum (s)***	***Carry (c)***
0	0	0	0
0	1	1	0
1	0	1	0
1	1	0	1

Now, write expression for sum and carry.

Sum Expression:

$$\text{Sum } (s) = \overline{A}B + A\overline{B} \qquad [\because \text{ Expression of Ex-OR gate}]$$

$$\boxed{S = A \oplus B}$$

Carry Expression:

$$\text{Carry } (c) = AB$$

$$\boxed{C = A \cdot B} \qquad [\because \text{ Expression of OR gate}]$$

Simple Representation of HALF Adder:

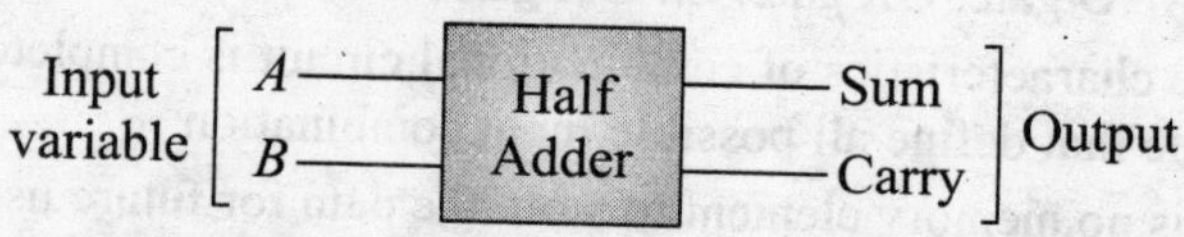

Circuit Diagram of Half Adder:

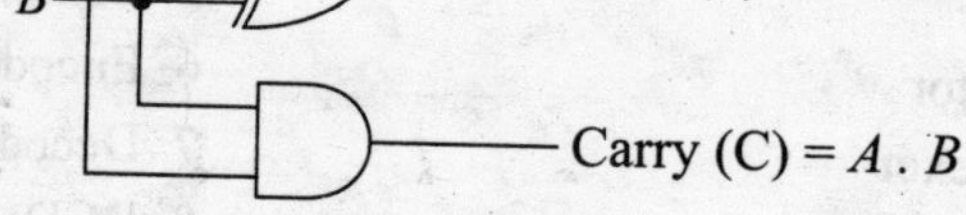

3.2.2 Full Adder

There is addition of 3 bits and these are taken as input variables.

Three Inputs = A, B, C_p

where C_p = Previous carry

Outputs = Sum (s), Carry (c)

Truth Table:

Input Variable			*Output*	
A	***B***	***C_p***	***S***	***C***
0	0	0	0	0
0	0	1	1	0
0	1	0	1	0
0	1	1	0	1
1	0	0	1	0
1	0	1	0	1
1	1	0	0	1
1	1	1	1	1

Expression for Sum (s):

$$\text{Sum } (s) = \bar{A}\,\bar{B}\,C_p + \bar{A}\,B\,\bar{C}_p + A\,\bar{B}\,\bar{C}_p + A\,B\,C_p$$

$$= C_p\ (AB + \bar{A}\,\bar{B}) + \bar{C}_p\ (\bar{A}\,B + A\,\bar{B})$$

$$= C_p\ (\overline{A \oplus B}) + \bar{C}_p\ (A \oplus B)$$

$$\boxed{\text{Sum} = C_p \oplus A \oplus B}$$

Expression for Carry:

$$\text{Carry } (c) = \bar{A}\,BC_p + A\,\bar{B}\,C_p + A\,B\,\bar{C}_p + A\,B\,C_p$$

$$= C_p\ (\bar{A}\,B + A\,\bar{B}) + AB\ (C_p + \bar{C}_p)$$

$$= C_p (A \oplus B) + AB \qquad [\because\ C_p + \bar{C}_p = 1]$$

$$\boxed{\text{Carry} = AB + C_p (A \oplus B)}$$

Circuit Diagram for full adder:

We will design a full adder using two Half adder.

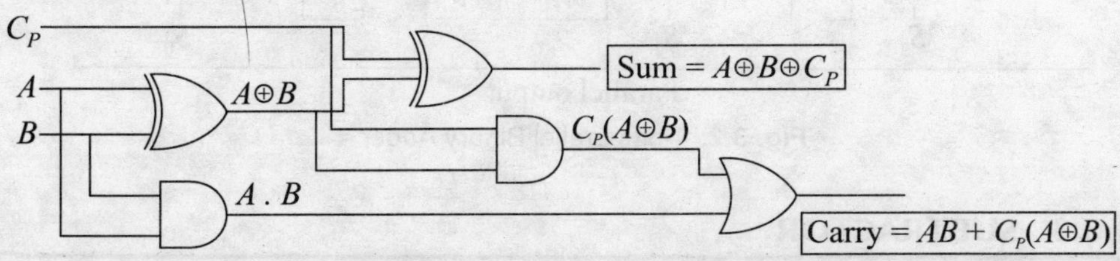

Fig. 3.1: Full adder using 2 Half adder

3.2.3 Parallel Binary Adder

A parallel adder is used to add two numbers in parallel form and to produce the sum-bits as parallel output.

A block diagram of 4-bit parallel adder is capable of adding two 4-bit number. *i.e.* $A_3 A_2 A_1 A_0$ and $B_3 B_2 B_1 B_0$ is shown in figure and resulting output bits are $S_3 S_2 S_1 S_0$ and C_{out} as carry bit.

$$\begin{array}{ccccc l} & C_3 & C_2 & C_1 & C_0 & \leftarrow \text{Carry} \\ & & A_3 & A_2 & A_1 & A_0 \\ & & B_3 & B_2 & B_1 & B_0 \\ \hline S_4 & S_3 & S_2 & S_1 & S_0 & \leftarrow \text{Sum} \end{array}$$

Example: $A_3 A_2 A_1 A_0 = 1011$
$B_3 B_2 B_1 B_0 = 1011$

A_0	B_0	C_{in}		S_0	C_{out}	
1 +	1 +	0	=	0	1	Least significant stage.
A_1	B_1	C_{in}		S_1	C_{out}	
1 +	1 +	1	=	1	1	
A_2	B_2	C_{in}		S_2	C_{out}	
0 +	0 +	1	=	1	0	
A_3	B_3	C_{in}		S_3	C_{out}	
1 +	1 +	0	=	0	1	Most significant bit

$S_4 \rightarrow$ Carry out

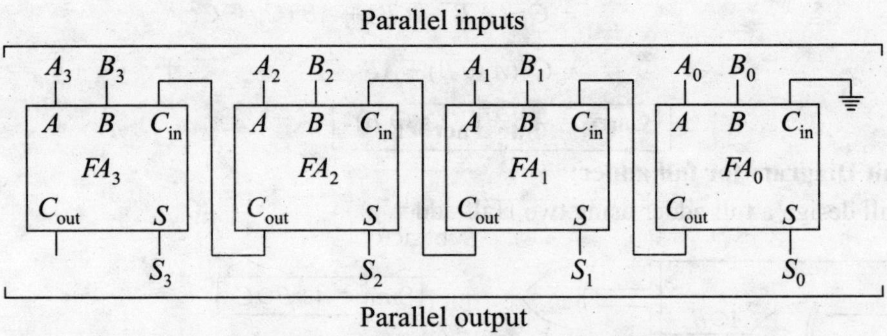

Fig. 3.2: 4-bit parallel Binary Adder

3.3 SUBTRACTOR

This is basically to perform the subtraction of bits.

The digit from which other digit is subtracted is called minuend.

The digit which is to be subtracted is called subtrahend. When minuend and subtrahend contains more significant digits, then borrow obtained from the subtraction of two bits is subtracted from next higher order pair significant bit.

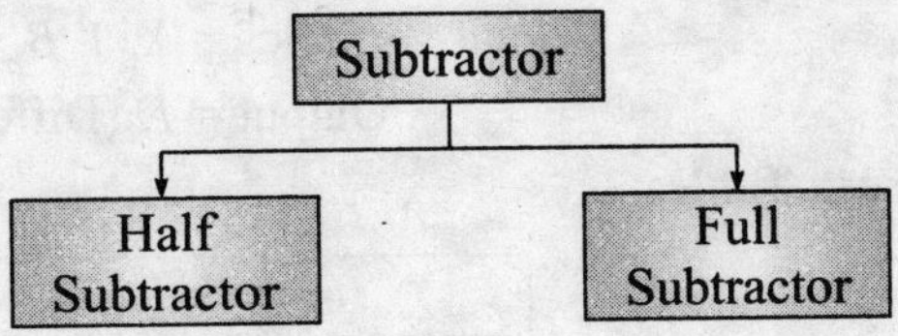

3.3.1 Half Subtractor

In half subtractor, the subtraction of two bits are taking place and two outputs are produced.

Two Inputs = X, Y

Two Outputs = Difference (D), Borrow (B).

Truth Table:

Input Variable		*Output*	
X	Y	D	B
0	0	0	0
0	1	(1)	(1)
1	0	(1)	0
1	1	0	0

Output Expression:

Expression for Difference (D):

$$D = X\bar{Y} + \bar{X}Y$$

$$\boxed{D = X \oplus Y} \qquad [\because X\bar{Y} + \bar{X}Y = Ex-OR]$$

Expression for Borrow (B):

$$\boxed{B = \bar{X}Y}$$

Simple Representation of Half Subtractor:

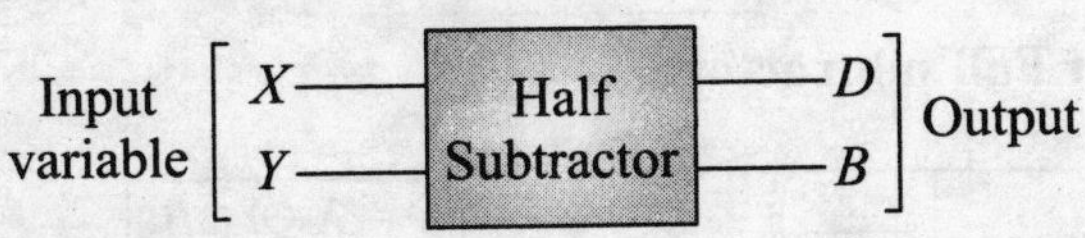

Circuit Diagram: Difference expression is similar to sum in Half adder but borrow expression is different that is one more invert/not gate is used in Half subtractor.

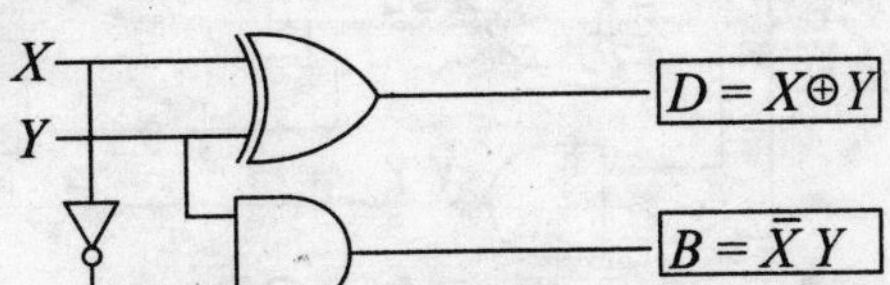

3.3.2 Full Subtractor

In full subtractor, the subtraction of 3 bits is taking place *i.e.* minuend, subtrahend and borrow are 3 bits and two outputs difference and borrow is produced.

Input variables = X, Y, B_p (Previous Borrow)

Output = D (Difference), B (Borrow)

Truth Table:

Input Variables			*Output*	
X	***Y***	***B_p***	***D***	***B***
0	0	0	0	0
0	0	1	1	1
0	1	0	1	1
0	1	1	0	1
1	0	0	1	0
1	0	1	0	0
1	1	0	0	0
1	1	1	1	1

Expression for Difference (*D*):

$$D = \overline{X}\,\overline{Y}\,B_p + \overline{X}\,y\,\overline{B}_p + X\,\overline{Y}\,\overline{B}_p + XY\,B_p$$

$$= B_p\,(X\,Y + \overline{X}\,\overline{Y}) + \overline{B}_p\,(X\,\overline{Y} + \overline{X}\,Y)$$

$$= B_p\,\overline{(X \oplus Y)} + \overline{B}_p\,(X \oplus Y)$$

$$D = X \oplus Y \oplus B_p \qquad [\because \text{ Same as sum of full adder}]$$

Expression for Borrow:

$$B = \overline{X}\,\overline{Y}\,B_p + \overline{X}\,Y\,\overline{B}_p + \overline{X}\,Y\,B_p + X\,Y\,B_p$$

$$= \overline{X}\,Y\,(B_p + \overline{B}_p) + \overline{X}\,\overline{Y}\,B_p + X\,Y\,B_p$$

$$= \overline{X}\,Y\,(1) + \overline{X}\,\overline{Y}\,B_p + X\,Y\,B_p \qquad [B_p + \overline{B}_p = 1]$$

$$= \overline{X}\,Y + B_p\,(XY + \overline{X}\,\overline{Y})$$

$$\boxed{B = \overline{X}\,Y + B_p\,\overline{(X \oplus Y)}}$$

Circuit Diagram for Full subtractor:

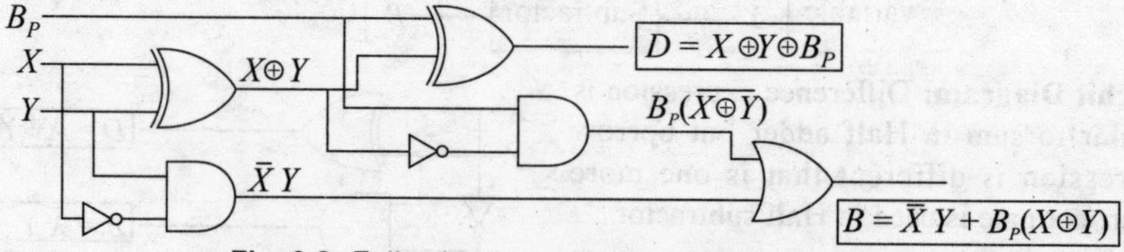

Fig. 3.3: Full subtractor using two Half Subtractor

3.4 MULTIPLEXER (DATA SELECTOR)

Multiplexer is also known as data selector, is a logic circuit allows the digital information from multi-inputs to a single output line. The selection of input data to be routed to output line is done by select terminals. The number of select terminals depends on number of input lines to be routed to output line, given by formula:

$$N = 2^k$$

where N = Number of input line.

K = Number of select terminals.

Multiplex means "many into one". Digital multiplex is a combinational circuit that selects binary information from one of many input channels and transmits to a single output line.

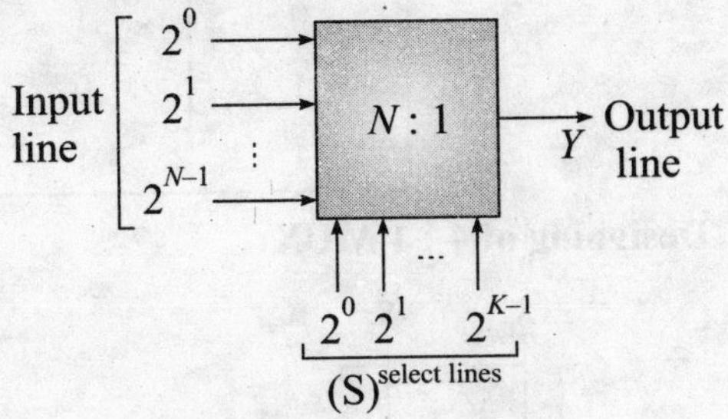

Different types of Multiplexer on basis of Input and select line:

Number of Input (Input) N : 1 (output)	Number of select line S (Select line)
1 : 1	0
2 : 1	1
4 : 1	2
8 : 1	3
16 : 1	4
32 : 1	5
64 : 1	6
128 : 1	7

Example: When $K = 2$ then input (N) 4

4 : 1 Multiplexer: A 4 to 1 line multiplexer is defined as the multiplexer consisting of four input channels and information of one of the channels can be selected and transmitted to output line according to select lines/inputs combinations. Selection of 4 inputs is possible with the help of two select lines.

4 Input lines are = X_0, X_1, X_2, X_3

2 Select lines are = S_0, S_1

Output = Y

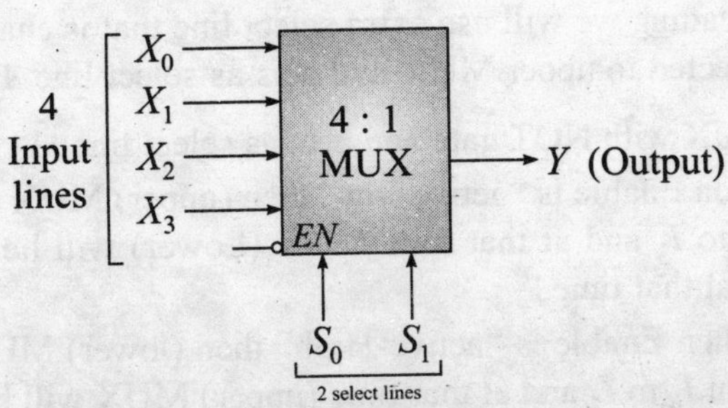

E_N = Enable or strobe pin is used for expand two or more multiplexer. It is always 'active-low'

Truth Table:

Select lines		Inputs	Output (Y)
S_0	S_1		
0	0	X_0	$X_0\,\bar{S}_0\,\bar{S}_1$
0	1	X_1	$X_1\,\bar{S}_0\,S_1$
1	0	X_2	$X_2\,S_0\,\bar{S}_1$
1	1	X_3	$X_3\,S_0\,S_1$

Designing of 4 : 1 MUX:

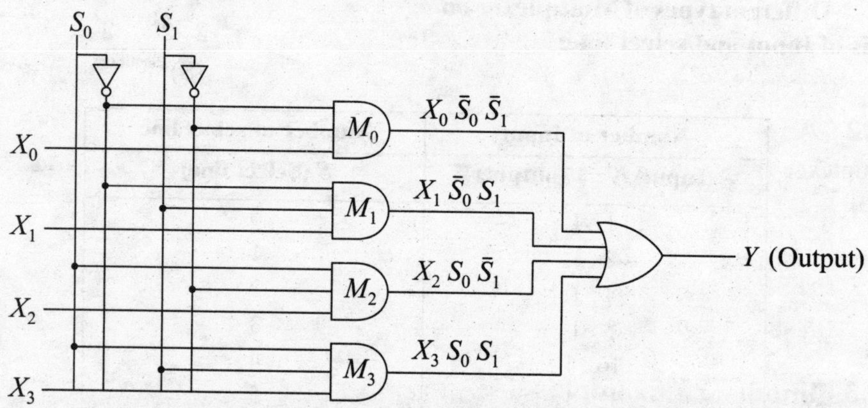

3.4.1 Cascading of Multiplexer

Cascading means combining two or more multiplexer together in order to have large number of inputs.

Example: Combining of two 16 : 1 multiplexer together in order to perform 32 : 1 multiplexer operation.

So, for the cascading of multiplexer additional select line, *i.e.* 'Enable' or 'strobe' pin is used.

Example: A 8 : 1 MUX is realized by two 4 : 1 line Multiplexer.

Operation: 4 : 1 MUX have two select line but 8 : 1 MUX having 3 select lines, so, to do cascading we will use extra select line that is enable pin or strobe pin. Enable pin is connected to upper MUX and acts as select line A and Enable pin connected to lower MUX with NOT gate and acts as select line $\bar{A}$.

Case I: When Enable is "active-low" then (upper) MUX will work and selects the inputs from I_0 to I_3 and at that time MUX (Lower) will be deactivated because Enable pin is high at that time.

Case II: When Enable is "active-High" then (lower) MUX will work and selects the inputs from I_4 to I_7 and at that time (upper) MUX will be deactivated.

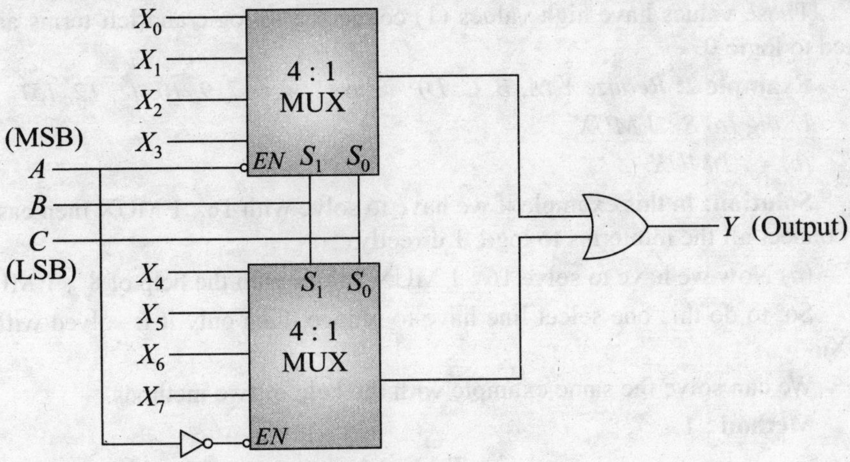

3.4.2 Applications of Multiplexer

Multiplexer have a variety of applications, in addition to their main use as data selectors.

Some other uses:

(*i*) Combination Logic Expression Generator.

(*ii*) Sequence Generator.

(*iii*) Parallel to serial Generator.

3.4.3 Important Example of Multiplexer

Example 1: *Use 8 : 1 MUX to realize following* function: $Y(A, B, C) = \Sigma m\ (0, 2, 4, 5, 7)$

Solution: Truth table of function:

A	*B*	*C*	*Minterm*	*Output (x)*
0	0	0	0	1
0	0	1	1	0
0	1	0	2	1
0	1	1	3	0
1	0	0	4	1
1	0	1	5	1
1	1	0	6	0
1	1	1	7	1

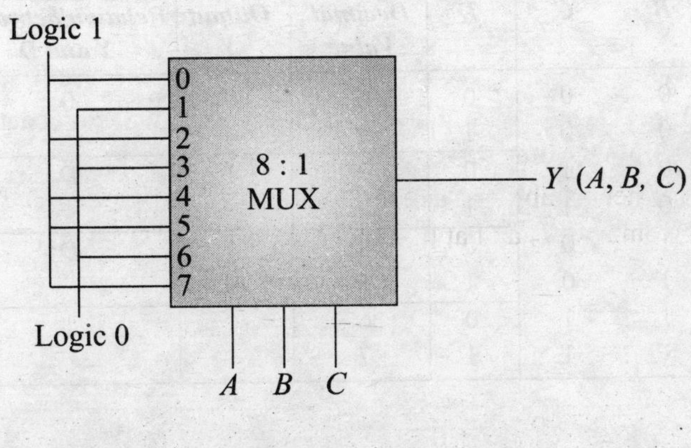

Those values have high values (1) connected to logic/and left terms are connected to logic 0.

Example 2: *Realize Y (A, B, C, D) = Σm (2, 4, 6, 7, 9, 10, 11, 12, 15)*
Using (a) 8 : 1 MUX
(b) 4 : 1 MUX.

Solution: In this example if we have to solve with 16 : 1 MUX then easily we can connect all the minterms to logic 1 directly.

(*a*) Now we have to solve 16 : 1 MUX values with the help of 8 : 1 MUX.

So, to do this one select line have to reduce, then only it is solved with 8 : 1 MUX.

We can solve the same example with the help of two methods.

Method : 1

Table 3.1

A	*B*	*C*	*D*	*Decimal Values*	*Y (output)*
0	0	0	0	0	0
0	0	0	1	1	0
0	0	1	0	2	1
0	0	1	1	3	0
0	1	0	0	4	1
0	1	0	1	5	0
0	1	1	0	6	1
0	1	1	1	7	1
1	0	0	0	8	0
1	0	0	1	9	1
1	0	1	0	10	1
1	0	1	1	11	1
1	1	0	0	12	1
1	1	0	1	13	0
1	1	1	0	14	0
1	1	1	1	15	1

Table 3.2

Now check the relation between *Y* and *D* to reduce *D* and make group of two values.

A	*B*	*C*	*D*	*Decimal Value*	*Output Y*	*Relation between Y and D*	*Input*
0	0	0	0	0	0	0	0
0	0	0	1	1	0		
0	0	1	0	2	1	$\overline{D}$	1
0	0	1	1	3	0		
0	1	0	0	4	1	$\overline{D}$	2
0	1	0	1	5	0		
0	1	1	0	6	1	1	3
0	1	1	1	7	1		

1	0	0	0	8	0	D	4
1	0	0	1	9	1		
1	0	1	0	10	1	1	5
1	0	1	1	11	1		
1	1	0	0	12	1	$\overline{D}$	6
1	1	0	1	13	0		
1	1	1	0	14	0	D	7
1	1	1	1	15	1		

Now 3 select lines are there.

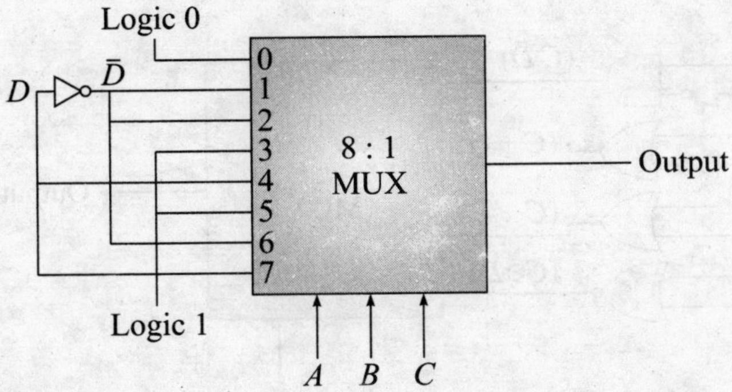

Method 2:

Input lines	*0*	*1*	*2*	*3*	*4*	*5*	*6*	*7*
$\overline{D}$	0	(2)	(4)	(6)	8	(10)	(12)	14
D	1	3	5	(7)	(9)	(11)	13	(15)
	0	$\overline{D}$	$\overline{D}$	1	D	1	$\overline{D}$	D

In this method we have to reduce one select line then if values are given in $\overline{D}$ then result is $\overline{D}$ if values is $(D + \overline{D})$ then 1

Now, again draw the same MUX.

(*b*) Now solve 16 : 1 MUX using 4 : 1 MUX, so, to do this minimize two select lines and make a relation between Y and C, D.

Method 1: Make group of four.

A	*B*	*C*	*D*	**Input line**	***Y* output**	**Relation between *Y* and C, *D***	**Input**
0	0	0	0	0	0	$Y = C\overline{D}$	0
0	0	0	1	1	0		
0	0	1	0	2	1		
0	0	1	1	3	0		
0	1	0	0	4	1	$Y = C + \overline{D}$	1
0	1	0	1	5	0		
0	1	1	0	6	1		
0	1	1	1	7	1		

1	0	0	0	8	0		
1	0	0	1	9	1	$Y = C + D$	2
1	0	1	0	10	1		
1	0	1	1	11	1		
1	1	0	0	12	1		
1	1	0	1	13	0	$Y = \overline{C \oplus D}$	3
1	1	1	0	14	0		
1	1	1	1	15	1		

Solve the relaiton between Y and C, D using K-Map.

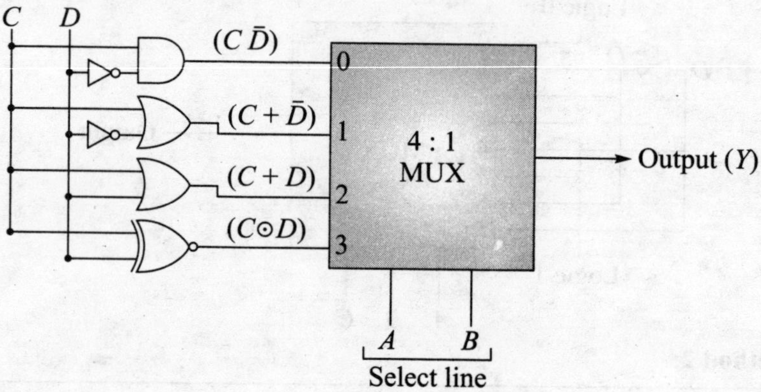

3.5 DEMULTIPLEXER (DATA DISTRIBUTOR)

Demultiplex means "one to many." Its function is to receive information an a single line and transmit this information to any one of 2^N possible output lines. Enable or strobe pin is used for cascading. The particular output line is selected by N-bit input address.

It is a process that receives information from one channel and distributes the data over several channels. So, that's why it is called *Data Distributor.*

It is reverse operation of Multiplexer.

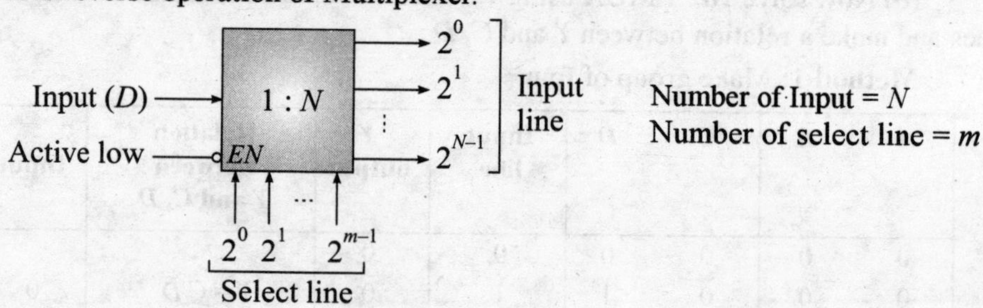

Formula is:

$$\boxed{1 : (N)2^m}$$

where 1 = Input

m = select line 0, 1, 2, 3,...

Select line	Types of DeMUX *Input : Output*	
$m = 0$	$2^0 = 1$	1 : 1
1	$2^1 = 2$	1 : 2
2	$2^2 = 4$	1 : 4
3	$2^3 = 8$	1 : 8
4	$2^4 = 16$	1 : 16
5	$2^5 = 32$	1 : 32

Example of Demultiplexer:

when $m = 2$

1 : 4 DeMUX:

Input = D

Select line = S_0, S_1

Output = Y_0, Y_1, Y_2, Y_3

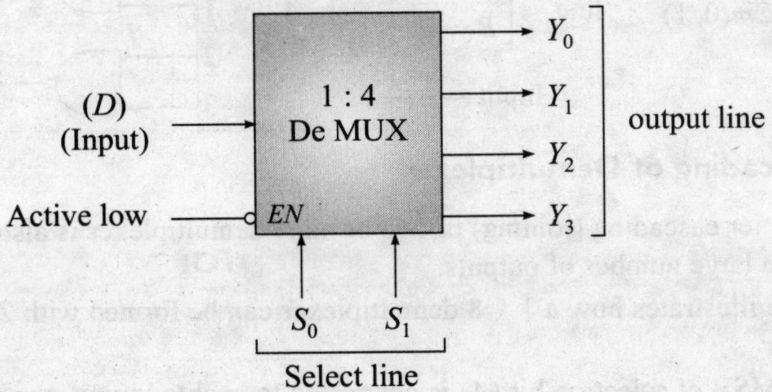

Truth Table:

Select line		***Output***			
S_0	S_1	Y_0	Y_1	Y_2	Y_3
0	0	D	0	0	0
0	1	0	D	0	0
1	0	0	0	D	0
1	1	0	0	0	D

$$Y_0 = \overline{S}_0\, \overline{S}_1\, D$$

$$Y_1 = \overline{S}_0\, S_1\, D$$

$$Y_2 = S_0\, \overline{S}_1\, D$$

$$Y_3 = S_0 S_1 D$$

Designing of 1 : 4 DeMUX:

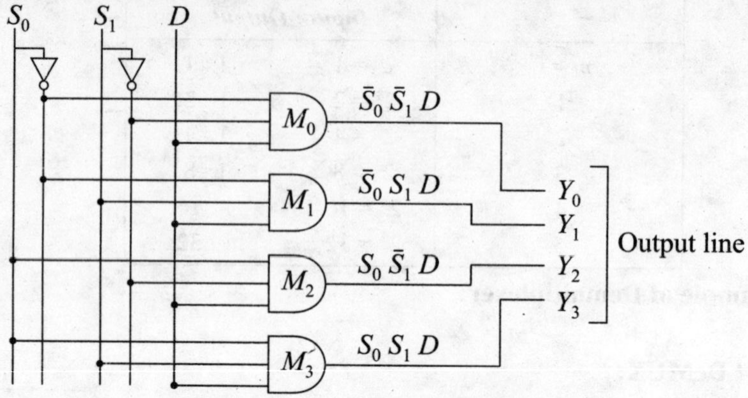

Example 1: Solve the function using Demultiplexer.

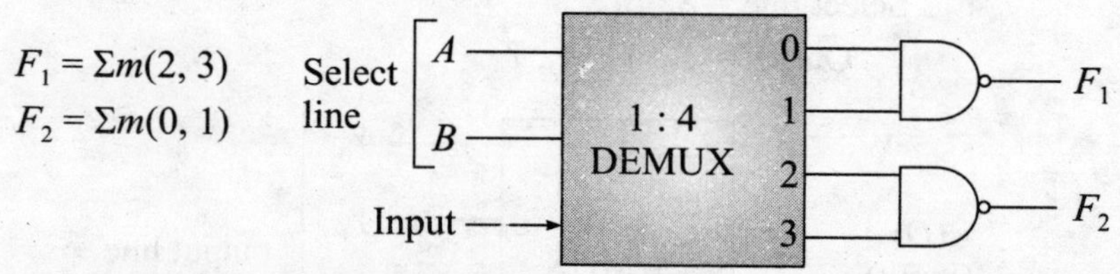

3.5.1 Cascading of Demultiplexer

Like multiplexer cascading (joining) of two or more demultiplexer is also possible in order to attain large number of outputs.

Figure illustrates how a 1 : 8 demultiplexer can be formed with 2 (two) 1 : 4 demultiplexer.

Here MSB of selection bit (A) is connected to enable inputs, one directly and one complemented.

When A *is logic '0'*, one of output lines $D_0 - D_3$ is selected according to select line B and C.

When A *is logic '1'* one of output lines $D_4 - D_7$ will be selected.

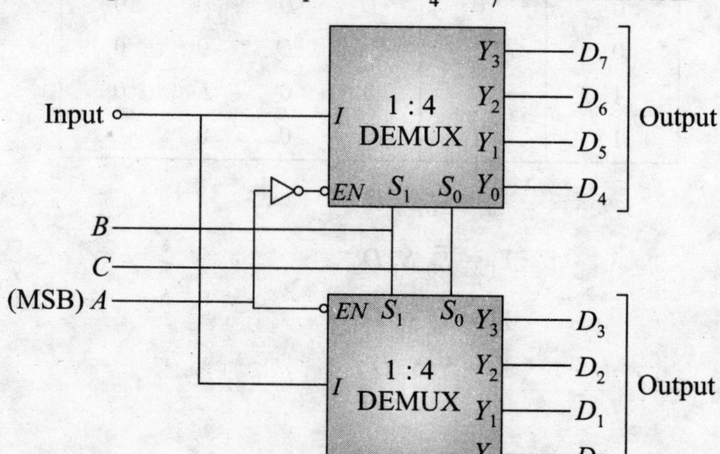

Fig. 3.4: Cascading of 1 : 8 Demultiplexer using two 1 : 4 Demultiplexer

3.5.2 Demultiplexer Applications

Demultiplexer is used for many applications in digital systems.

Application is:

(*i*) Serial to parallel conversion.

(*ii*) Binary to decimal conversion

3.5.3 Comparison Between MUX and DEMUX

S. No.	*Multiplexer*	*Demultiplexer*
1.	It means "many to one" *i.e.* multiple input and single output.	It means "one to many" *i.e.* one Input and multiple output.
2.	It is also known as "Data Selector."	It is also known as "Data Distributor."
3.	Function Realized using it does not required additional gates.	Function realized using it requires additional gates. *i.e.* NAND gate
4.	Used in data selection data routing, parallel to serial conversion.	used in serial to parallel conversion, binary to decimal conversion.
5.	Examples are: 4 : 1, 8 : 1, 16 : 1, 32 : 1	Examples are: 1 : 4, 1 : 8, 1 : 16, 1 : 32

3.6 MAGNITUDE COMPARATOR

It is also called as magnitude digital (or binary) comparator. It compares the two digits (A, B) in three function: $\boxed{A < B, A = B, A > B}$

One function = A less than B. $(A < B)$

Second function = A equal to B $(A = B)$

Third function = A greater than B $(A > B)$

Circuit Diagram:

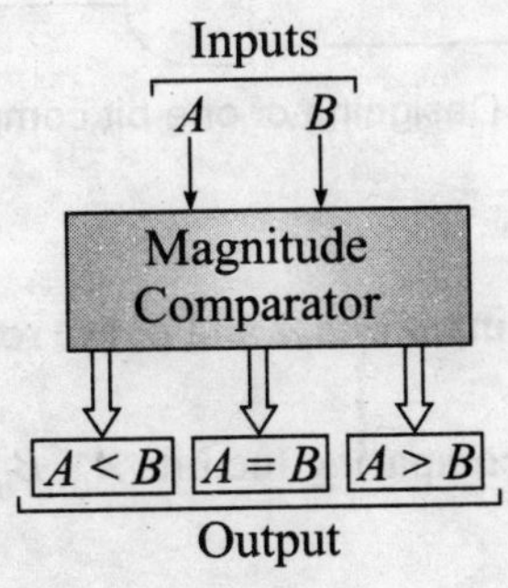

Types of Comparator:

1. 1-bit comparator
2. 2-bit comparator

⋮

till n bit comparator.

3.6.1 1-bit Comparator

For 1-bit comparator circuit there are two bits only *i.e.*, A, B and compare these two bits according to three function.

Truth Table:

Inputs		*Output*		
A	B	$A < B$	$A = B$	$A > B$
0	0	0	(1)	0
0	1	(1)	0	0
1	0	0	0	(1)
1	1	0	(1)	0

Expression for 1-bit comparator:

For $A < B$: $\boxed{\bar{A}B}$

For $A = B$: $\bar{A}\bar{B} + AB$ or $\boxed{A \odot B}$ → Ex-Nor gate output

For $A > B$: $\boxed{A\bar{B}}$

Designing of 1-bit Comparator:

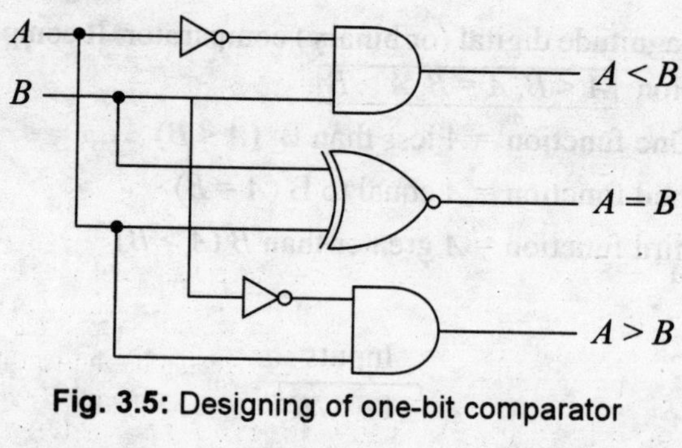

Fig. 3.5: Designing of one-bit comparator

3.6.2 2-Bit Comparator

For 2-bit comparator again two input bits A and B are represented by A is A_0, A_1 and B is B_0, B_1.

Output are obtained by comparing the bits A_0, B_0, A_1, B_1.

Truth Table:

Decimal	Inputs						Outputs		
Value	A_0	A_1	A	B_0	B_1	B	$A<B$	$A=B$	$A>B$
0	0	0	0	0	0	0	0	1	0
1	0	0	0	0	1	1	1	0	0
2	0	0	0	1	0	2	1	0	0
3	0	0	0	1	1	3	1	0	0
4	0	1	1	0	0	0	0	0	1
5	0	1	1	0	1	1	0	1	0
6	0	1	1	1	0	2	1	0	0
7	0	1	1	1	1	3	1	0	0
8	1	0	2	0	0	0	0	0	1
9	1	0	2	0	1	1	0	0	1
10	1	0	2	1	0	2	0	1	0
11	1	0	2	1	1	3	1	0	0
12	1	1	3	0	0	0	0	0	1
13	1	1	3	0	1	1	0	0	1
14	1	1	3	1	0	2	0	0	1
15	1	1	3	1	1	3	0	1	0

For the expression we will solve *k*-map.

K-Map for $A < B$:

A_0A_1 \ B_0B_1	$\bar{B}_0\bar{B}_1$ 00	$\bar{B}_0B_1$ 01	B_0B_1 11	$B_0\bar{B}_1$ 10
$\bar{A}_0\bar{A}_1$ 00	0	1 (1)	1 (3)	1 (2)
$\bar{A}_0A_1$ 01	(4)	(5)	1 (7)	1 (6)
A_0A_1 11	(12)	(13)	(15)	(14)
$A_0\bar{A}_1$ 10	(8)	(9)	1 (11)	(10)

$$\boxed{A<B} = \bar{A}_0\bar{A}_1B_1 + B_0\bar{A}_0 + B_0B_1\bar{A}_1$$

Designing of $A < B$:

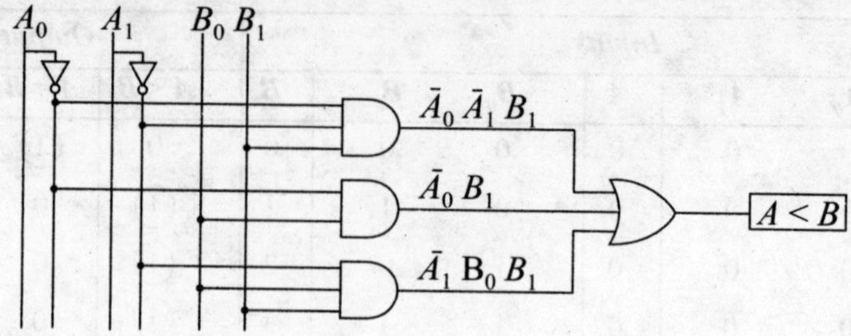

K-Map for $A = B$:

A_0A_1 \ B_0B_1	$\bar{B}_0\bar{B}_1$ 00	$\bar{B}_0B_1$ 01	B_0B_1 11	$B_0\bar{B}_1$ 10
$\bar{A}_0\bar{A}_1$ 00	1 (0)	(1)	(3)	(2)
$\bar{A}_0A_1$ 01	(4)	1 (5)	(7)	(6)
A_0A_1 11	(12)	(13)	1 (15)	(14)
$A_0\bar{A}_1$ 10	(8)	(9)	(11)	1 (10)

$$\boxed{A = B} = \bar{A}_0\bar{A}_1\bar{B}_0\bar{B}_1 + \bar{A}_0A_1\bar{B}_0B_1 + A_0A_1B_0B_1 + A_0\bar{A}_1B_0\bar{B}_1$$

$$= \bar{A}_0\bar{B}_0[A_1B_1 + \bar{A}_1\bar{B}_1] + A_0B_0[A_1B_1 + \bar{A}_1\bar{B}_1]$$

$$= \bar{A}_0\bar{B}_0[A_1 \odot B_1] + A_0B_0[A_1 \odot B_1]$$

$$= (A_1 \odot B_1)\,(A_0B_0 + \bar{A}_0\bar{B}_0]$$

$$= \boxed{(A_1 \odot B_1)\,(A_0 \odot B_0)}$$

Designing for $A = B$:

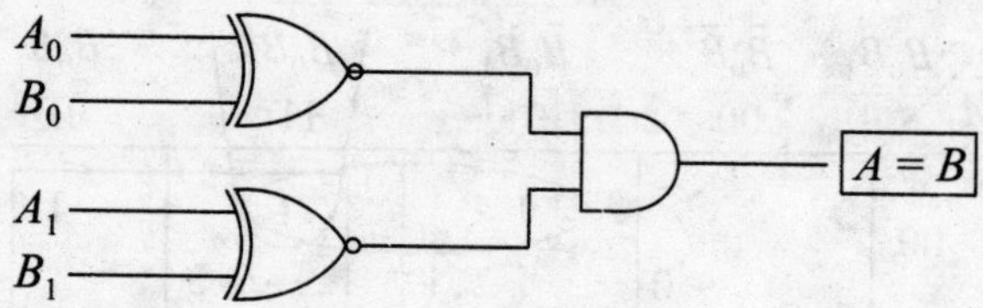

K-Map for $A > B$:

A_0A_1 \ B_0B_1	$\bar{B}_0\bar{B}_1$ 00	$\bar{B}_0B_1$ 01	B_0B_1 11	$B_0\bar{B}_1$ 10
$\bar{A}_0\bar{A}_1$ 00	(0)	(1)	(3)	(2)
$\bar{A}_0A_1$ 01	1 (4)	(5)	(7)	(6)
A_0A_1 11	1 (12)	1 (13)	(15)	1 (14)
$A_0\bar{A}_1$ 10	1 (8)	1 (9)	(11)	(10)

$$\boxed{A > B} = A_1\,\bar{B}_0\bar{B}_1 + A_0A_1\bar{B}_1 + A_0\bar{B}_0$$

Designing for $A > B$:

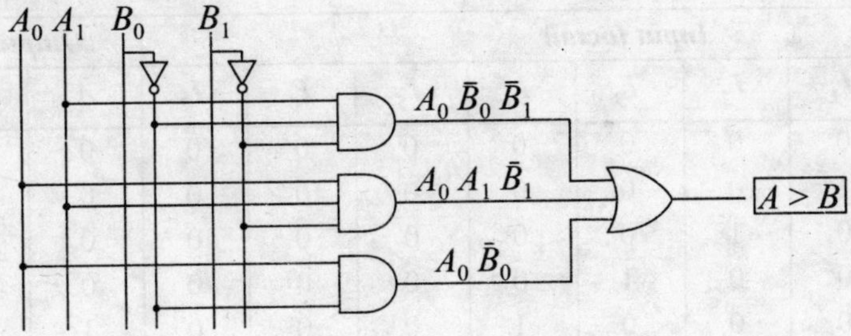

3.7 ENCODER

Encoding is the process of converting from familiar symbols or numbers to a coded format. Example: Binary coding for decimal digits.

A typical encoder has N-bit wide input port and a k-bit wide output port, where N is in the vicinity of 2^k. One and only one input line can take a value 1 at a time and encoder will send the code on k-output line.

N = Input line.

K = Output line.

Circuit Diagram:

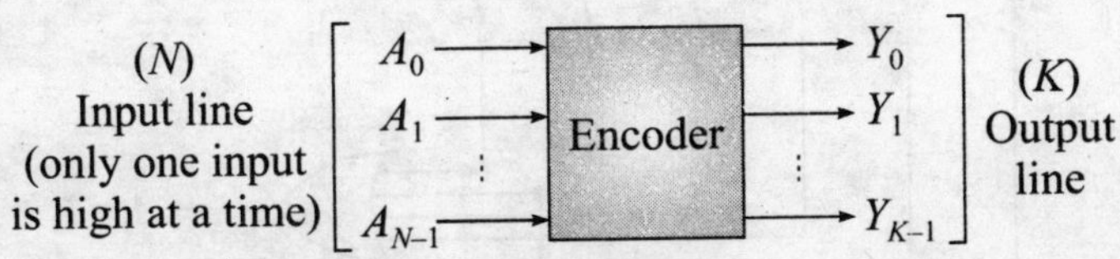

3.7.1 Types of Encoder

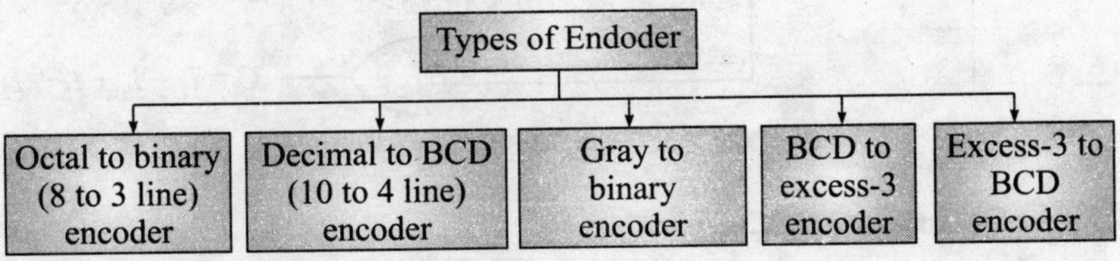

3.7.2 Octal to Binary Encoder

It is 8 line ($I_0 - I_1$) to 3 line (A, B, C) Encoder. It accepts 8 inputs lines and produces a 3-bit output code corresponding to activated code.

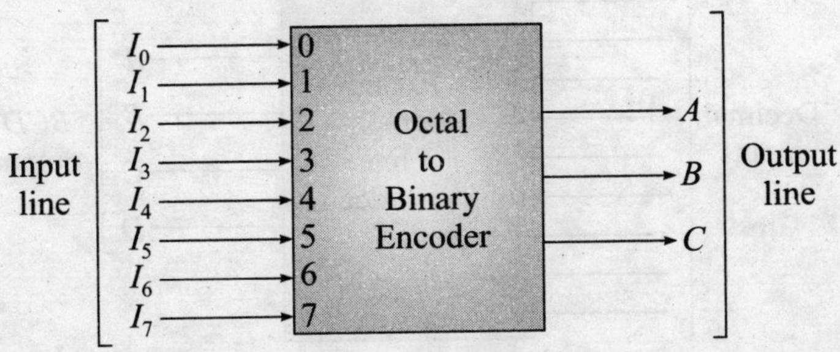

Truth Table:

Input (octal)								Output (Binary)		
I_0	I_1	I_2	I_3	I_4	I_5	I_6	I_7	A	B	C
1	0	0	0	0	0	0	0	0	0	0
0	1	0	0	0	0	0	0	0	0	1
0	0	1	0	0	0	0	0	0	1	0
0	0	0	1	0	0	0	0	0	1	1
0	0	0	0	1	0	0	0	1	0	0
0	0	0	0	0	1	0	0	1	0	1
0	0	0	0	0	0	1	0	1	1	0
0	0	0	0	0	0	0	1	1	1	1

Reduced expression for A, B and C is given as:

$$A = I_4 + I_5 + I_6 + I_7$$
$$B = I_2 + I_3 + I_6 + I_7$$
$$C = I_1 + I_3 + I_5 + I_7$$

Designed Procedure:

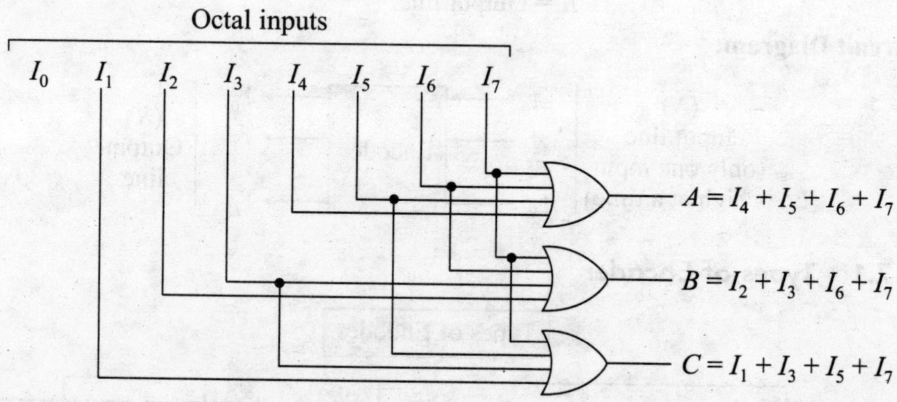

Fig. 3.6: Designing of 8 to 3 line Encoder

3.7.3 Decimal to BCD Encoder

It is 10 line to 4 line encoder. This type of encoder have ten inputs, one for each decimal digit and 4 outputs. So, block diagram is:

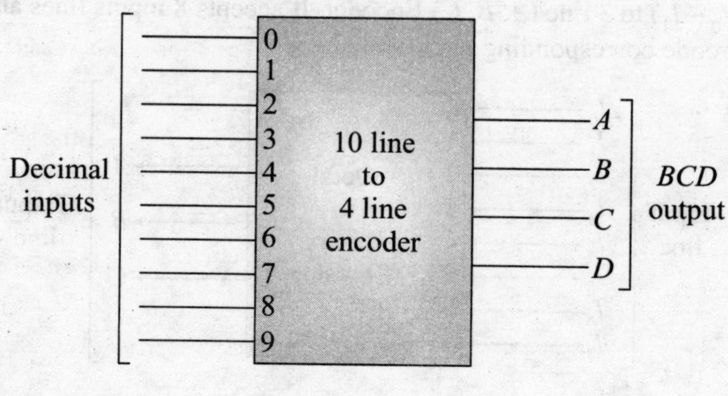

Truth Table:

Decimal Inputs		Binary Output			
		D	*C*	*B*	*A*
0	D_0	0	0	0	0
1	D_1	0	0	0	1
2	D_2	0	0	1	0
3	D_3	0	0	1	1
4	D_4	0	1	0	0
5	D_5	0	1	0	1
6	D_6	0	1	1	0
7	D_7	0	1	1	1
8	D_8	1	0	0	0
9	D_9	1	0	0	1

Designing:

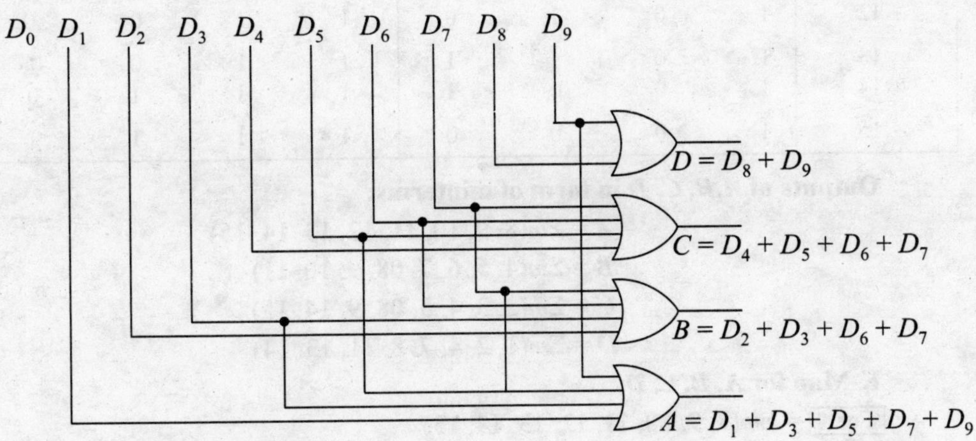

3.7.4 Gray to Binary Encoder

In this encoder Gray is taking as inputs line and Binary values are taken as output line.

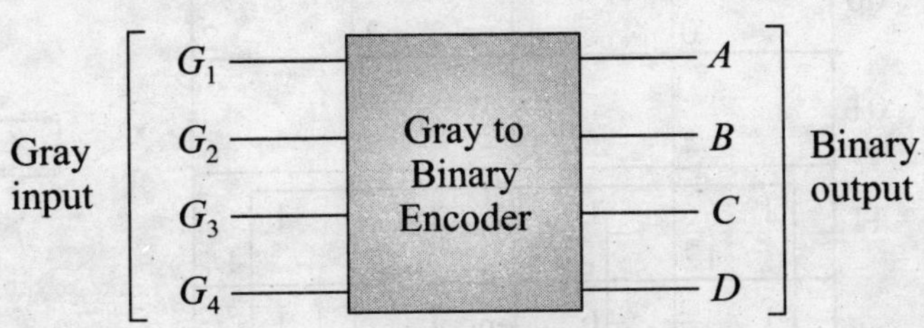

Truth Table:

Decimal No.	*Gray Code*				*Binary Values*			
	G_1	G_2	G_3	G_4	*A*	*B*	*C*	*D*
0	0	0	0	0	0	0	0	0
1	0	0	0	1	0	0	0	1
2	0	0	1	1	0	0	1	0
3	0	0	1	0	0	0	1	1
4	0	1	1	0	0	1	0	0
5	0	1	1	1	0	1	0	1
6	0	1	0	1	0	1	1	0
7	0	1	0	0	0	1	1	1
8	1	1	0	0	1	0	0	0
9	1	1	0	1	1	0	0	1
10	1	1	1	1	1	0	1	0
11	1	1	1	1	1	0	1	1
12	1	0	1	0	1	1	0	0
13	1	0	1	1	1	1	0	1
14	1	0	0	1	1	1	1	0
15	1	0	0	0	1	1	1	1

Outputs of *A,B, C, D* in form of minterms:

$$A = \Sigma m(8, 9, 10, 11, 12, 13, 14, 15)$$
$$B = \Sigma m(4, 5, 6, 7, 08, 9, 10, 11)$$
$$C = \Sigma m(2, 3, 4, 5, 08, 9, 14, 15)$$
$$D = \Sigma m(1, 2, 4, 7, 8, 11, 13, 14)$$

K-Map for *A, B, C, D*:

For A $= \Sigma m(8, 9, 10, 11, 12, 13, 14, 15)$

G_1G_2 \ G_3G_4	$\bar{G}_3\bar{G}_4$ 00	$\bar{G}_3G_4$ 01	G_3G_4 11	$G_3\bar{G}_4$ 10
$\bar{G}_1\bar{G}_2$ 00	(0)	(1)	(3)	(2)
$\bar{G}_1G_2$ 01	(4)	(5)	(7)	(6)
G_1G_2 11	1 (12)	1 (13)	1 (15)	1 (14)
$G_1\bar{G}_2$ 10	1 (8)	1 (9)	1 (11)	1 (10)

$A = G_1$

$\boxed{A = G_1}$

$\boxed{\text{For } B} = \Sigma m(4, 5, 6, 7, 08, 9, 10, 11)$

$G_1G_2 \backslash G_3G_4$	$\bar{G}_3\bar{G}_4$ 00	$\bar{G}_3G_4$ 01	G_3G_4 11	$G_3\bar{G}_4$ 10
$\bar{G}_1\bar{G}_2$ 00	(0)	(1)	(3)	(2)
$\bar{G}_1G_2$ 01	1 (4)	1 (5)	1 (7)	1 (6)
G_1G_2 11	(12)	(13)	(15)	(14)
$G_1\bar{G}_2$ 10	1 (8)	1 (9)	1 (11)	1 (10)

$$B = \bar{G}_1G_2 + G_1\bar{G}_2$$

$$\boxed{B = G_1 \oplus G_2}$$

$\boxed{\text{For } C} = \Sigma m(2, 3, 4, 5, 08, 9, 14, 15)$

$G_1G_2 \backslash G_3G_4$	$\bar{G}_3\bar{G}_4$ 00	$\bar{G}_3G_4$ 01	G_3G_4 11	$G_3\bar{G}_4$ 10
$\bar{G}_1\bar{G}_2$ 00	(0)	(1)	1 (3)	1 (2)
$\bar{G}_1G_2$ 01	1 (4)	1 (5)	(7)	(6)
G_1G_2 11	(12)	(13)	1 (15)	1 (14)
$G_1\bar{G}_2$ 10	1 (8)	1 (9)	(11)	(10)

$$C = \bar{G}_1\bar{G}_2G_3 + \bar{G}_1G_2\bar{G}_3 + G_1G_2G_3 + G_1\bar{G}_2\bar{G}_3$$

$$= \bar{G}_1(\bar{G}_2G_3 + G_2\bar{G}_3) + G_1(G_2G_3 + \bar{G}_2\bar{G}_3)$$

$$= \bar{G}_1(G_2 \oplus G_3) + G_1(\overline{G_2 \oplus G_3})$$

$$\boxed{C = G_1 \oplus G_2 \oplus G_3}$$

For D $=\Sigma m(1, 2, 4, 7, 8, 11, 13, 14)$

G_1G_2 \ G_3G_4	$\bar{G}_3\bar{G}_4$ 00	$\bar{G}_3G_4$ 01	G_3G_4 11	$G_3\bar{G}_4$ 10
$\bar{G}_1\bar{G}_2$ 00	0	① 1	3	① 2
$\bar{G}_1G_2$ 01	① 4	5	① 7	6
G_1G_2 11	12	① 13	15	① 14
$G_1\bar{G}_2$ 10	① 8	9	① 11	10

$D = G_1 \oplus G_2 \oplus G_3 \oplus G_4$

Designing of Circuit:

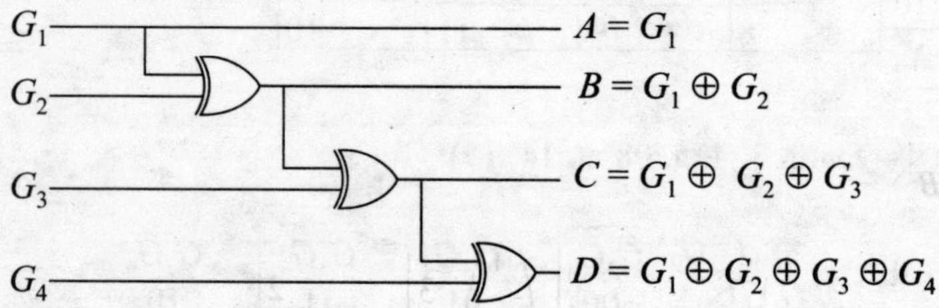

3.7.5 BCD to Excess-3 Encoder

In this Binary coded decimal are taken at input terminal and Excess-3 code of BCD values are at output terminal.

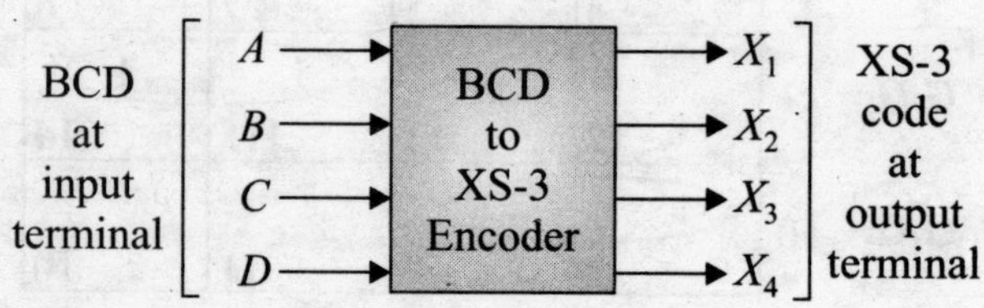

Truth Table:

Decimal Value	*BCD (Input)* A	B	C	D	*Excess-3 (output)* X_4	X_3	X_2	X_1
0	0	0	0	0	0	0	1	1
1	0	0	0	1	0	1	0	0
2	0	0	1	0	0	1	0	1
3	0	0	1	1	0	1	1	0
4	0	1	0	0	0	1	1	1
5	0	1	0	1	1	0	0	0
6	0	1	1	0	1	0	0	1
7	0	1	1	1	1	0	1	0
8	1	0	0	0	1	0	1	1
9	1	0	0	1	1	1	0	0
10-15	All are don't care values							

K-map for X_1, X_2, X_3, X_4

For X_1

$AB \backslash CD$	$\bar{C}\bar{D}$ 00	$\bar{C}D$ 01	CD 11	$C\bar{D}$ 10
$\bar{A}\bar{B}$ 00	1 (0)	(1)	(3)	1 (2)
$\bar{A}B$ 01	1 (4)	(5)	(7)	1 (6)
AB 11	× (12)	× (13)	× (15)	× (14)
$A\bar{B}$ 10	1 (8)	(9)	× (11)	× (10)

$$X_1 = \bar{D}$$

For X_2

$AB \backslash CD$	$\bar{C}\bar{D}$ 00	$\bar{C}D$ 01	CD 11	$C\bar{D}$ 10
$\bar{A}\bar{B}$ 00	1 (0)	(1)	1 (3)	(2)
$\bar{A}B$ 01	1 (4)	(5)	1 (7)	(6)
AB 11	× (12)	× (13)	× (15)	× (14)
$A\bar{B}$ 10	1 (8)	(9)	× (11)	× (10)

$$X_2 = \bar{C}\bar{D} + CD$$

$$X_2 = C \odot D$$

For X_3

$AB \backslash CD$	$\bar{C}\bar{D}$ 00	$\bar{C}D$ 01	CD 11	$C\bar{D}$ 10
$\bar{A}\bar{B}$ 00	1 (0)	1 (1)	1 (3)	1 (2)
$\bar{A}B$ 01	1 (4)	(5)	1 (7)	(6)
AB 11	× (12)	× (13)	× (15)	× (14)
$A\bar{B}$ 10	1 (8)	1 (9)	× (11)	× (10)

$$X_3 = D\bar{B} + C\bar{B} + B\bar{C}\bar{D}$$

$$X_3 = \bar{B}(C + D) + B\bar{C}\bar{D}$$

For X_4

AB \ CD	$\bar{C}\bar{D}$ 00	$\bar{C}D$ 01	CD 11	$C\bar{D}$ 10
$\bar{A}\bar{B}$ 00	0	1	3	2
$\bar{A}B$ 01	4	1 (5)	1 (7)	1 (6)
AB 11	× (12)	× (13)	× (15)	× (14)
$A\bar{B}$ 10	1 (8)	1 (9)	× (11)	× (10)

$$X_4 = BC + DB + A$$

$$X_4 = B(C + D) + A$$

Designing of Circuit:

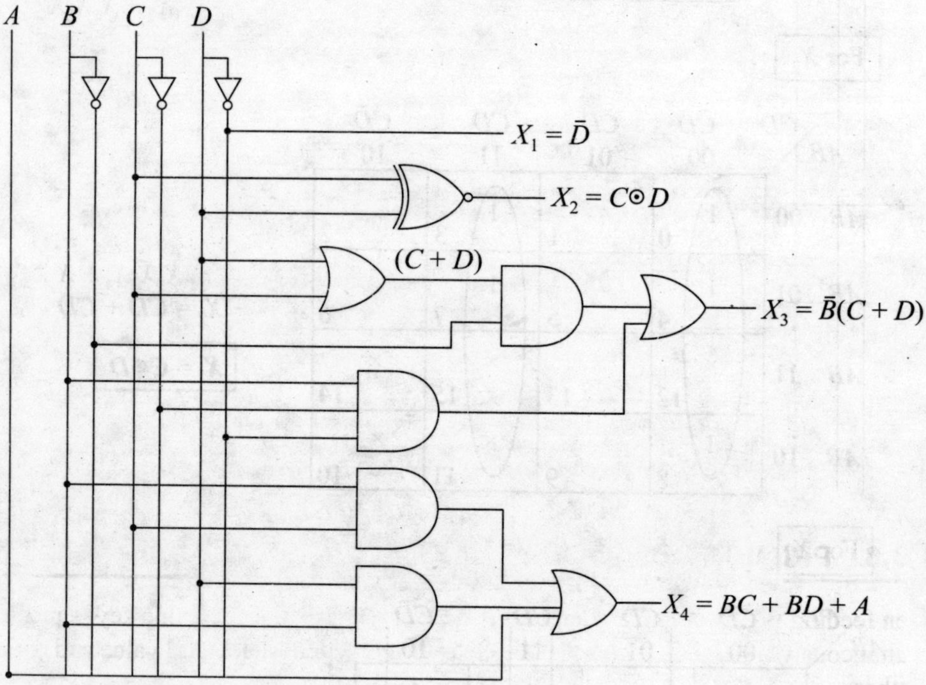

3.7.6 Excess-3 to BCD Encoder

In this Excess-3 code at input and BCD code at output terminal. In this decimal number 0, 1, 2, 13, 14, 15 will not appear in *XS*-3 code, these terms are used as don't care condition.

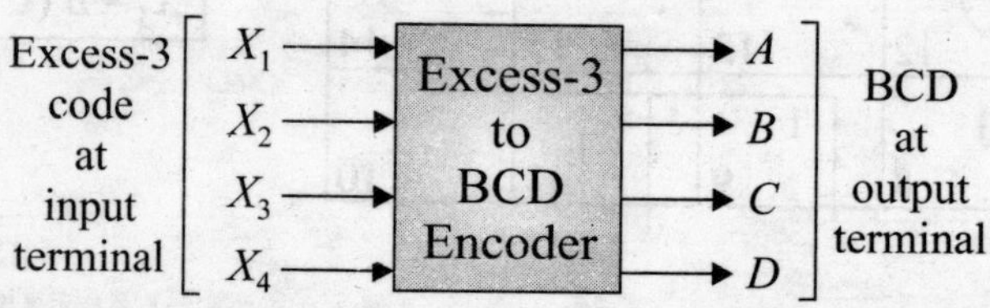

Excess-3 to BCD encoder is opposite of BCD to Excess-3 code Encoder.

Output Expression:

$$A = X_1 X_2 + X_1 X_3 X_4$$

$$B = \bar{X}_2 \bar{X}_4 + X_1 \bar{X}_3 X_4 + X_2 X_3 X_4$$

$$C = \bar{X}_3 X_4 + X_3 \bar{X}_4 \Rightarrow X_3 \oplus X_4$$

$$D = \bar{X}_4$$

Designing of circuit:

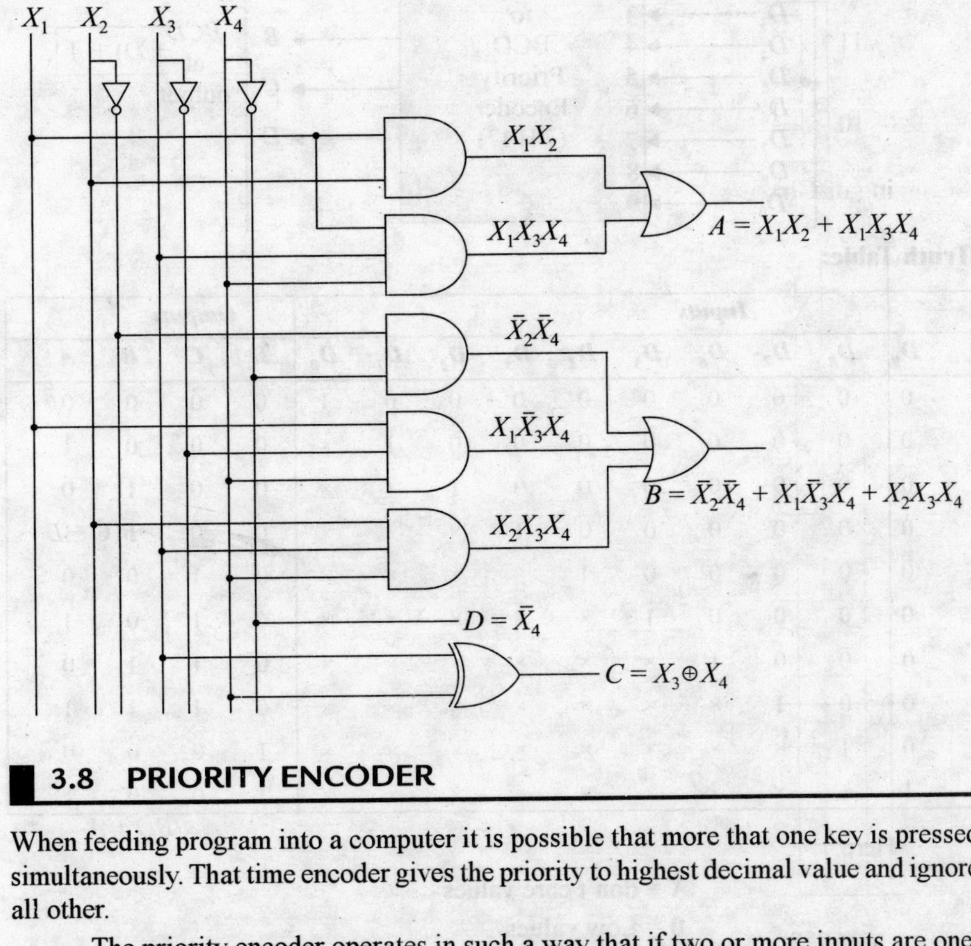

3.8 PRIORITY ENCODER

When feeding program into a computer it is possible that more that one key is pressed simultaneously. That time encoder gives the priority to highest decimal value and ignore all other.

The priority encoder operates in such a way that if two or more inputs are one, the input having the higher priority will take precedence.

Example: In priority encoder, decimal 7 and 3 are together and both are high means 1, the encoder will send 7 at output and ignore 3.

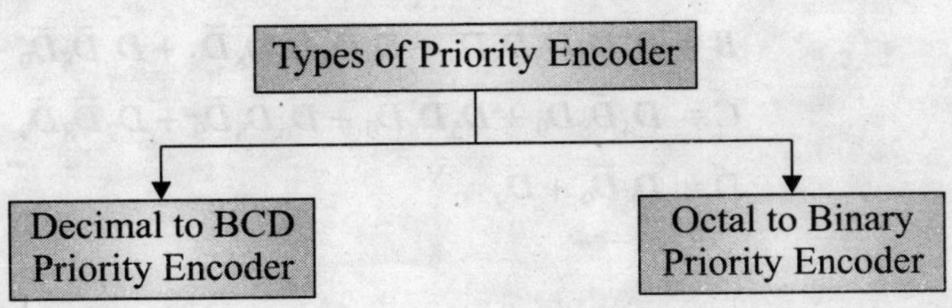

3.8.1 Decimal to BCD Priority Encoder

In this at input total 10 (0-9) values are taken then encoder will check priority when more than one value are high at one time. It will send the value at output which have high value and BCD codes are at output terminal.

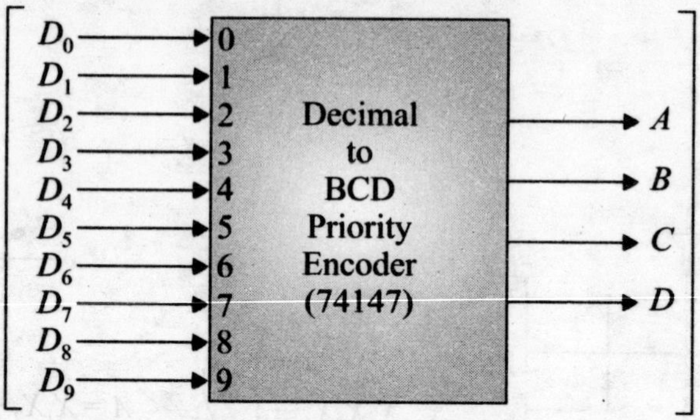

Truth Table:

Inputs										*Outputs*			
D_9	D_8	D_7	D_6	D_5	D_4	D_3	D_2	D_1	D_0	D	C	B	A
0	0	0	0	0	0	0	0	0	1	0	0	0	0
0	0	0	0	0	0	0	0	1	×	0	0	0	1
0	0	0	0	0	0	0	1	×	×	0	0	1	0
0	0	0	0	0	0	1	×	×	×	0	0	1	1
0	0	0	0	0	1	×	×	×	×	0	1	0	0
0	0	0	0	1	×	×	×	×	×	0	1	0	1
0	0	0	1	×	×	×	×	×	×	0	1	1	0
0	0	1	×	×	×	×	×	×	×	0	1	1	1
0	1	×	×	×	×	×	×	×	×	1	0	0	0
1	×	×	×	×	×	×	×	×	×	1	0	0	1

Here:

X = don't care values

0 = Low values

1 = High values

Expression For *A, B, C, D*:

$$A = D_1\bar{D}_2\bar{D}_4\bar{D}_6\bar{D}_8 + D_3\bar{D}_4\bar{D}_6\bar{D}_8 + D_5\bar{D}_6\bar{D}_8 + D_7\bar{D}_8 + D_9$$

$$B = D_2\bar{D}_4\bar{D}_5\bar{D}_8\bar{D}_9 + D_3\bar{D}_4\bar{D}_5\bar{D}_8\bar{D}_9 + D_6\bar{D}_8\bar{D}_9 + D_7\bar{D}_8\bar{D}_9$$

$$C = D_4\bar{D}_8\bar{D}_9 + D_5\bar{D}_8\bar{D}_9 + D_6\bar{D}_8\bar{D}_9 + D_7\bar{D}_8\bar{D}_9$$

$$D = D_8\bar{D}_9 + D_9$$

Designing for Circuit:

For A

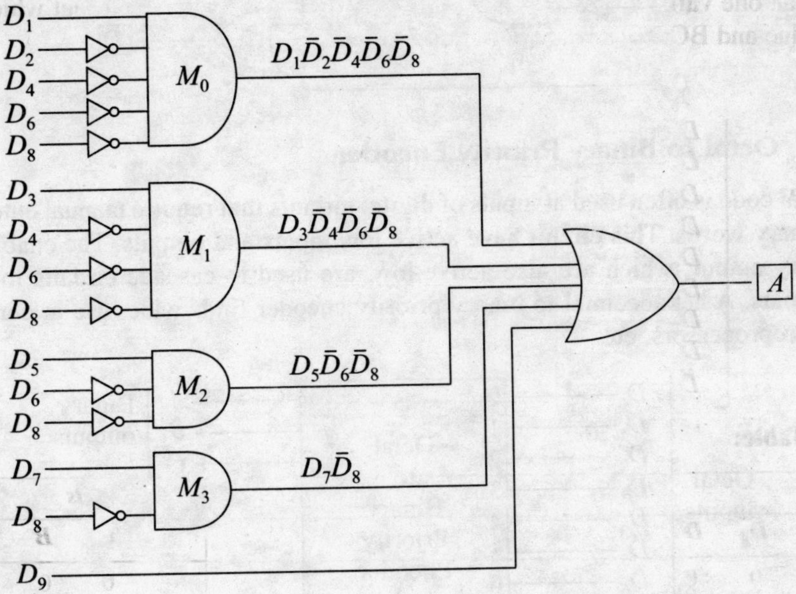

For B

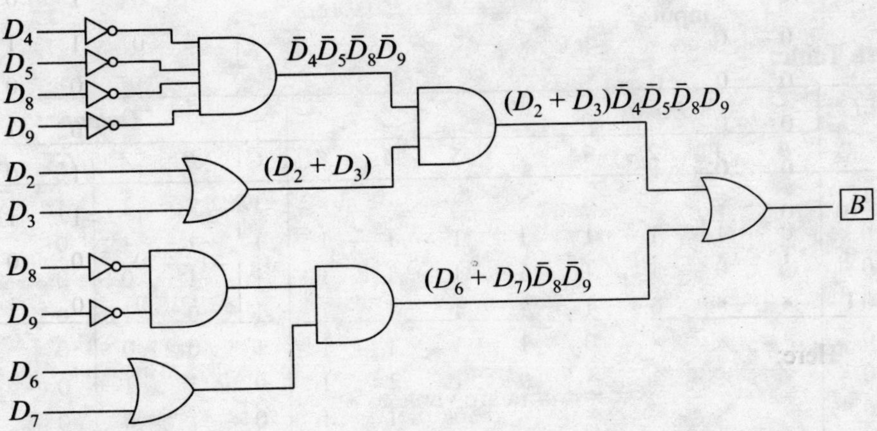

For C

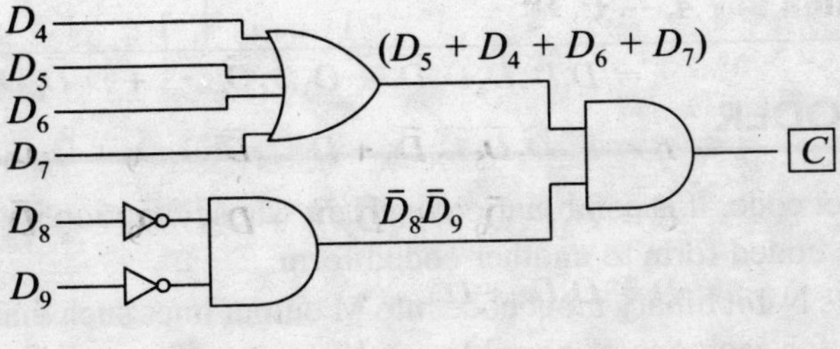

For D

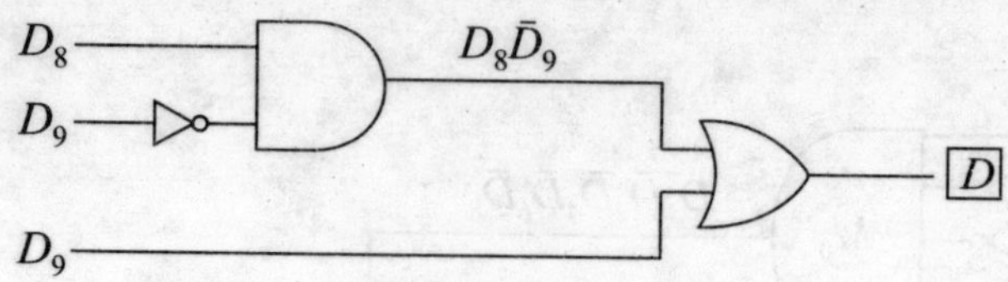

3.8.2 Octal to Binary Priority Encoder

The octal code is often used at inputs of digital circuits that require manual entering of long binary words. This circuit have active low inputs and outputs. The enable input and carry output; which are also active-low, are used to cascade circuits to handle more inputs. A hexadecimal to binary priority encoder finds wide use in computers and microprocessors, etc.

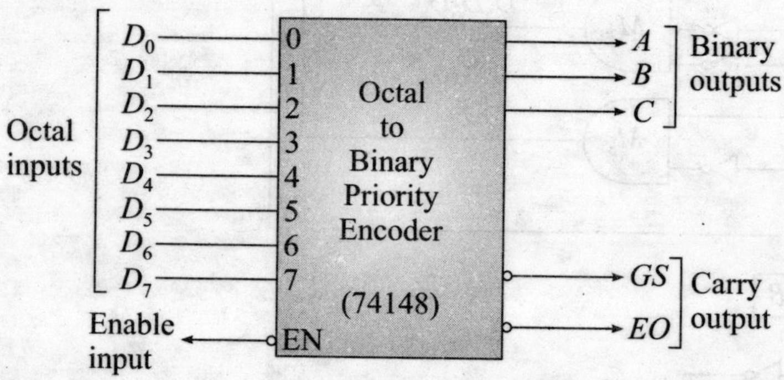

Truth Table:

EI	*Inputs*								*Outputs*				
	0	*1*	*2*	*3*	*4*	*5*	*6*	*7*	*C*	*B*	*A*	*GS*	*EO*
1	×	×	×	×	×	×	×	×	1	1	1	1	1
0	0	1	1	1	1	1	1	1	1	1	1	0	1
0	×	0	1	1	1	1	1	1	1	1	0	0	1
0	×	×	0	1	1	1	1	1	1	0	1	0	1
0	×	×	×	0	1	1	1	1	1	0	0	0	1
0	×	×	×	×	0	1	1	1	0	1	1	0	1
0	×	×	×	×	×	0	1	1	0	1	0	0	1
0	×	×	×	×	×	×	0	1	0	0	1	0	1
0	×	×	×	×	×	×	×	0	0	0	0	0	1
0	1	1	1	1	1	1	1	1	1	1	1	1	0

3.9 DECODER

It is used to detect code. It is combinational circuit whose function is to convert binary strings from one coded form to another coded form.

It converts N-bit binary input code into M output lines such that only one output line is activated for each one of possible combinations of inputs. Such as decoder is called N to $2^N(M)$ line decoder.

Block Diagram: Since each of N inputs can be 0 or 1, there are 2^N possible input combinations or codes. For each of these input combinations, only one of M outputs will be active (High) all other outputs will remain inactive (Low). Some decoder are designed to produce active low output, while all the other outputs remain high.

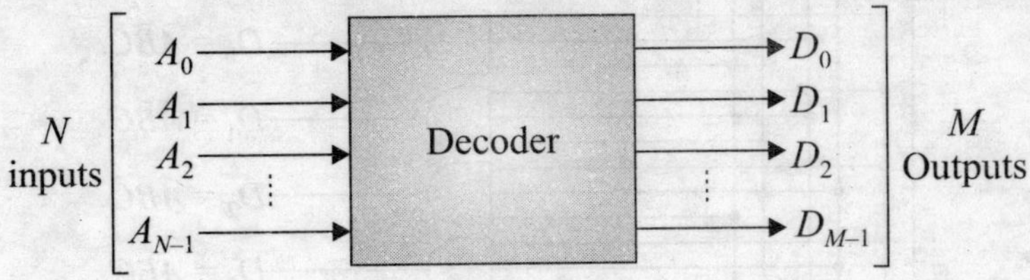

Types of Decoder:

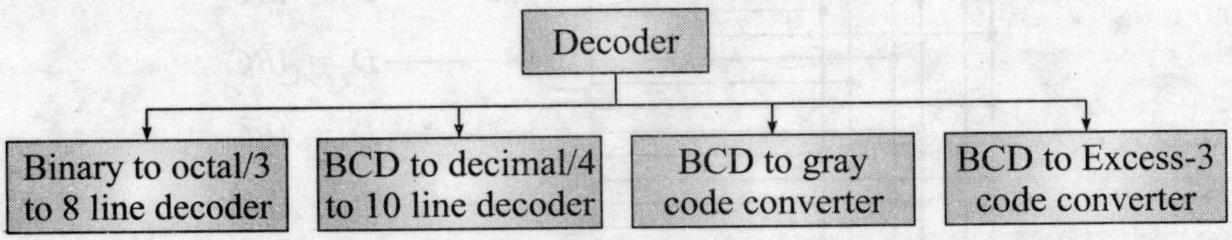

3.9.1 Binary to Octal (3 to 8 line decoder)

In this decoder 3 inputs and 8 outputs are there. It uses all AND gate, therefore the outputs are active high. For active low outputs NAND gates are used. This is also known as Binary to octal decoder.

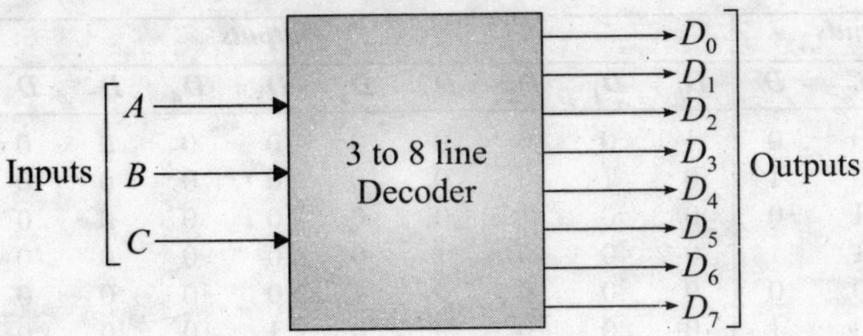

Truth Table:

Inputs			*Outputs*							
A	***B***	***C***	D_0	D_1	D_2	D_3	D_4	D_5	D_6	D_7
0	0	0	(1)	0	0	0	0	0	0	0
0	0	1	0	(1)	0	0	0	0	0	0
0	1	0	0	0	(1)	0	0	0	0	0
0	1	1	0	0	0	(1)	0	0	0	0
1	0	0	0	0	0	0	(1)	0	0	0
1	0	1	0	0	0	0	0	(1)	0	0
1	1	0	0	0	0	0	0	0	(1)	0
1	1	1	0	0	0	0	0	0	0	(1)

Designing of Circuit:

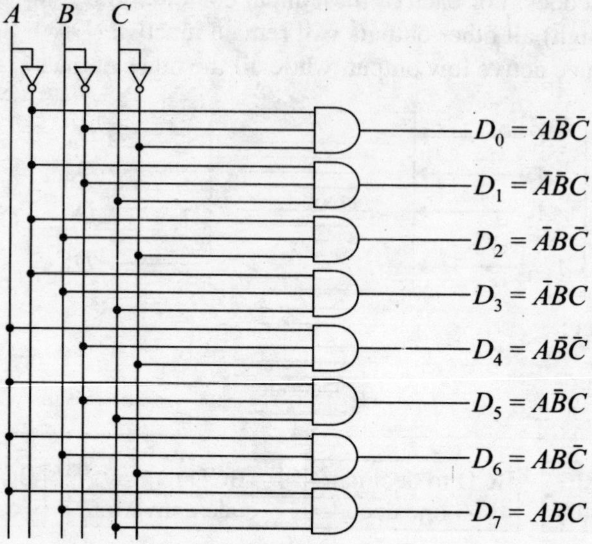

3.9.2 BCD to Decimal Decoder: (4 to 10 line)

In this decoder BCD codes are taken at input and decimal value 10 (0-9) are taken at output. It is also known as 4 to 10 line decoder.

Truth Table:

Inputs				Outputs									
A	*B*	*C*	*D*	D_0	D_1	D_2	D_3	D_4	D_5	D_6	D_7	D_8	D_9
0	0	0	0	1	0	0	0	0	0	0	0	0	0
0	0	0	1	0	1	0	0	0	0	0	0	0	0
0	0	1	0	0	0	1	0	0	0	0	0	0	0
0	0	1	1	0	0	0	1	0	0	0	0	0	0
0	1	0	0	0	0	0	0	1	0	0	0	0	0
0	1	0	1	0	0	0	0	0	1	0	0	0	0
0	1	1	0	0	0	0	0	0	0	1	0	0	0
0	1	1	1	0	0	0	0	0	0	0	1	0	0
1	0	0	0	0	0	0	0	0	0	0	0	1	0
1	0	0	1	0	0	0	0	0	0	0	0	0	1

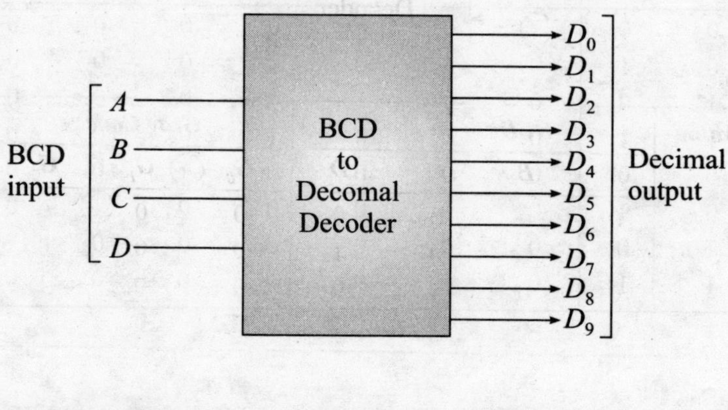

Circuit Diagram:

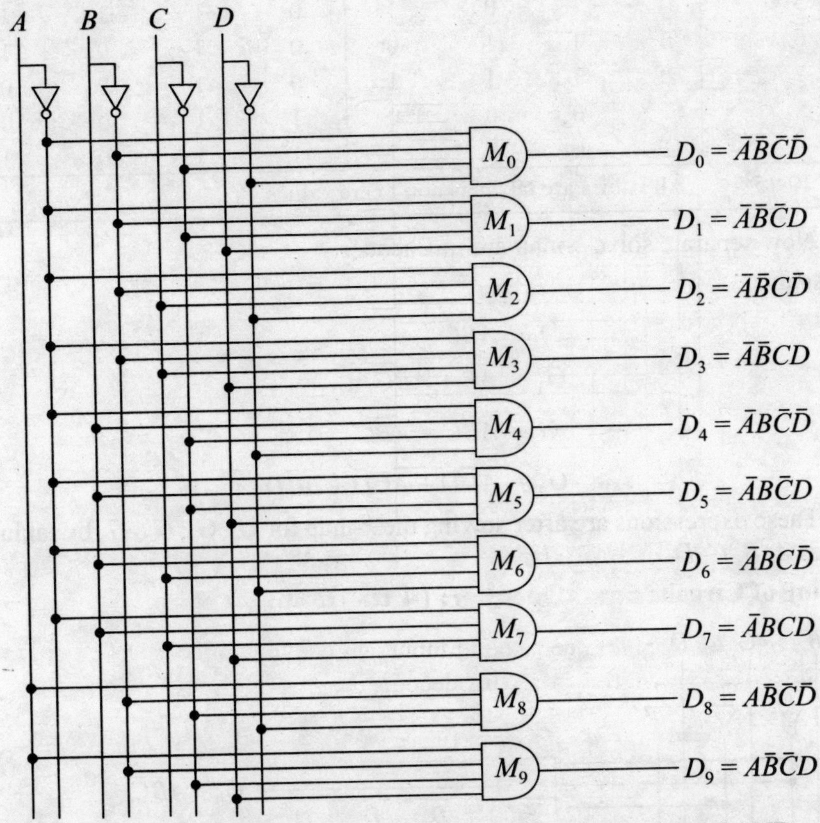

This is designing of BCD to decimal decoder with the help of AND gate. and output expression is the expression of decimal values from D_0 to D_9.

3.9.3 BCD TO GRAY CODE DECODER

In this BCD is taken as input and gray code are at output.

Truth Table:

Decimal Value	*BCD*				*Gray Code*			
	A	*B*	*C*	*D*	G_0	G_1	G_2	G_3
0	0	0	0	0	0	0	0	0
1	0	0	0	1	0	0	0	1
2	0	0	1	0	0	0	1	1
3	0	0	1	1	0	0	1	0

4	0	1	0	0	0	1	1	0
5	0	1	0	1	0	1	1	1
6	0	1	1	0	0	1	0	1
7	0	1	1	1	0	1	0	0
8	1	0	0	0	1	1	0	0
9	1	0	0	1	1	1	0	1
10-15	All values are taken as don't care values.							

Now separate solve *k*-map are in Encoder:

Expression:

$$G_0 = A\bar{B}\bar{C}$$

$$G_1 = \bar{A}B + A\bar{B}\bar{C}$$

$$G_2 = \bar{A}B\bar{C} + A\bar{B}C$$

$$G_3 = \bar{A}\bar{C}D + \bar{A}C\bar{D} + \bar{B}\bar{C}D$$

These expressions are after solving the *k*-map for G_0, G_1, G_2, G_3 by taking four variables *A*, *B*, *C*, *D*.

Designing of Circuit:

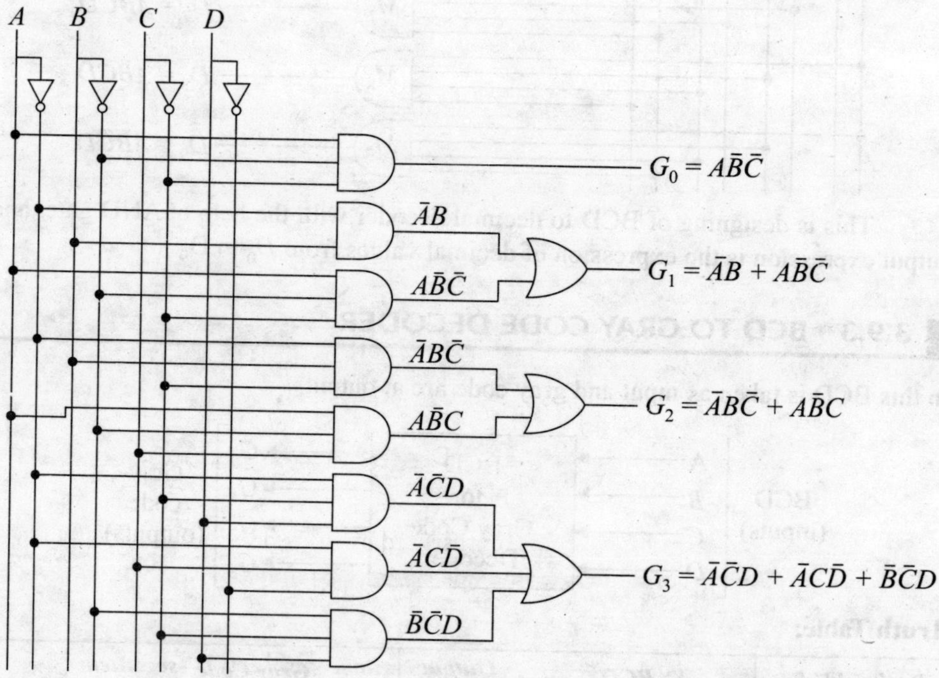

3.10 BCD TO 7-SEGMENT DISPLAY DEVICES DECODER

It is most popular device used in digital system. For displaying data using this device, the data have to be converted from BCD to 7-segment code. The seven output of decoder will drive the seven segments on a corresponding display.

The *BCD* system is used to represent the decimal numbers from 0 to 9 in binary format suitable for digital devices. A 4-bit code is required for the decimal number 0 to 9 represented by binary number 0000 to 1001. But values from 10 to 15 are not used as these are invalid codes.

The input to decoder is a number from o to 9 in *BCD* and outputs provides the seven signals required to drive 7-segment display.

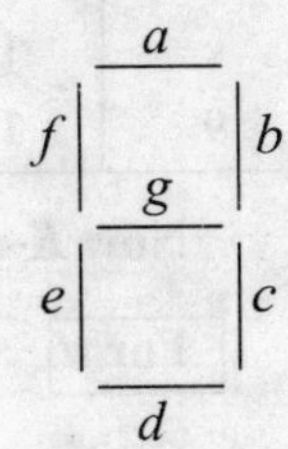

7 segment unit

The block diagram of 7 segment display is shown in figure.

For its shape, each of LED can be controlled individually, this display is called 7-segment display. It is used to represent BCD number that is from 0 to 9 by LED's.

a, b, c, d, e, f, g are 7 segments of display device which are driven by 7 outputs of decoder.

Decimal digits 0 to 9 can be displayed by glowing some particular LED segments. Now representation of all digit from 0 to 9 by glowing the segments is shown in figure:

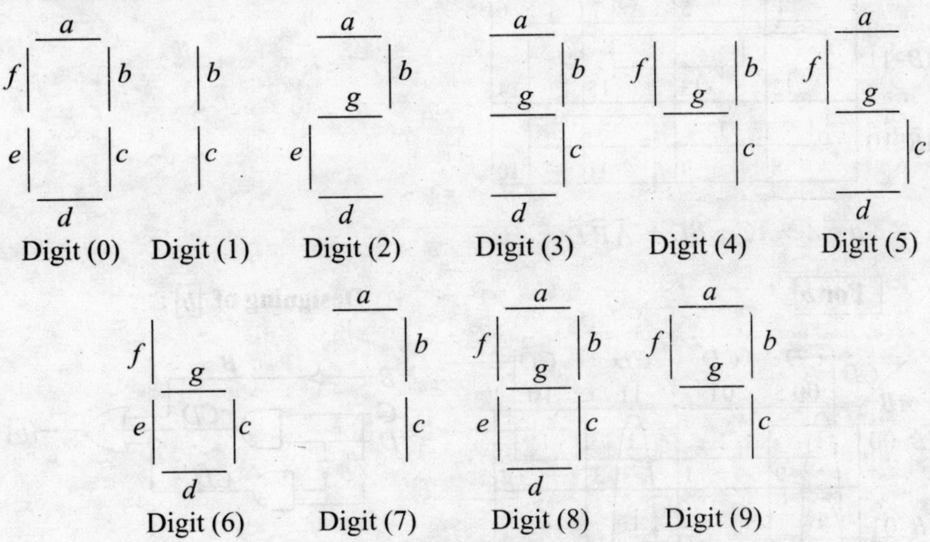

Figure

Every decimal digit of 0 to 9 is represented by BCD data, consisting of 4 input variables *A, B, C, D*. Truth table can be made for each of the LED segment.

Truth Table:

Decimal Number	*Input Variables*				*Output variable Display as 7-segement*						
	A	*B*	*C*	*D*	*a*	*b*	*c*	*d*	*e*	*f*	*g*
0	0	0	0	0	1	1	1	1	1	1	0
1	0	0	0	1	0	1	1	0	0	0	0
2	0	0	1	0	1	1	0	1	1	0	1
3	0	0	1	1	1	1	1	1	0	0	1

4	0	1	0	0	0	1	1	0	0	1	1
5	0	1	0	1	1	0	1	1	0	1	1
6	0	1	1	0	0	0	1	1	1	1	1
7	0	1	1	1	1	1	1	0	0	0	0
8	1	0	0	0	1	1	1	1	1	1	1
9	1	0	0	1	1	1	1	0	0	1	1

Now *K*-map for *a, b, c, d, e, f, g*:

For *a* 10 to 15 are don't care values

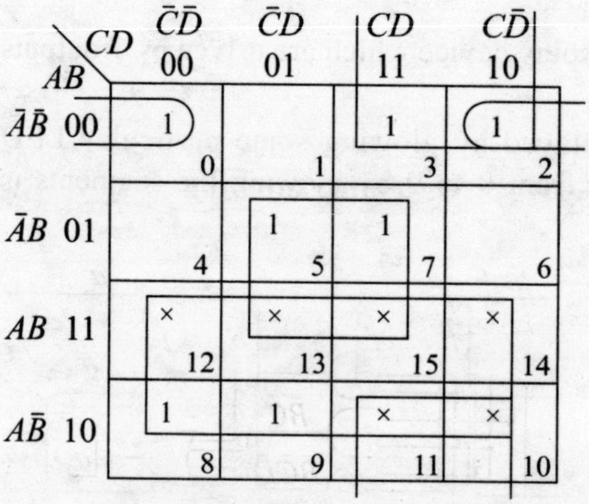

$a = A + AC + BD + \bar{A}\bar{B}\bar{D}$

Designing of $\boxed{a}$:

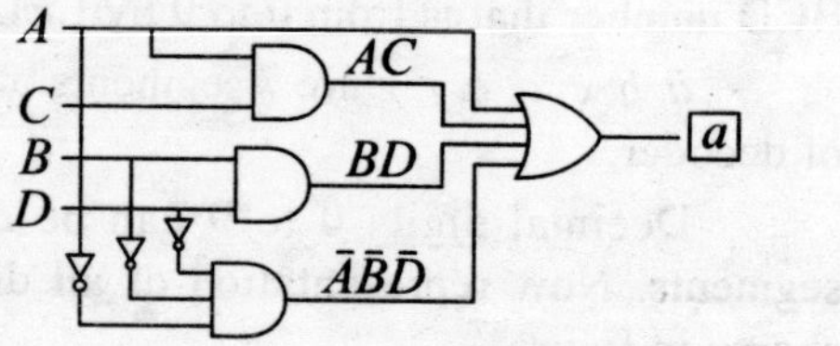

For *b*

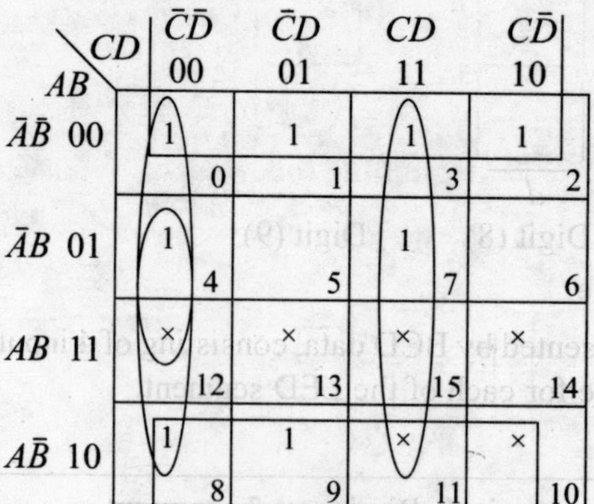

$b = \bar{B} + \bar{C}\bar{D} + CD$

Designing of $\boxed{b}$:

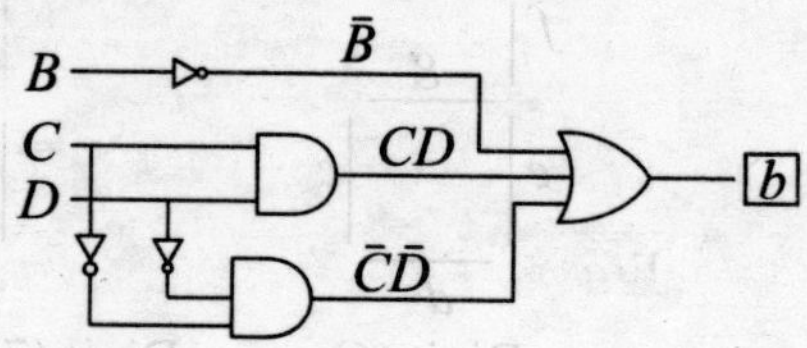

For c

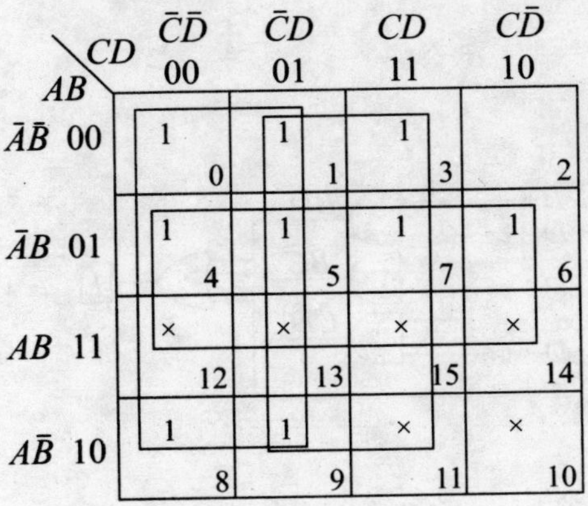

$c = D + B + \bar{C}$

Designing for $\boxed{c}$:

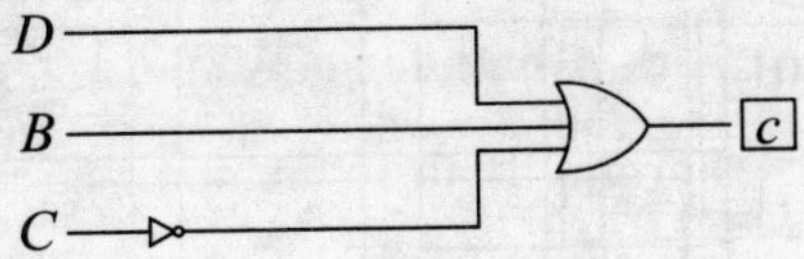

For d

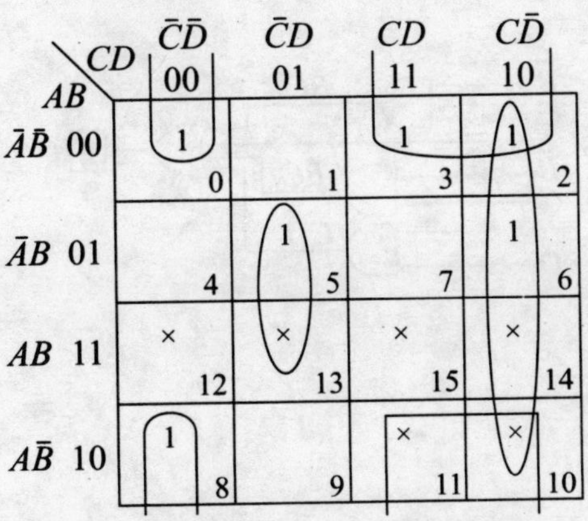

$d = B\bar{C}D + \bar{B}C + \bar{C}\bar{D}\bar{B} + C\bar{D}$

Designing for $\boxed{d}$:

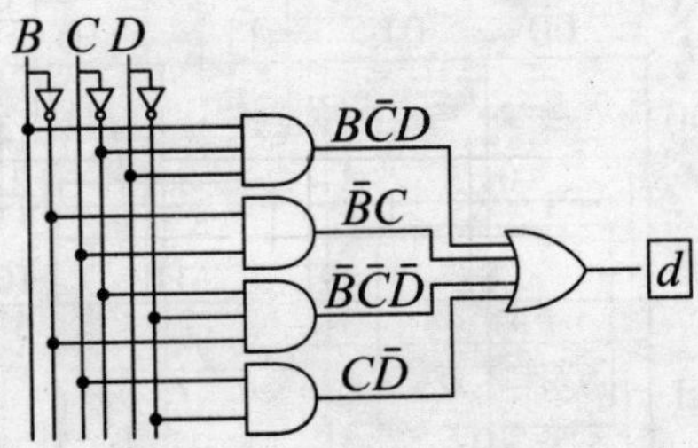

For e

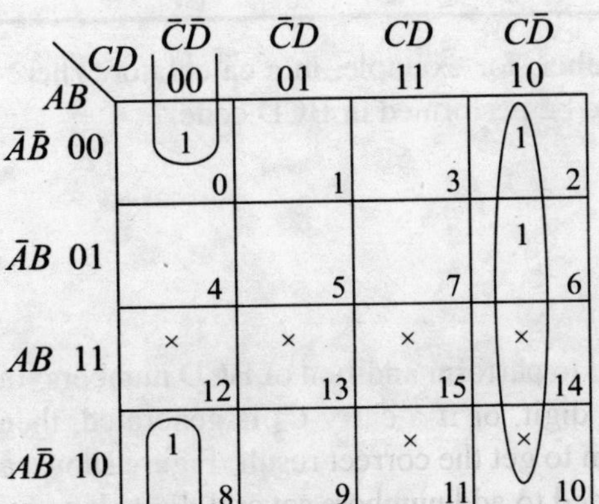

$e = \bar{B}\bar{C}\bar{D} + C\bar{D}$

Designing for $\boxed{e}$:

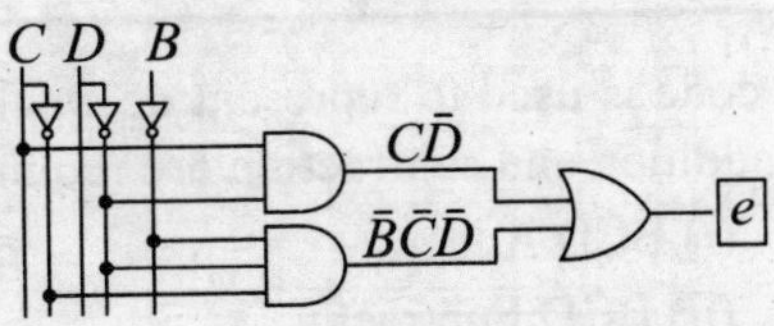

For f

$AB \backslash CD$	$\bar{C}\bar{D}$ 00	$\bar{C}D$ 01	CD 11	$C\bar{D}$ 10
$\bar{A}\bar{B}$ 00	1 (0)	(1)	1 (3)	(2)
$\bar{A}B$ 01	1 (4)	1 (5)	1 (7)	1 (6)
AB 11	× (12)	× (13)	× (15)	× (14)
$A\bar{B}$ 10	1 (8)	1 (9)	× (11)	× (10)

$$f = B\bar{D} + B\bar{C} + A + \bar{C}\bar{D}$$

Designing for f:

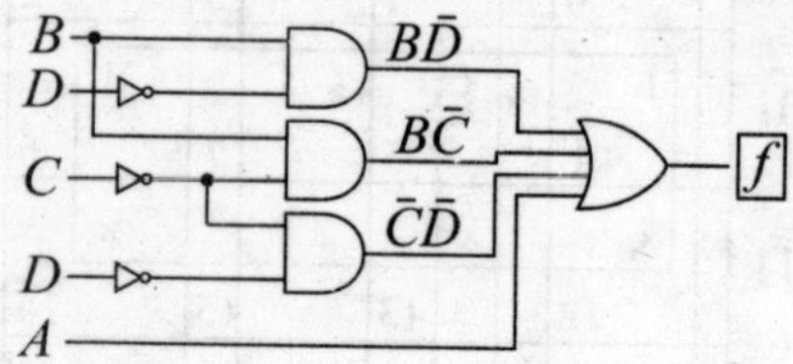

For g

$AB \backslash CD$	$\bar{C}\bar{D}$ 00	$\bar{C}D$ 01	CD 11	$C\bar{D}$ 10
$\bar{A}\bar{B}$ 00	(0)	(1)	1 (3)	1 (2)
$\bar{A}B$ 01	1 (4)	1 (5)	(7)	1 (6)
AB 11	× (12)	× (13)	× (15)	× (14)
$A\bar{B}$ 10	1 (8)	1 (9)	× (11)	× (10)

$$g = B\bar{C} + A + \bar{B}C + C\bar{D}$$

Designing for g:

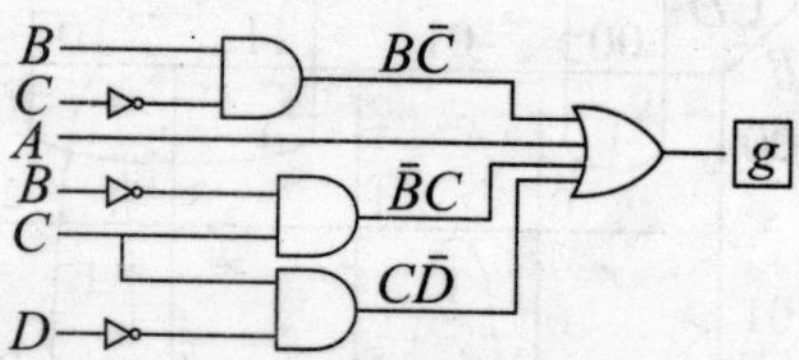

3.11 BCD ARITHMETIC ELEMENT

BCD code is used to represent decimal number, for example, in a calculator. Therefore, addition and subtraction are required to be performed in BCD code.

(*i*) BCD Adder.

(*ii*) BCD Subtractor.

3.11.1 BCD Adder

The 4-bit binary adder IC (7483) can be used to perform addition of BCD numbers. In this, if 4-bit sum output is not a valid BCD digit, or if a carry C_3 is generated, then decimal 6 (0110 binary) is to be added to sum to get the correct result. Figure shows a 1-digit BCD adder BCD adder can be cascaded to add numbers several digits long by connecting the carry-out of a stage to the carry in of next stage.

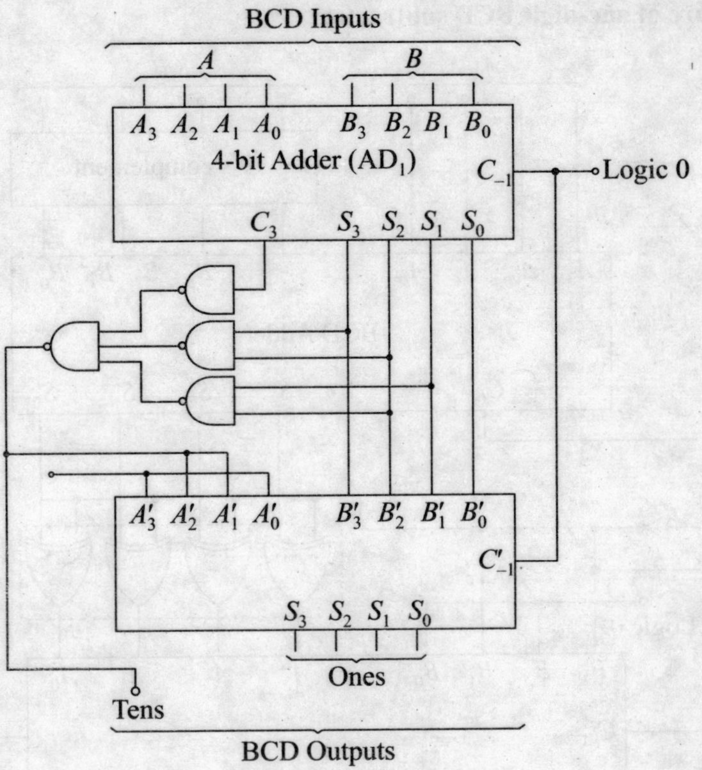

One digit BCD Adder

3.11.2 BCD Subtractor

For BCD Subtraction, 9'*s* complement of subtrahend is added to minuend. The 9'*s* complement of BCD number is given by 9 minus that number.

For example : 9'*s* complement of 7 $= 9 - 7 \Rightarrow 2$.

In BCD code we say that 9'*s* complement of 0111 is 0010.

In figure one digit BCD subtractor is given.

Figure of 9'*s* complement:

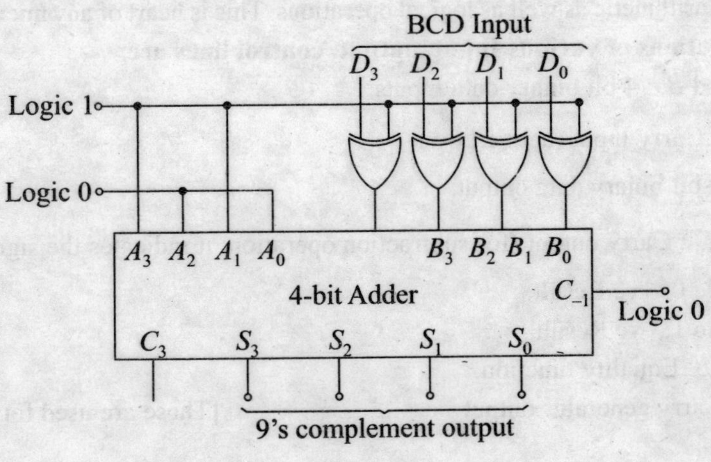

Figure of one-digit BCD subtractor:

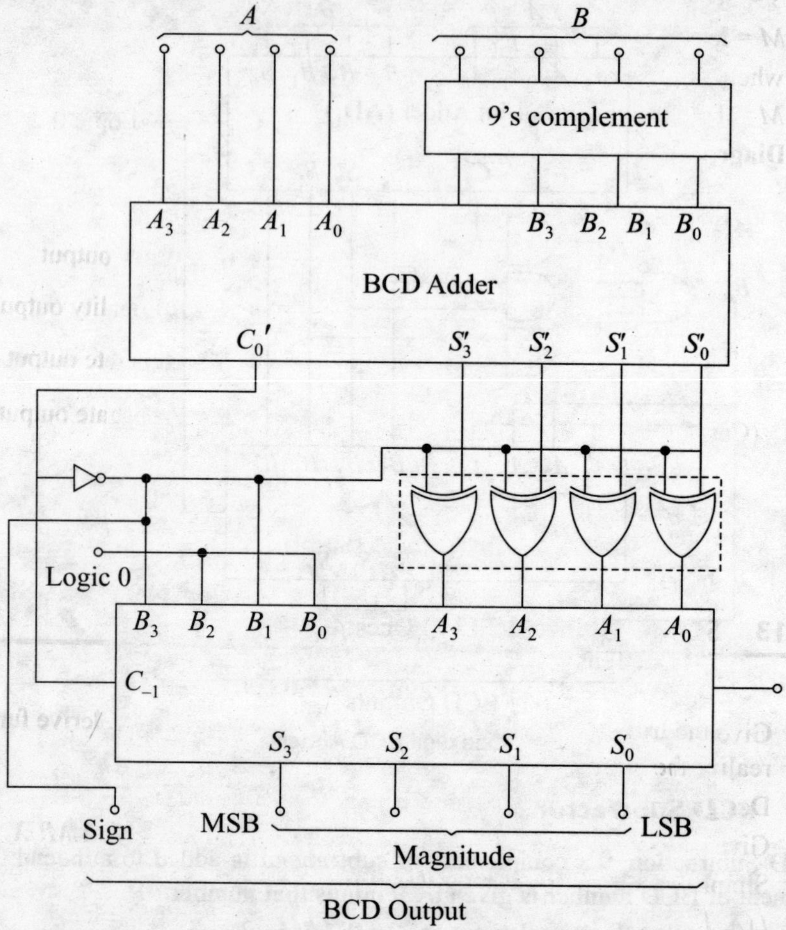

3.12 ARITHMETIC LOGIC UNIT (ALU)

A very popular and widely used combinational circuit is ALU which is capable of performing arithmetic as well as logical operations. This is heart of any microprocessor.

Functions of various Input, output, control lines are:

A and B : 4 bit binary data Inputs.

$\overline{C}_n$: Carry Input (active Low)

F: 4-bit binary data output.

$\overline{C}_{n+4}$: Carry output (for subtraction operation, it indicates the sign of output)

Logic 0: +ve Result.

Logic 1: –ve Result.

$A = B$: Equality function.

G: Carry generates output [These are used for cascading]

P: Carry propagate output

S = Select Input [Used to select any operation]

M = Mode Control

when $M = 0$ Arithmetic operation

$M = 1$ Logic operation.

Block Diagram:

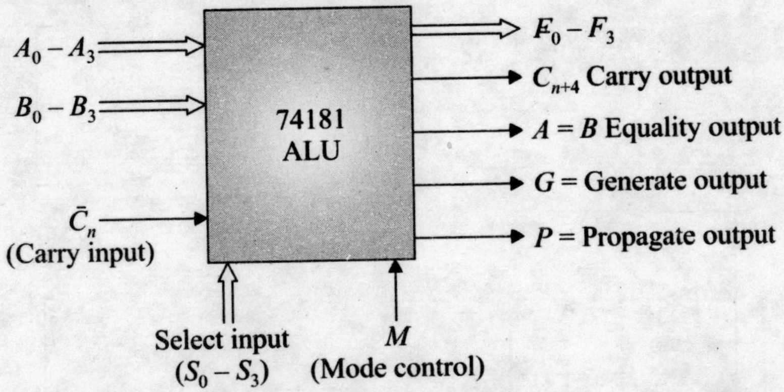

3.13 SOME UNSOLVED QUESTIONS

(Combinational Circuit)

1. Give the truth table for XOR and XNOR gates and hence derive function and realize the circuit diagram in SOP and POS form.
2. Design the circuit for full adder using two Half adder.
3. Give truth table for a full adder and design circuit using 4 : 1 *MUX*.
4. Simplify the function using *MUX* :
   $f(A, B, C, D) = \Sigma m\ (0, 2, 4, 7, 8, 9, 12, 15)$
   (*i*) Using 4 : 1 *MUX*. (*ii*) Using 8 : 1 *MUX*.
5. Design a combinational circuit which generates output 1 if a 4 bit number contains even number of 1's otherwise output is 0.
6. Implement the following function using decoder:
   (*a*) $m\ (2, 3, 5, 7) + m\ (1, 3, 6)$
   (*b*) $m\ (0, 3, 5, 10, 13) + m\ (2, 4, 6, 8, 11)$
7. Write difference between:
   (*a*) Multiplexer and Demultiplexer.
   (*b*) Decoder and Encoder.
   (*c*) Decoder and Demultiplexer.

UNIT 4

SEQUENTIAL CIRCUITS

Topics are:

- Introduction
- Flip-Flop
- Conversion of one Flip-Flop into Another Flip- Flop
- Register
- Counter
- Application of Counter
- Sequence Generator
- Designing of Sequential Circuits (Counters)

CHAPTER 4

Sequential Circuits

4.1 INTRODUCTION

The output of a sequential logic circuit depends on the present input and past outputs. So, we need memory element to store the past output. Figure 4.1 Shows the block diagram of sequential logic circuit. The combinational logic circuits operates on these inputs to produce various outputs. The outputs of memory elements, is given to inputs of combinational logic circuits.

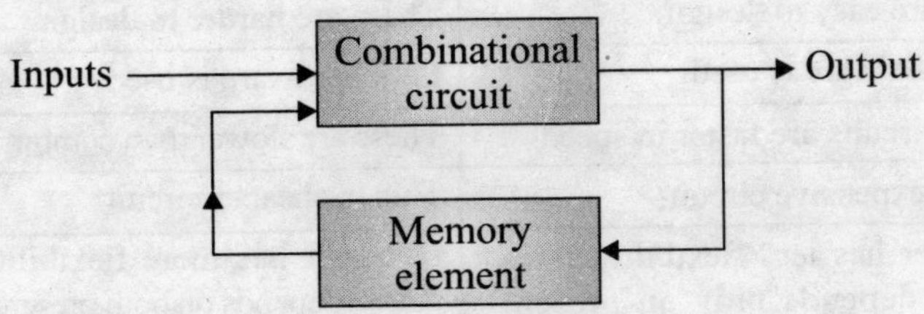

Fig. 4.1: Block diagram of sequential logic circuit.

The sequential circuit receives binary information from external inputs. These inputs and present state (past outputs) of memory elements determine the binary value of output circuit. They also determine the condition for changing the state in memory elements. Thus the next state of memory elements is a function of the external inputs, and the present state.

A sequential circuit is specified by a time sequence of inputs, outputs and internal states. Sequential circuits classified on basis of timing of their signals.

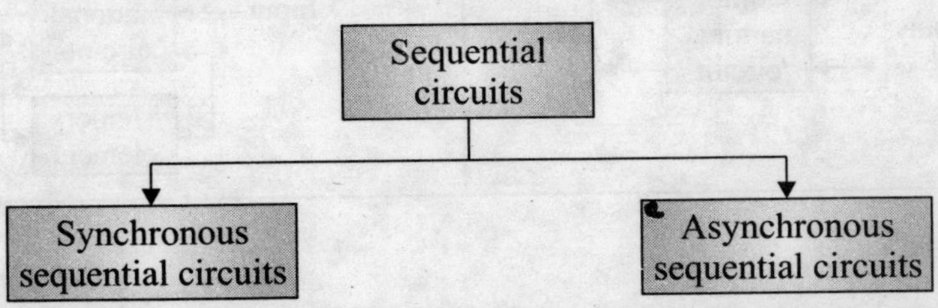

Synchronous sequential circuits: It is a system whose behaviour can be defined by the knowledge of its signals at discrete instants of time *i.e.*, signals can affect the memory element only at discrete instants of time. Also known as **clocked sequential circuit.**

Asynchronous Sequential circuits: It is a system in which change in input signals can affect memory element at any instant of time, *i.e.*, the behaviour of an asynchronous sequential circuits depends upon the order in which its input signals change and can affected at any instant of time. It is also known as **unclocked sequential circuits.**

Different sequential circuits:

1. Flip-Flops
2. Registers
3. Counters

4.1.1 Comparison between Combinational and Sequential

S.No.	*Combinational Circuits*	*Sequential Circuits*
1.	A circuit whose output is dependent only on inputs at that instant or the circuit that does not have feedback.	A circuit whose output depends not only on the present inputs, but also on the past history of inputs or circuit have feedback.
2.	No need of memory element.	Memory element is needed.
3.	These are easy to design.	These are harder to design.
4.	More hardware is used.	Less hardware is used.
5.	These circuits are faster in speed.	These are slower than combinational circuits.
6.	This is expensive circuit.	This is cheaper circuit.
7.	Designer has less flexibility since output depends only on present inputs.	Designer has more flexibility because the output depends on both present input and past history of input.
8.	The behaviour is defined by set of output functions only.	The behaviour is defined by set of output function and next state functions.
9.	Examples: Adders, Subtractor, Multiplexer, Demultiplexer.	Examples: Flip-Flops, counters, Registers, etc.
10.	Circuit Diagram: 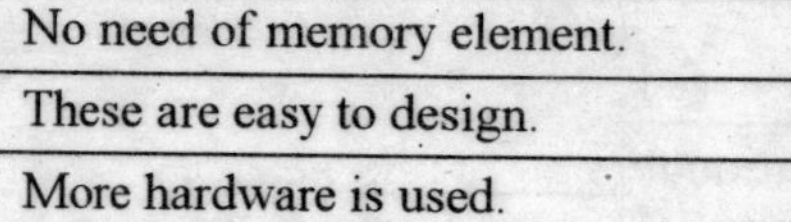	Circuit Diagram: 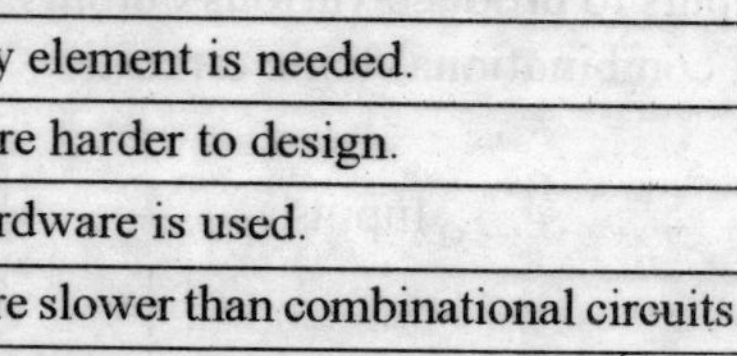

4.1.2 Latch

A flip-flop may be used to store or "Lock" one bit of information. This locking of information also known as Latching. Latches are useful for storing binary information and for the design of asynchronous sequential circuits. Shown in.

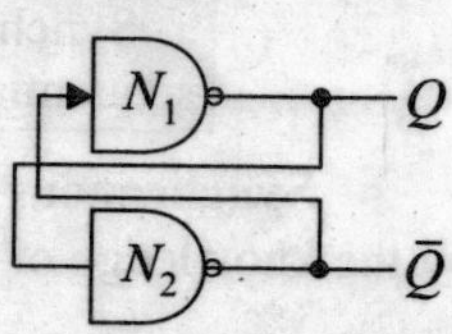

Fig. 4.2: Latch Circuit

4.2 FLIP-FLOP

It is basic memory element used to store one bit of information. It is also called Bistable multivibrator.

It have two states = '1' and '0'

Flip-flops have 2 output signals $(Q, \bar{Q})$ and one is always complement of other.

A flip-flop is a semiconductor device that has a digital output which can be toggled between two stable states by providing it appropriate digital input signals. Once output is put in one state, it remains there until a change in input causes it to toggle again.

Basic circuit diagram of flip-flop is shown by fig. 4.3.

Both Latch and flip-flop are circuit element whose output depends not only on current input, but also depends on previous input and output.

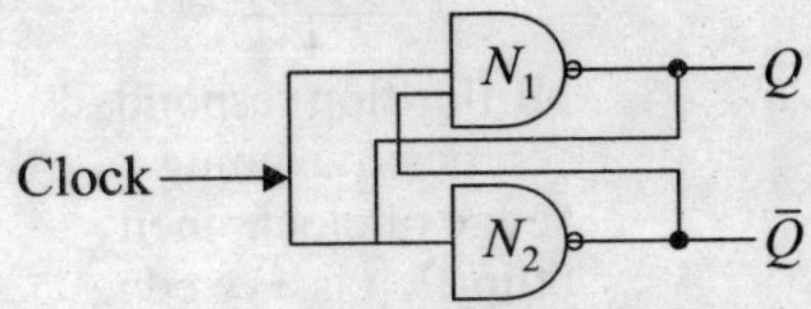

Fig. 4.3: Flip-flop

4.2.1 Comparison between Latch and Flip-Flop

S.No.	*Latch*	*Flip-Flop*
1.	It stores one bit of information.	It stores one bit of information.
2.	It do not have clock input.	It always have clock input.
3.	It is a level sensitive device.	It is edge sensitive device.
4.	These are asynchronous that output changes very soon after input changes.	These are synchronous version of Latch.
5.	These takes less gates to implement.	These takes more gates to implement.
6.	It takes less power.	It takes more power.
7.	These are faster.	These are slower.
8.	It have less propagation delay.	It have more propagation delay.

Flip-Flop is modified form of Latch with additional control input (clock/Enable). In this way, we can better control the output.

Clock is basically used to control flip-flop.

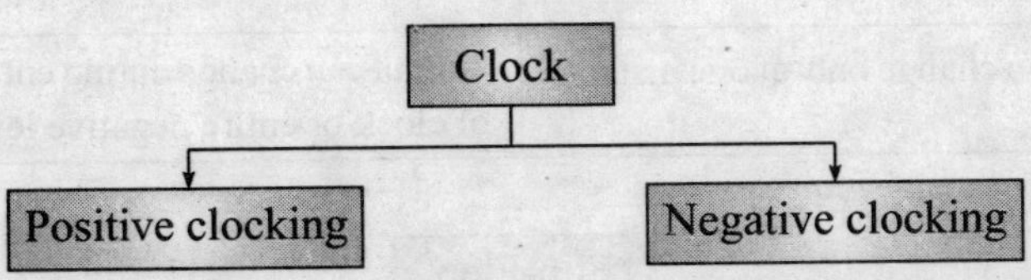

(*a*) **Positive Clocking:** If flip-flop changes its states at high clock pulse, then flip-flop is said to have positive clocking.

(*b*) **Negative Clocking:** If flip-flop changes its states, then it is negative clocking.

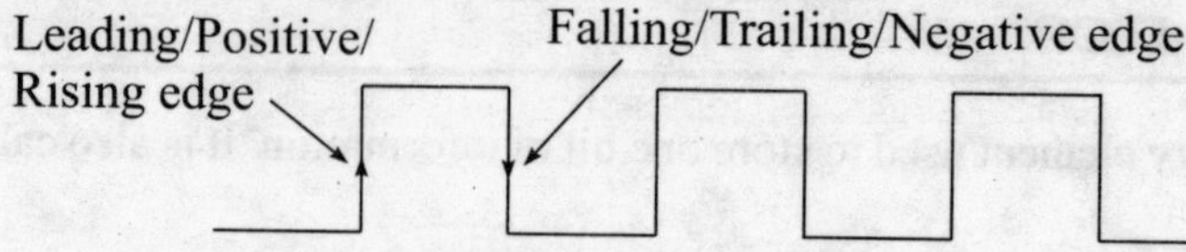

Gates are activated on positive spike and hence flip-flop responds according to applied input. This process is most commonly known as *edge triggering.*

Edge trigerring

- **Positive**: If flip-flop responds on positive going edge of clock, then flip-flop is +ve edge triggering.
- **Negative**: If flip-flop responds on negative going edge of clock, then flip-flop is said to have –ve edge triggering.

For negative clocking we can pass clock from the *R-C* circuit.

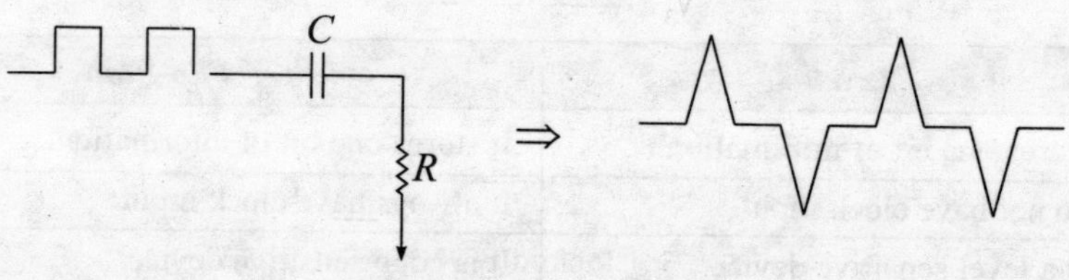

R-C differentiator circuit

Comparison between Edge Triggering and Level Triggering.

S.No.	*Edge Triggering*	*Level Triggering*
1.	Clock change in output takes place only on rising or falling edge of clock pulse. 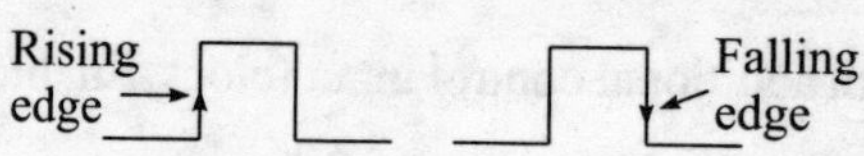	Clock change in output takes place on high clock or low clock. 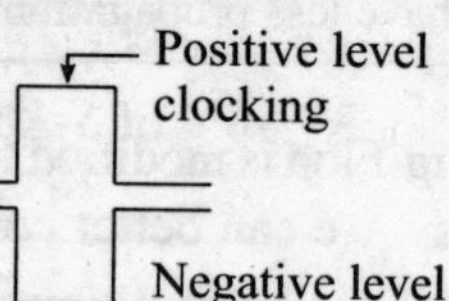
2.	Output can change only at one instant.	Output can change during entire positive level of clock or entire negative level of clock.

4.2.2 Types of Flip-Flop

Flip-Flop

- S-R Flip-Flop
- D Flip-Flop
- T Flip-Flop
- J-K Flip-Flop

There are basically four types of flip-flop and all the flip-flop have a particular circuit diagram and operation of flip-flop is shown by truth-Table.

4.2.3 S-R Flip-Flop

Any flip-flop in **clocked form.**

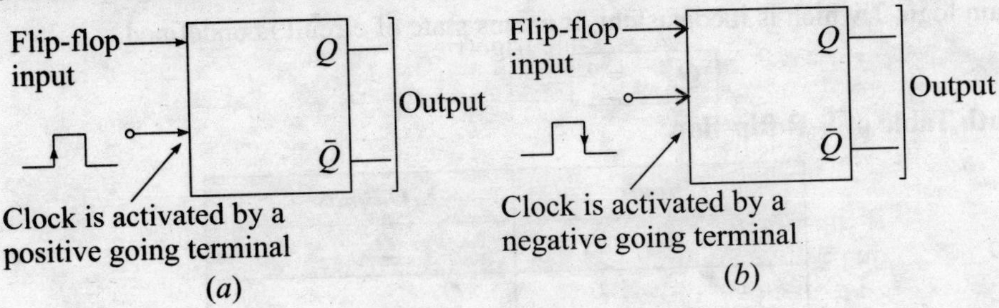

Fig. 4.4: Clocked Flip-flop

S-R is known as Set-Reset flip-flop. Figure 4.5 shows the clocked set-reset (*S-R*) flip-flop.

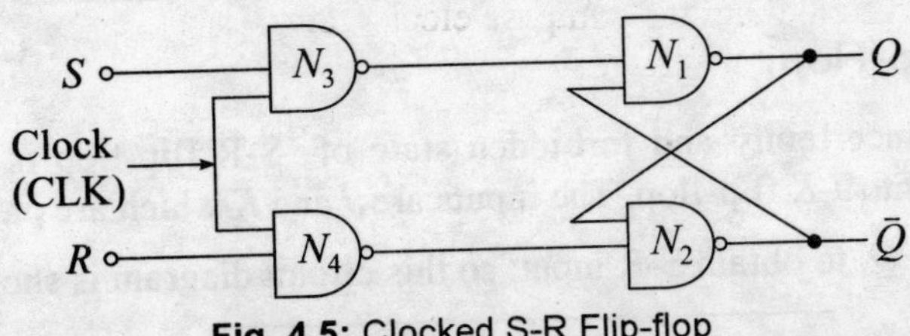

Fig. 4.5: Clocked S-R Flip-flop

Flip-flop will work when clock is given. In this two inputs *S* and *R*, two outputs Q, $\bar{Q}$. Then two inputs have four cases in which complete operation of *S-R* flip-flop is shown and finally given by truth table. In truth table we have

S_n and R_n denotes the inputs.

Q_n = Output during bit time *n*.

Q_{n+1} = Output Q after the pulse passes, *i.e.*, in the bit time $(n + 1)$.

Logic symbol of *S-R* flip-flop is:

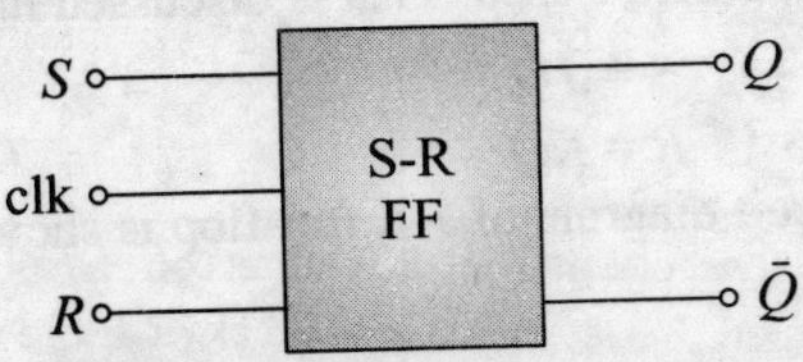

Case I: When $\boxed{S_n = R_n = 0}$ and clock pulse is applied, the output at the end of clock pulse is same as the output before the clock pulse,

$$\boxed{Q_{n+1} = Q_n.}$$

Case II: When $\boxed{S_n = 0, R_n = 1}$ and in this case output will be:

$$\boxed{Q_{n+1} = 0 \text{ (Reset)}}$$

Case III: When $S_n = 1, R_n = 0$ and in this case output will be:

$$Q_{n+1} = 1(\text{set})$$

Case IV: When $S_n = R_n = 1$, and clock is present then output of gates N_3 and N_4 are both 0, making one of the input of N_1 and N_2 NAND gate 0 and Q, $\bar{Q}$ both will attain logic 1 which is inconsistent. It means state of circuit is undefined.

$$Q_{n+1} = ?$$

Truth Table of S-R flip-flop:

Inputs		*Outputs*
S_n	R_n	Q_{n+1}
0	0	Q_n
0	1	0 (Reset)
1	0	1 (Set)
1	1	? (Forbidden state)

4.2.4 J-K Flip-Flop

The undefined uncertainty and forbidden state of S-R flip-flop is eliminating by converting S-R into *J-K* flip-flop. The inputs are *J* and *K* which are passed from AND gate with $\bar{Q}$ and Q, to obtain S-R input; so this circuit diagram is shown by Fig. 4.6.

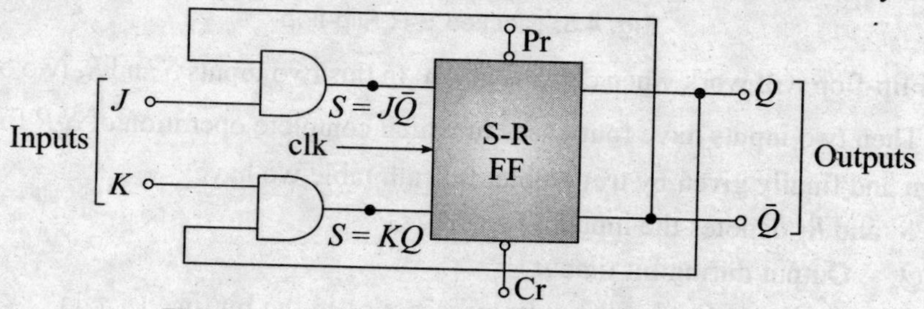

Fig. 4.6: S-R flip-flop converted into *J-K* flip-flop

Here two New terms Cr, Pr used. That is discussed in next section.

$$S = J\bar{Q}$$
$$R = KQ$$

Now the basic circuit diagram of *J-K* flip-flop is shown in Fig. 4.7.

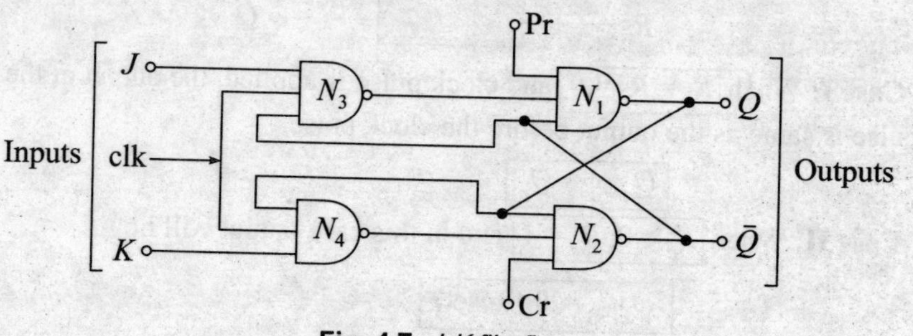

Fig. 4.7: *J-K* flip-flop

A symbol of *J-K* flip-flop is:

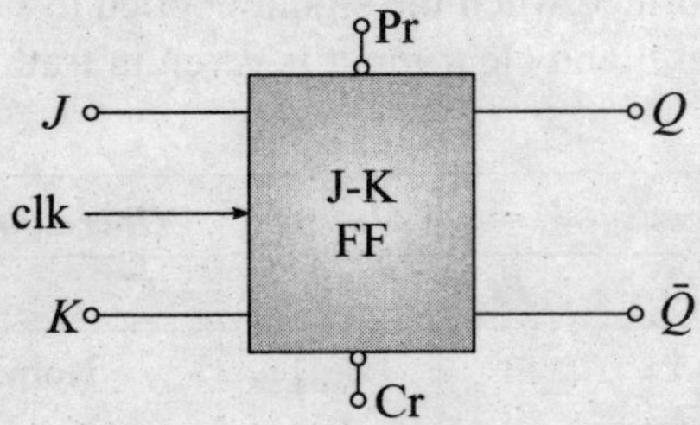

After this complete operation of *J-K* flip-flop according to $S = J\bar{Q}$, $R = KQ$ is given in truth table.

Data Inputs		***Output***		***S-R flip-flop***		***Output***	
J	***K***	Q_n	$\bar{Q}_n$	S_n	R_n	Q_{n+1}	
0	0	0	1	0	0	0	$= Q_n$
0	0	1	0	0	0	1	
0	1	0	1	0	0	0	= 0 (Reset)
0	1	1	0	0	1	0	
1	0	0	1	1	0	1	= 1(Set)
1	0	1	0	0	0	1	
1	1	0	1	1	0	1	$= \bar{Q}_n$
1	1	1	0	0	1	0	

Final truth Table of *J-K* flip-flop:

Inputs		***Outputs***
J_n	K_n	Q_{n+1}
0	0	Q_n
0	1	0 (Reset)
1	0	1 (Set)
1	1	$\bar{Q}_n$ → Race Around Condition.

4.2.5 Preset and Clear Signal

In fig. 4.5 when the power is switched on, the state of circuit is uncertain. It may come set ($Q = 1$) or reset ($Q = 0$) state. These inputs may be applied at any time between clock pulses and are not in synchronism with the clock. So, we can say these are asynchronous inputs.

An *S-R* flip-flop with preset (Pr) and clear(Cr) is shown in Fig. 4.8.

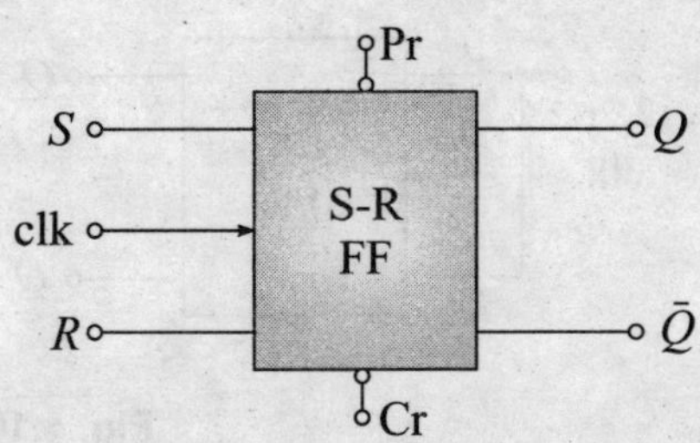

Fig. 4.8: S-R flip-flop with Preset and clear

Bubbles are used for Pr and Cr inputs, which means these are active-Low. *i.e.*, intended function is performed when the signal applied to *Pr* and *Cr* is low operation of *S-R* flip-flop using preset and clear input is given in truth table.

Truth Table:

Inputs			*Output*	*Operation Performed*
Clk	*Cr*	*Pr*	*Q*	
1	1	1	Q_{n+1}	Normal flip-flop
0	0	1	0	Clear (Cr)
0	1	0	1	Preset (Pr)

4.2.6 D-Type Flip-Flop

Here D stands for Data input. If we use two middle rows of *S-R* flip-flop truth table then that is truth table of D-type flip-flop. It is shown in Fig. 4.9.

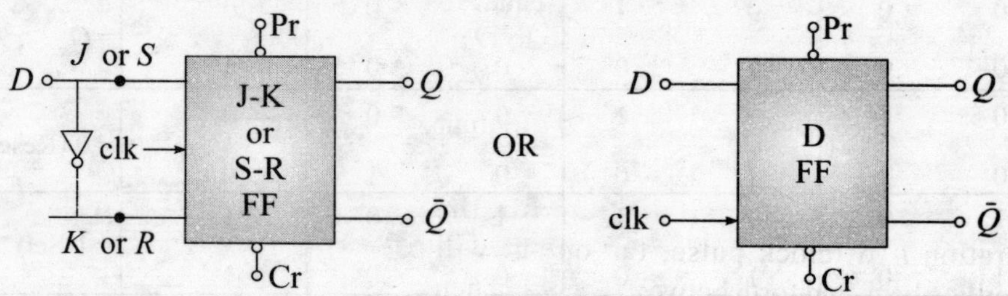

Fig. 4.9: D-Type flip-flop

Truth Table:

Input	*Output*
D	Q_{n+1}
0	0
1	1

In D-type flip-flop what is input that is output but the values of output will be given by some delay. So, that's why we can say it **Delay flip-flop** also.

- **T-Type Flip-Flop**

In a *J-K* flip-flop, if *J* = *K*, then resulting flip-flop is known as T-flip-flop. It has one input (*T*). It means it is a **toggle flip-flop.** The circuit diagram for this is shown in Fig. 4.10.

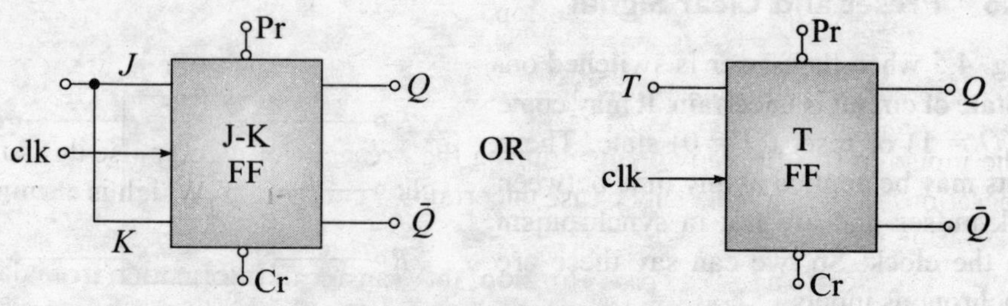

Fig. 4.10: T-type flip-flop

Operation of the flip-flop is shown by truth table when *J* = *K*.

Inputs	*Outputs*
T	Q_{n+1}
0	Q_n
1	$\bar{Q}_n$

S-R flip-flop cannot give *T*- type flip-flop because $S = R = 1$ will not determine the state.

4.2.7 Race Around Condition

In case of *J-K* flip-flop when $\boxed{J = K = 1,}$ then output $Q_{n+1} = \bar{Q}_n$ *i.e.*, output is the invert of previous output and this yields to continuous change in value of Q_{n+1} from zero to one and one to zero in a single clock pulse is term as **Race around condition.**

When inputs $J = K = 1$ and $Q = 0$ and *a* pulse shown in Fig. 4.11 is applied at clock input. After a time interval Δt equal to propagation delay through two NAND gates in series, the output will change to $Q = 1$.

When inputs $J = K = 1$ and $Q = 1$, and after another time interval of Δt, the output will change back to $Q = 0$.

Hence, we conclude that for the duration t_p of clock pulse, the output will oscillate back and forth between 0 and 1. Then it is Race around condition and this equation must satisfied.

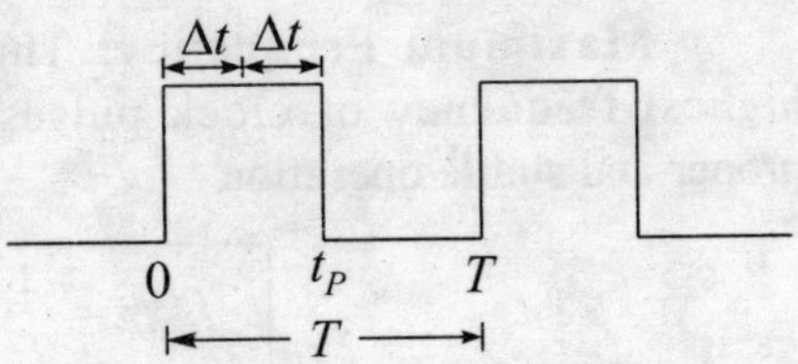

Fig. 4.11: Clock pulse

$$\boxed{0 < \Delta t < t_p}$$

t_p = Time period where pulse is high.

4.2.7.1 Methods to Remove Race-Around Condition

There are three methods to remove race-Around conditions, those are:

(*i*) Flip-flop must follow the equation:

$$\boxed{t_p < \Delta t < T}$$

That is not possible.

(*ii*) Use Edge-Triggered flip-flop.

(*iii*) Use of Master-Solve flip-flop.

4.2.8 Edge-Triggered Flip-Flop

The inputs to flip-flop any change during the presence of clock pulse due to certain operations in the system. This cause uncertainily in flip-flop. Which is eliminated by edge-triggered flip-flop.

In case of edge-triggered flip-flop, the transfer of information from data input to the output of flip-flop occurs at positive (or negative) edge of clock pulse. The only time the outputs can change state is during the brief interval of time when the clock signal is making a transition from 0 to 1 (↑), or in some circuits, from 1 to 0 (↓) states. A flip-flop which responds only to rising (or falling) edge is referred to as positive-edge triggering (or negative-edge triggering). The data lock-out occurs at end of edge.

The logic symbol for this is shown in Fig. 4.12.

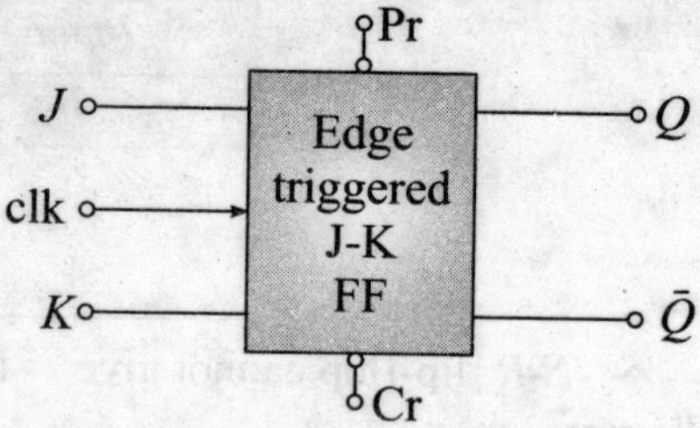

Fig. 4.12: Ege triggered flip-flop

Timing specifications of edge-triggered flip-flop is illustrated in Fig. 4.13 and explained below.

Set-up time: It is time required for the input data to settle in before the triggering edge of clock.

Hold Time: It is time for which the data must remain stable after triggering pulse of clock.

Maximum Frequency: This is highest frequency of clock pulses for proper and stable operation.

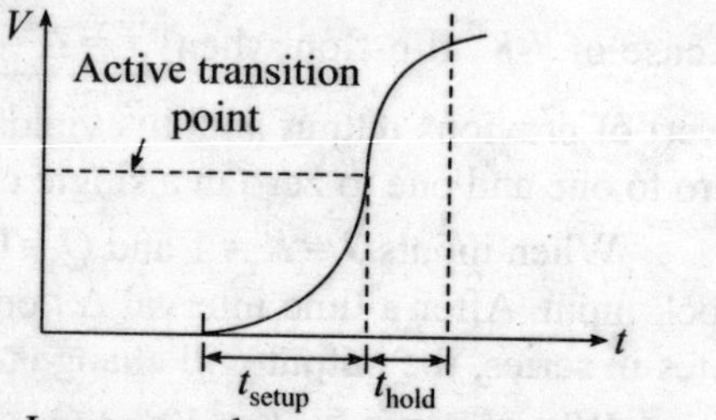

Fig. 4.13: Timing specification of edge-Triggered FF

$$f_{max} = \frac{1}{T_c}$$

4.2.9 Master-Slave Flip-Flop

A master-slave *J-K* flip-flop is the cascade of two *S-R* flip-flops, with feedback from the outputs of seconds to inputs of first is shown in Fig. 4.14. Positive clock pulse is applied to first flip-flop and clock pulses are inverted before these are applied to second flip-flop.

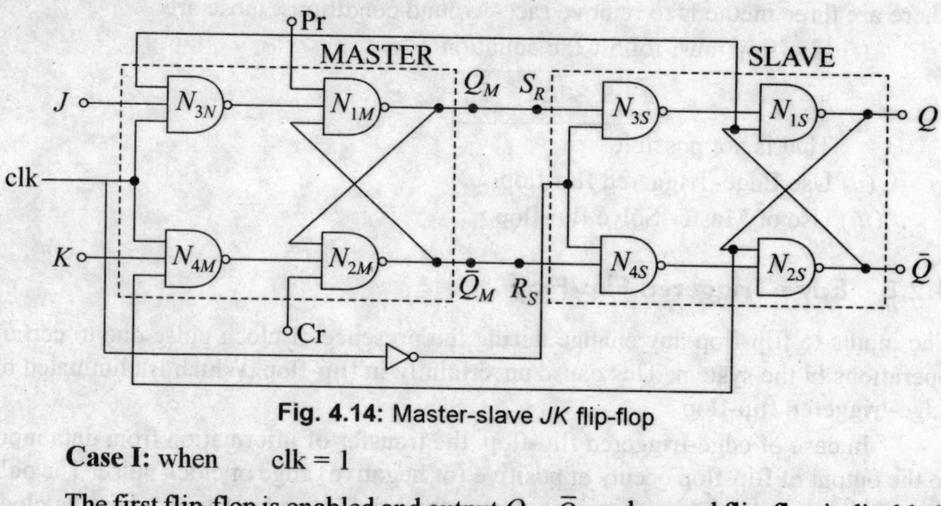

Fig. 4.14: Master-slave *JK* flip-flop

Case I: when clk = 1

The first flip-flop is enabled and output Q_m, $\bar{Q}_m$ and second flip-flop is disabled.

Case II: when clk = 0

The first flip-flop is disabled and second flip-flop is enabled and output Q and $\bar{Q}$ follows the output of first flip-flop Q_m, $\bar{Q}_m$.

It simple mean second flip-flop follows the first flip-flop.

So, first flip-flop is master.

Second flip-flop is slave.

Hence, it is referred as master-slave flip-flop. In this circuit N_{3M}, N_{4M} does not change during the clock pulse, therefore Race-around condition does not exist. State of master-slave flip-flop changes at negative transition (trailing edge) of pulse.

Symbol of master slave flip-flop:

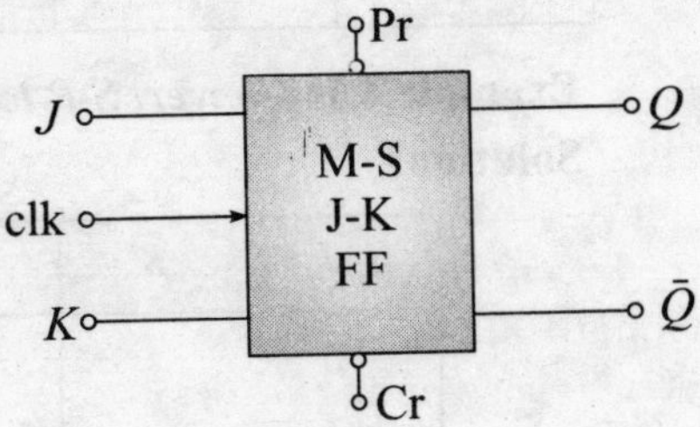

4.2.10 Application of Flip-Flop

(*i*) Parallel data storage.

(*ii*) It can utilized in digital counters.

(*iii*) It also used in switch debouncer circuit.

(*iv*) We can divide the frequency of periodic waveform by using flip-flop.

- It means it is "Divide by 2"
- Divide by 2^2

⋮

- Divide by 2^N.

4.3 CONVERSION OF ONE FLIP-FLOP INTO ANOTHER FLIP- FLOP

Steps of converting one flip-flop into another flip-flop:

1. Write the excitation of that flip-flop which is output.
2. After this present state (Q_n) and next state (Q_{n+1}) table.
3. Next write excitation table of input.
4. Next is to solve that *k*-map for inputs *i.e.*, J, K, Q_n (3-variable *k*-map) for two output (S, R).
5. Design the circuit by using the expressions obtained from *k*-map.

4.3.1 Excitation Table of Flip-Flops

Excitation table defines the procedure of getting inputs from outputs (*i.e.*, present and next state output). It is used in order to conversion of flip-flop and to design synchronous counter.

Excitation Table of all flip-flop

Present State (Q_n)	*Next State* (Q_{n+1})	*S Flip*	*R Flop*	*J Flip*	*K Flop*	*T FF*	*D FF*
0	0	0	×	0	×	0	0
0	1	1	0	1	×	1	1
1	0	0	1	×	1	1	0
1	1	×	0	×	0	0	1

Example 4.1: *Convert S-R to J-K flip-flop.*

Solution:

J	*K*	Q_n	Q_{n+1}	*S*	*R*
0	×	0	0	0	⊗
1	×	0	1	1	0
×	1	1	0	0	①
×	0	1	1	⊗	0

The *k*-map for *J*, *K*, Q_n is:

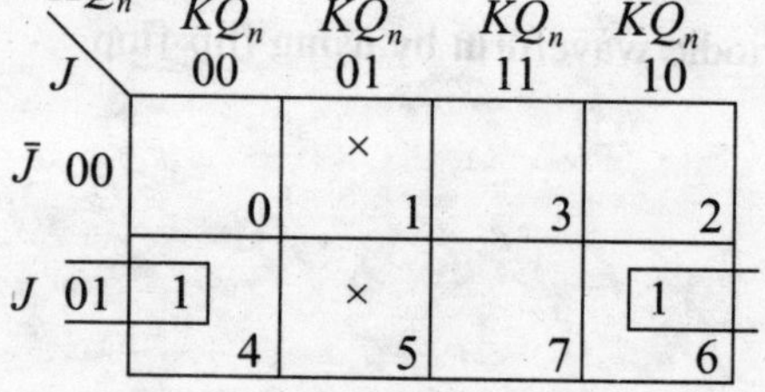

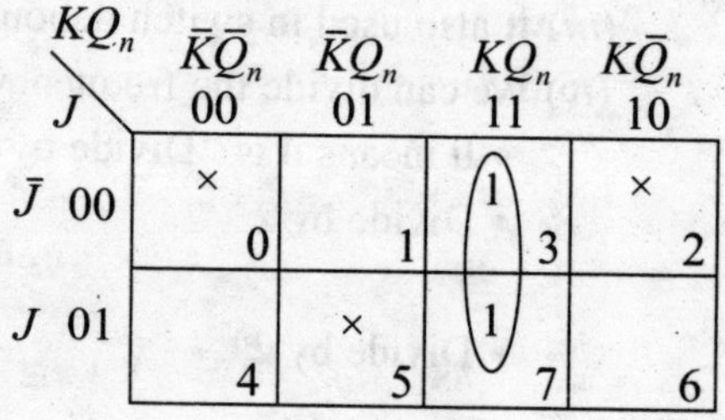

$S = J\bar{Q}_n$ $R = KQ_n$

Designing is given:

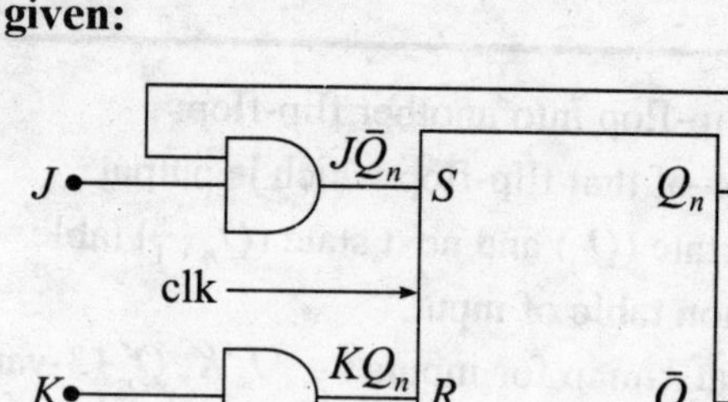

Example 4.2: *Convert S-R to T flip-flop.*

Solution:

T	Q_n	Q_{n+1}	*S*	*R*
0	0	0	0	⊗
1	0	1	①	0
1	1	0	0	①
0	1	1	⊗	0

The k-map for T, Q_n is:

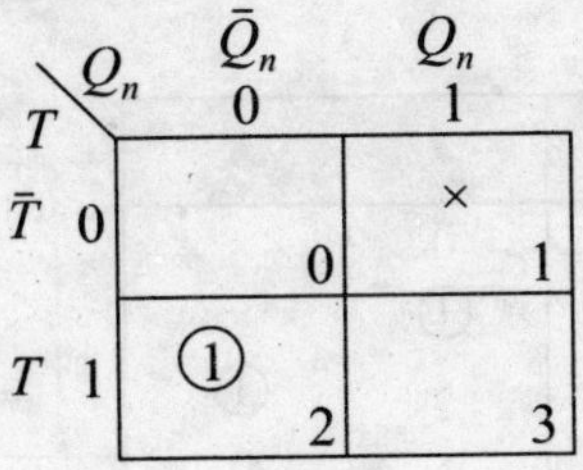

$\boxed{S = T\bar{Q}_n}$

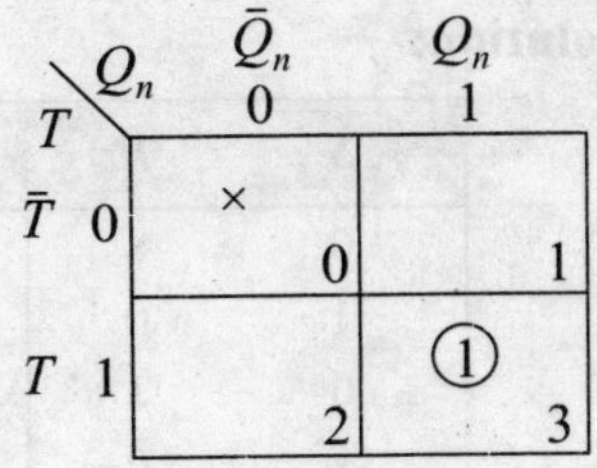

$\boxed{R = TQ_n}$

Designing is given:

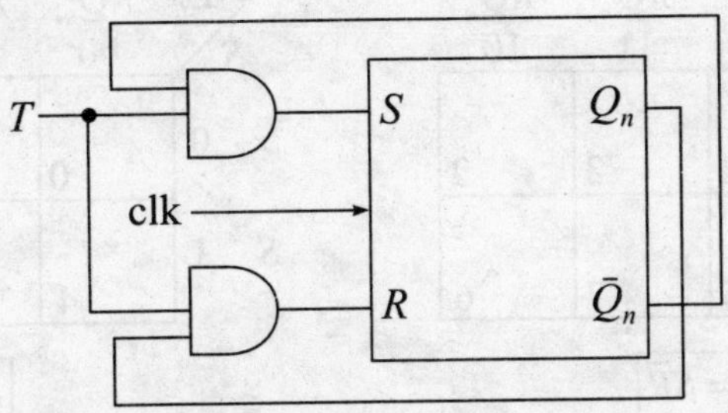

Example 4.3: *Convert J-K to D flip-flop*

Solution:

D	Q_n	Q_{n+1}	S	R
0	0	0	0	×
1	0	1	①	×
0	1	0	×	①
1	1	1	×	0

k-map for D, Q_n

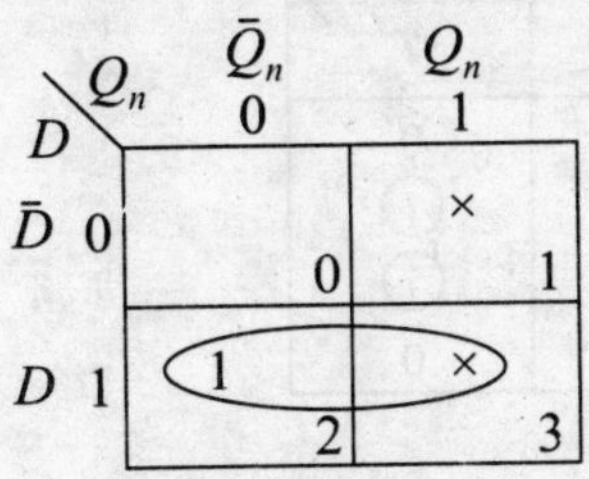

$\boxed{J = D}$

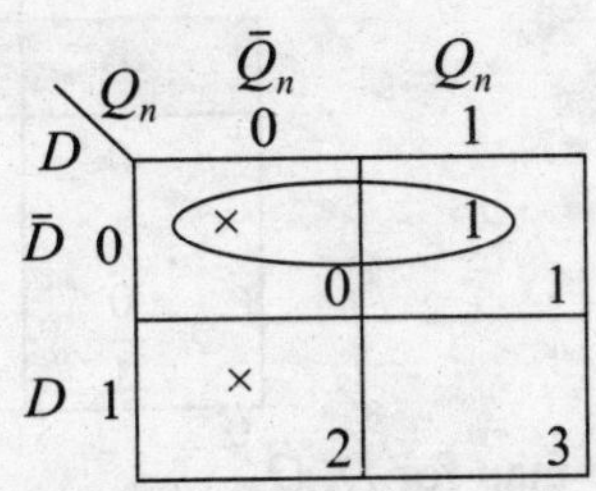

$\boxed{K = \bar{D}}$

Designing is given:

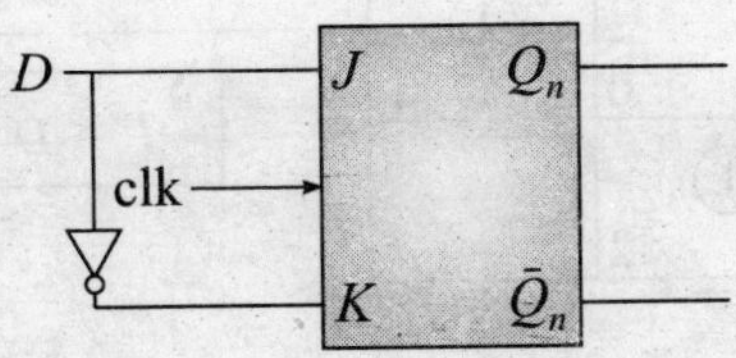

Example 4.4: *Convert J-K to SR flip-flop.*

Solution:

S	R	Q_n	Q_{n+1}	J	K
0	×	0	0	0	×
1	0	0	1	①	×
0	1	1	0	×	①
×	0	1	1	×	0

k-map for S, R, Q_n:

$S \backslash RQ_n$	$\bar{R}\bar{Q}_n$ 00	$\bar{R}Q_n$ 01	RQ_n 11	$R\bar{Q}_n$ 10
$\bar{S}$ 0	(0)	× (1)	× (3)	(2)
S 1	1 (4)	× (5)	(7)	(6)

$$\boxed{J = S\bar{R}}$$

$S \backslash RQ_n$	$\bar{R}\bar{Q}_n$ 00	$\bar{R}Q_n$ 01	RQ_n 11	$R\bar{Q}_n$ 10
$\bar{S}$ 0	× (0)	(1)	1 (3)	× (2)
S 1	× (4)	(5)	(7)	(6)

$$\boxed{K = \bar{S}R}$$

Designing is given:

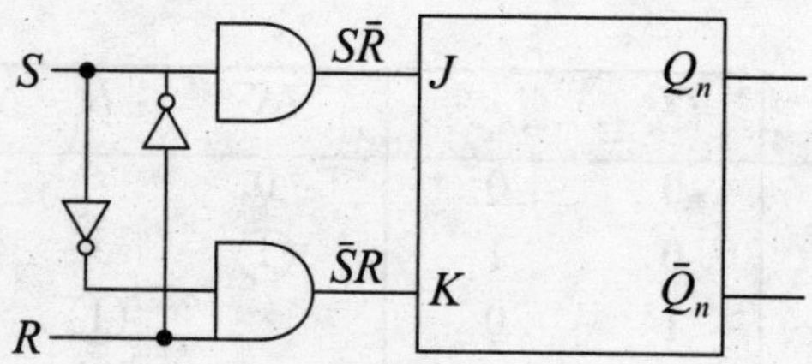

Example 4.5: *Convert T to D flip-flop.*

Solution:

D	Q_n	Q_{n+1}	T
0	0	0	0
1	0	1	①
0	1	0	①
1	1	1	0

k-map for D, Q_n:

$D \backslash Q_n$	$\bar{Q}_n$ 0	Q_n 1
$\bar{D}$ 0	(0)	① (1)
D 1	① (2)	(3)

$$T = \bar{D}Q_n + D\bar{Q}_n$$

$$\boxed{T = D \oplus Q_n}$$

Designing is given:

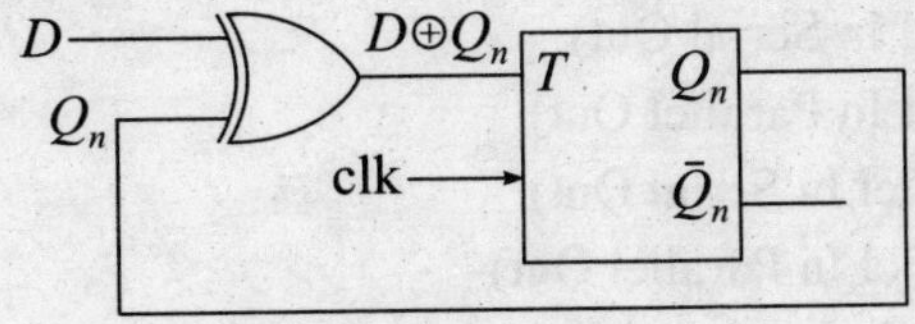

Example 4.6: *Convert D to J-K flip-flop.*

Solution:

J	K	Q_n	Q_{n+1}	D
0	×	0	0	0
1	×	0	1	①
×	1	1	0	0
×	0	1	1	①

k-map for J, K, Q_n:

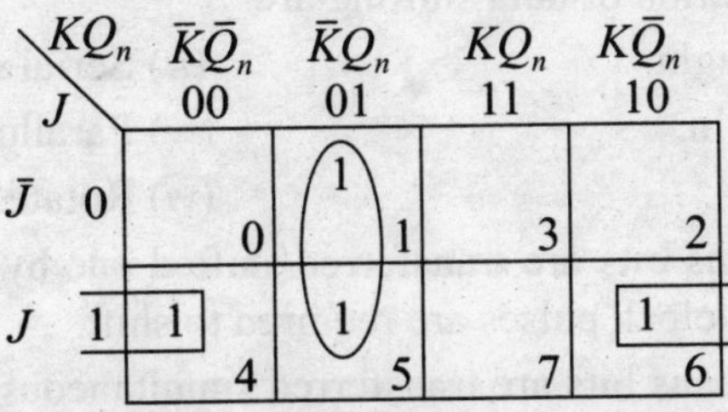

$$D = J\bar{Q}_n + \bar{K}Q_n$$

Designing is given:

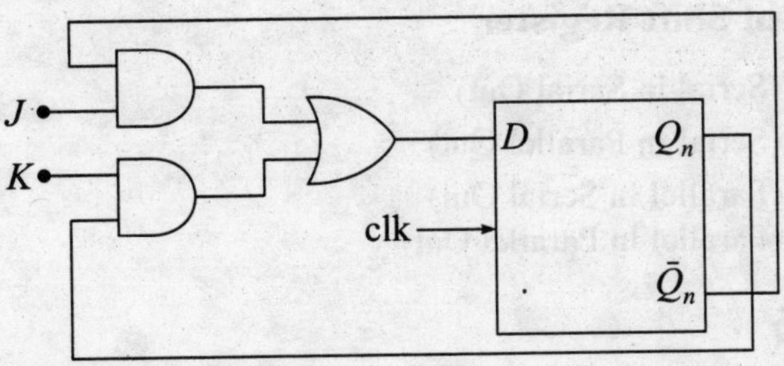

4.4 REGISTER

An array of flip-flop is required to store information, the number of flip-flop required being equal to the number of bits in binary word (one flip-flop for each bit) and referred as Register.

The data can be entered in serial (One bit at a time) or in parallel form (all bits simultaneously) can be retrieved in serial or parallel from.

Register can classified depending upon the way in which data are entered and retrieved.

There are 4 possible mode of operation:

(*a*) SISO (Serial In Serial Out)

(*b*) SIPO (Serial In Parallel Out)

(*c*) PISO (Parallel In Serial Out)

(*d*) PIPO (Parallel In Parallel Out)

Register in which data are entered or out in serial form are referred as **Shift Register** since bits are shifted in flip-flop with occurrence of clock pules either in Right direction (Right shift Register) or in left direction (left-shift Register).

1. Right-shift Register.
2. Left-shift Register.
3. Bi-directional shift Register.
4. Universal Register.

4.4.1 Shift Register

These are commonly used for temporary storage of data with in digital system. The shifting of data is the process of movement of bits into or out of register with the help of clock input. Classification of data shifting are:

(*i*) Serial shift Right.

(*ii*) Serial shift Left.

(*iii*) Parallel shift In.

(*iv*) Parallel shift Out.

(*v*) Rotate Left.

(*vi*) Rotate Right.

Serial Shift means bits are transferred/shifted one by one according to clock pulse *e.g.*, for 4 bits, '4' clock pulses are required to shift.

Parallel Shift means bits are transferred simultaneous means if '4' bits are to be transferred then all '4' bits are passed at same time, so only '1' clock pulse is used to transmit the data.

4.4.2 Types of Shift Register

(*i*) SISO (Serial In Serial Out)

(*ii*) SIPO (Serial In Parallel Out)

(*iii*) PISO (Parallel In Serial Out)

(*iv*) PIPO (Parallel In Parallel Out)

4.4.2.1 SISO

In this data bits to be entered serially and data bits are outed in serial order.

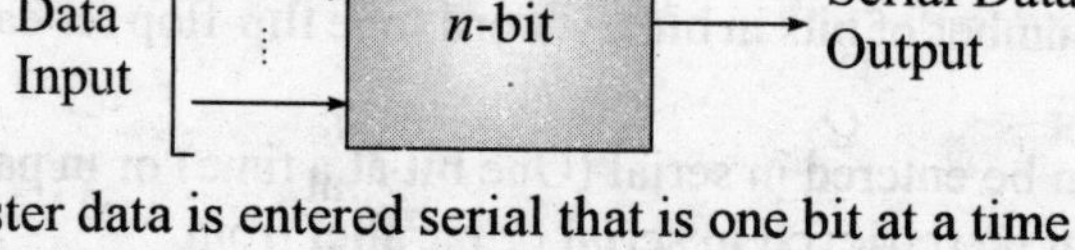

In this register data is entered serial that is one bit at a time and then next bit at next clock pulse. Generally one clock pulse is consumed when one bit of data is stored. The circuit using *D* flip-flop is shown as in Fig. 4.15.

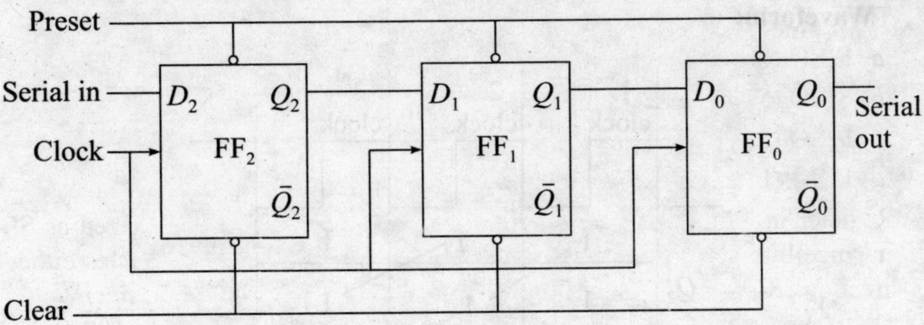

Fig. 4.15: SISO

Data is entered serial but LSB is stored first and then next and at last MSB is stored.

But at serial out LSB is taken out firstly that next bit and at last MSB is taken out.

Suppose we want to store 110 in SISO register

Serial IN:

Clock	Q_2	Q_1	Q_0
Initial	0	0	0
1st	0	0	0
2nd	1	0	0
3rd	1	1	0

Value stored in 3rd clock

Wave form:

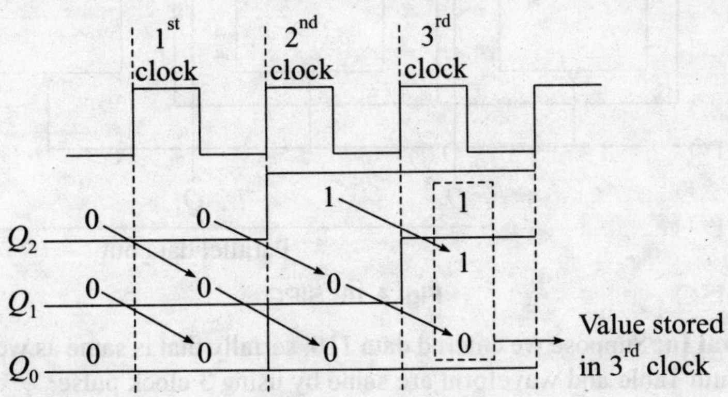

Serial Out:

Clock	Q_2	Q_1	Q_0 (Serial output)
1st	1	1	0
2nd	1	1	1
3rd	1	1	1

LSB, Serial output, MSB

Waveform:

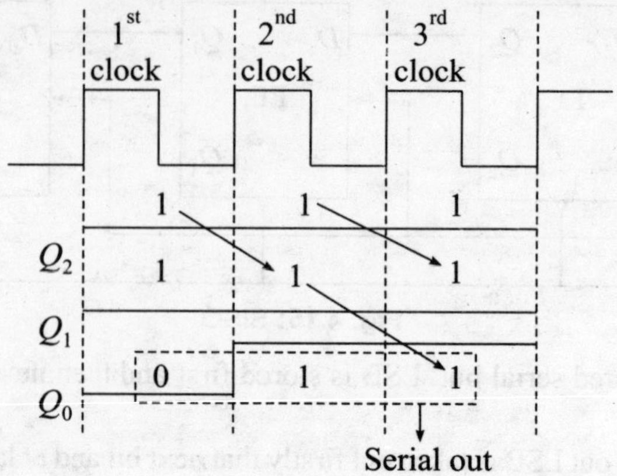

4.4.2.2 SIPO

In serial in parallel out the data is entered serially using three clock pulse but data is out parallely means once the data is stored, each bit appears on its respective output line and all the bits are available simultaneously, rather than bit by bit. It is shown in Fig. 4.16.

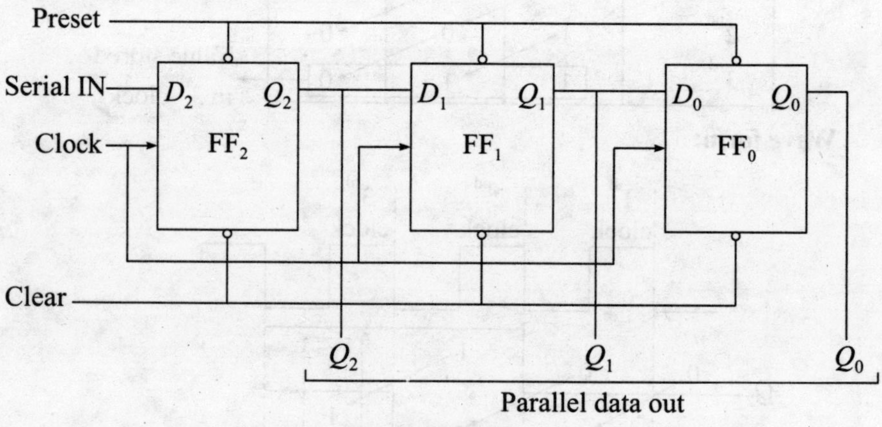

Fig. 4.16: SIPO

Serial In: Suppose we entered data 110, serially that is same as we did in SISO register. Truth Table and waveform are same by using 3 clock pulse.

Parallel Out: For the parallel output only one clock pulse is required to arrive data at output terminal.

Clock	Q_2	Q_1	Q_0
Ist clock	1	1	0

→ Parallel out

Waveform

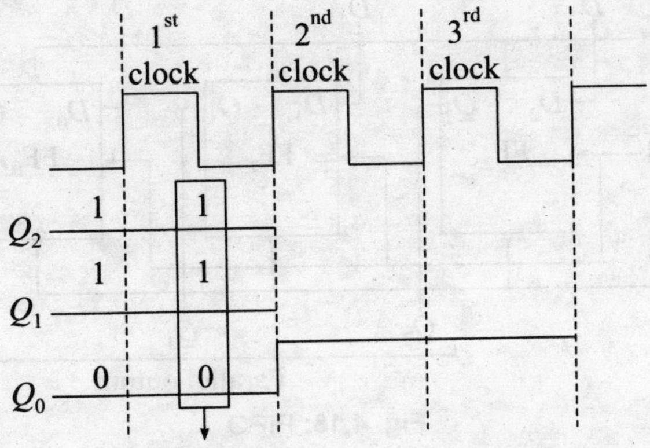

4.4.2.3 PISO

This parallel in serial out works as reverse operation of SIPO (serial In Parallel out). In PISO data is entered parallely but data is out serially. In this for parallel in only single bit is required but three clock pulse are required for serial out data. It is shown in Fig. 4.17.

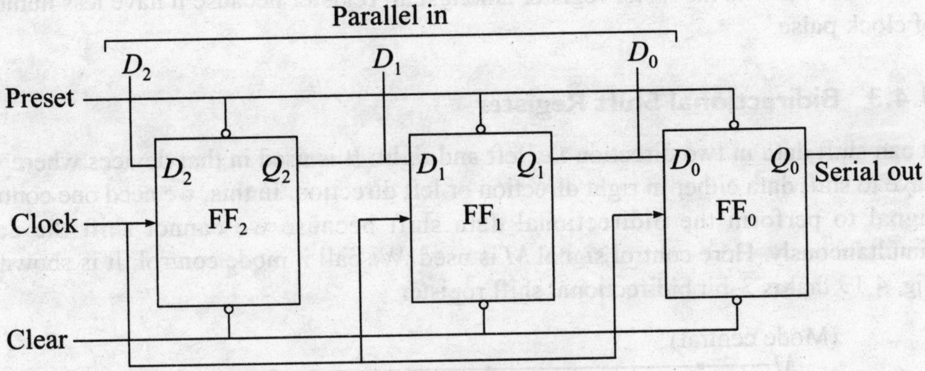

Fig. 4.17: PISO

Parallel IN: If we want to entered data 110 parallely, there is single clock pulse is used.

Clock	D_2	D_1	D_0
Initial	0	0	0
Ist	1	1	0

Wave form is same as parallel out in SIPO.

Serial Out: It is similar to serial out in SISO.

4.4.2.4 PIPO (Buffer Register)

It is Parallel in Parallel out in which data is entered parallely and data is out parallely. It is shown in Fig. 4.18.

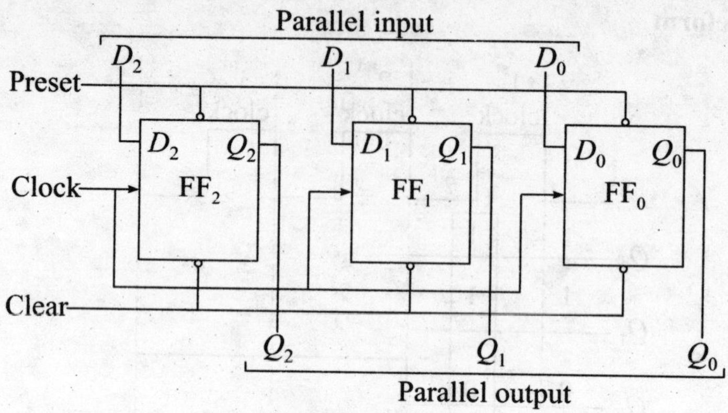

Fig. 4.18: PIPO

Parallel IN: It is same as parallel in PISO.

Parallel out: It is same as parallel out in SIPO.

Total Clock pulse = 1 for parallel Input.

= 1 for parallel output.

= 1 + 1

= 2

So, PIPO is the faster register amongst all register because it have less number of clock pulse.

4.4.3 Bidirectional Shift Register

It can shift data in two direction *i.e.* left and right: It is used in that devices where we have to shift data either in right direction or left direction. In this, we need one control signal to perform the bidirectional data shift because we cannot shift the data simultaneously. Here control signal M is used. We call it mode control. It is shown is Fig. 4.19 that is 3-bit bidirectional shift register.

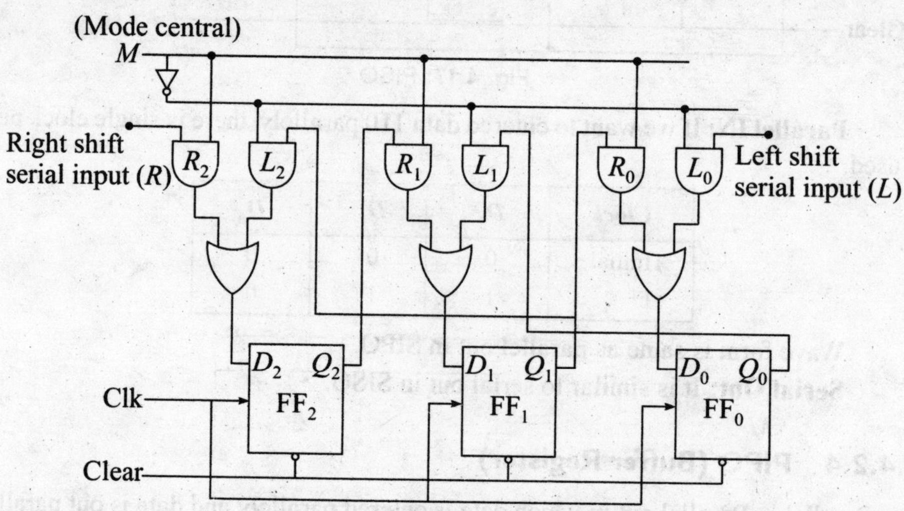

Fig. 4.19: Bidirectional Shift Register

Working of this depend upon mode control (M)

Case I : When $M = 1$, then all (R_2, R_1, R_0) gates are enable and hence the data applied at input R is shift towards right as soon as the positive edge of clock pulses are arrived. The output is taken from the terminal Q_0 of flip-flop (FF_0).

Suppose we want to shift right is '1'.

$M = 1$	Shift Right Input (R)	Q_2	Q_1	Q_0
	1	1	0	0
	1	1	1	0
	1	1	1	1

In this table, everytime, the shift right input (R) is feeded with logic '1'.

Case II: When $M = 0$, then all (L_2, L_1, L_0) gates are enabled and all (R_2, R_1, R_0) are giving output 0, *i.e.* they are disabled. Here output is taken from the terminal Q_2 of flip-flop (FF_2).

Here data will be shift towards left.

$M = 0$	Shift Leftt Input (L)	Q_2	Q_1	Q_0
	1	0	0	1
	1	0	1	1
	1	1	1	1

Here, everytime the shift left input (L) is feeded with Logic '1'.

Waveform:

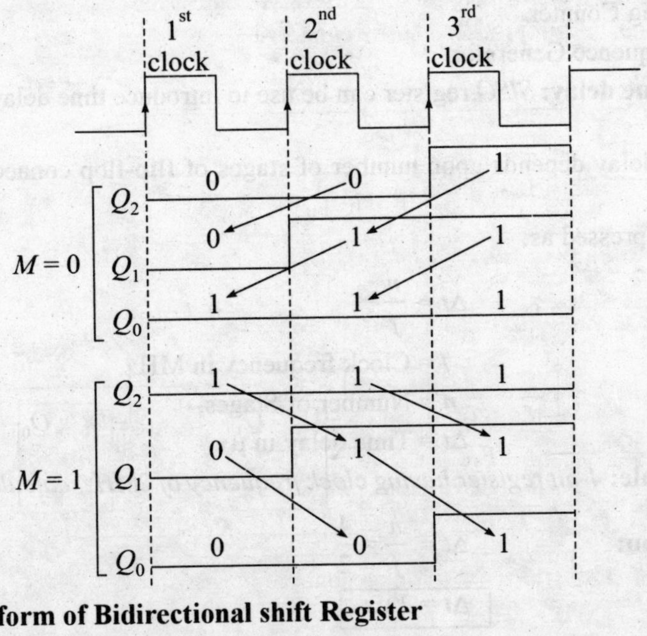

Waveform of Bidirectional shift Register

4.4.4 Universal Shift Register

This register operate in all the possible modes *i.e. SISO, SIPO, PISO, PIPO,* Bidirectional (Left and Right Shift) register, then it is known as Universal shift register. It can be used at place of any of above explained register. Its working depends on which type of shift register we want to use. It is shown in Fig. 4.20.

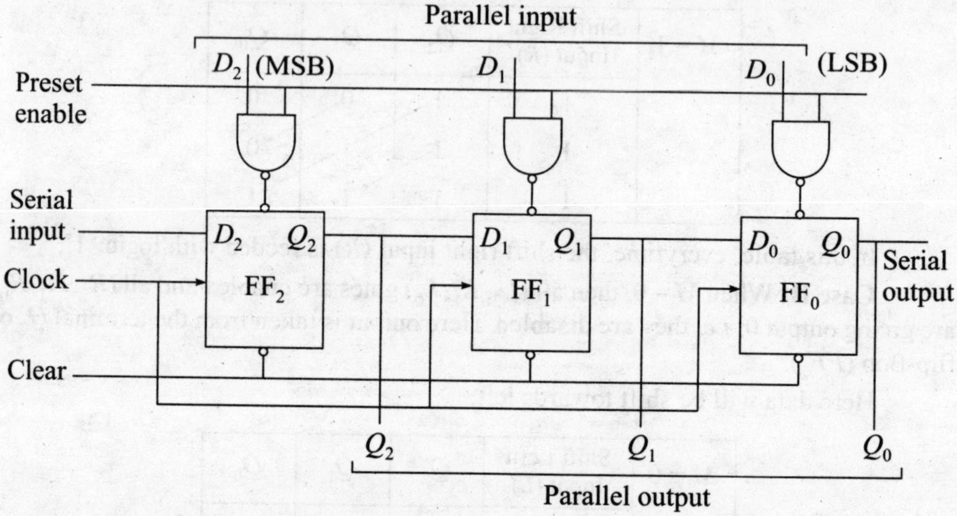

Fig. 4.20: Universal shift Register

Its working is similar to shift register.

4.4.5 Application of Shift Register

(*a*) Serial to parallel data conversion.

(*b*) Parallel to serial data conversion.

(*c*) Ring Counter.

(*d*) Sequence Generator.

(*e*) **Time delay:** *SISO* register can be use to introduce time delay from input to output.

Time delay depends upon number of stages of flip-flop connected and clock frequency.

It is expressed as:

$$\Delta t = \frac{n}{f}$$

where f = Clock frequency in MHz.

n = Number of Stages.

Δt = Time delay in μs.

Example: *4-bit register having clock frequency of 2MHz, calculate time delay.*

Solution:

$$\Delta t = \frac{n}{f} = \frac{4}{2}$$

$$\boxed{\Delta t = 2\ \mu s}$$

4.5 COUNTER

It is special sequential device which is used to count the clock pulses. A counter is a sequential machine that is constructed with the help of flip-flop and it changes it state according to state diagram. The counting of counter may be ascending, descending or in any manner defined by designer.

The counter may be up-counter and down-counter.

Up-Counter is a counter which counts in upward direction *i.e.* 0, 1, 2, ... $N-1, N$

Down-Counter is a counter which counts in the downward direction *i.e.* N, $N-1$, ... 2, 1, 0

Each count of counter is called the state of counter.

Modulus Counter: The number of state through which the counter passes before returning to starting state or initial state is called modulus counter.

A 2-bit counter has 4-states, it is called Mod-4 counter. It divides input clock signal frequency by 4, then it is called Divide by-4 counter

$$MOD \text{ Number} = 2^n$$

where n is the number of flip-flop.

4.5.1 Classification of Counter

Asynchronous/Ripple/Serial Counter	*Synchronous/Parallel Counter*
1. In this, first flip-flop is clocked by external clock pulse and then each successive flip-flop is clocked from previous flip-flop output.	1. In this, the common clock input is connected to all the flip-flop and thus they are clocked simultaneously.
2. Simpler to design	2. Difficult to design.
3. Frequency of operation is lesser than synchronous.	3. Frequency of operation can be much higher than asynchronous.
4. Minimum Number of Logic device needed.	4. Number of logic devices required more than asynchronous.
5. Less costly.	5. More costly.
6. Example: Divide by N counter Decade counter.	6. Example: Ring counter Johnson counter.
7. In this unwanted spikes are produced.	7. No unwanted spikes are produced.
8. 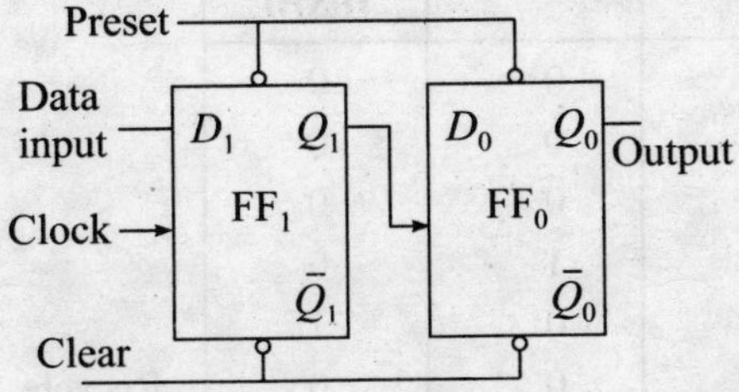	8. 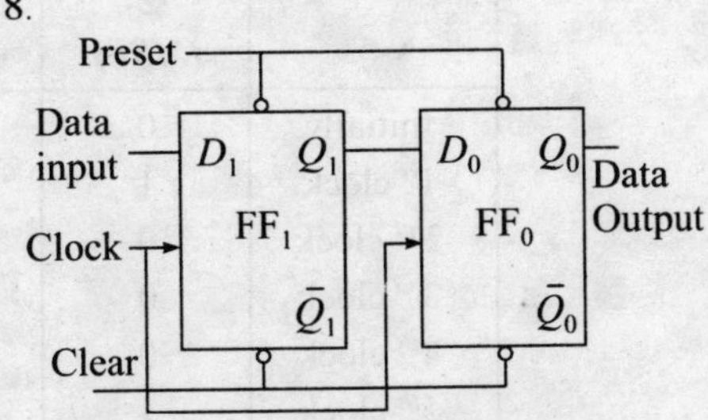

4.5.2 Basic Concept of Counter

4.5.2.1 State Diagram: It is graphical representation. States are represented by circles and the transition between states is indicated by directed lines connecting the circles. the binary number inside each circle identifies the state of logic circuit. The directed lines are labeled with two binary numbers separated by a symbol (/). For example x/y is indicated in direct line where x is input and y is output. Figure 4.21 shows the simple state diagram.

S_i, S_k = State of digital system.

x = Input representation.

y = Output representation.

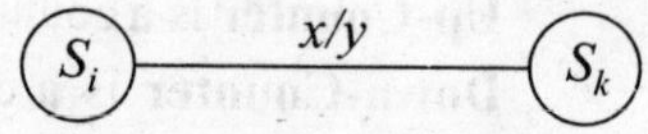

Fig. 4.21: State Diagram

4.5.2.2 State Table: It represents the state diagram in tabular form. The state table consists of **3 sections:**

Present state section.

Next state section.

Flip-Flop inputs.

Present State	Next State	Flip-Flop Inputs

4.5.3 Ring Counter

It is synchronous counter in which last output (*i.e.* LSB) is feeded back to the input of first flip-flop as shown in Fig. 4.22. In this clock pulse of all flip-flop is same.

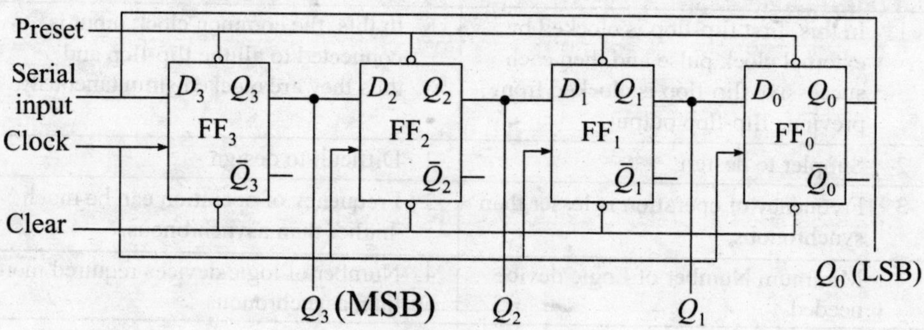

Fig. 4.22: Ring Counter

In this bit are at output will form ring and back connected to input data and according to clock pulse data shift and out.

	Q_3 *(MSB)*	Q_2	Q_1	Q_0 *(LSB)*	
Initially	0	0	0	0	
I[st] clock	1	0	0	0	
2nd clock	0	1	0	0	
3rd clock	0	0	1	0	
4th clock	0	0	0	1	
5th clock	1	0	0	0	Recycle

Waveform:

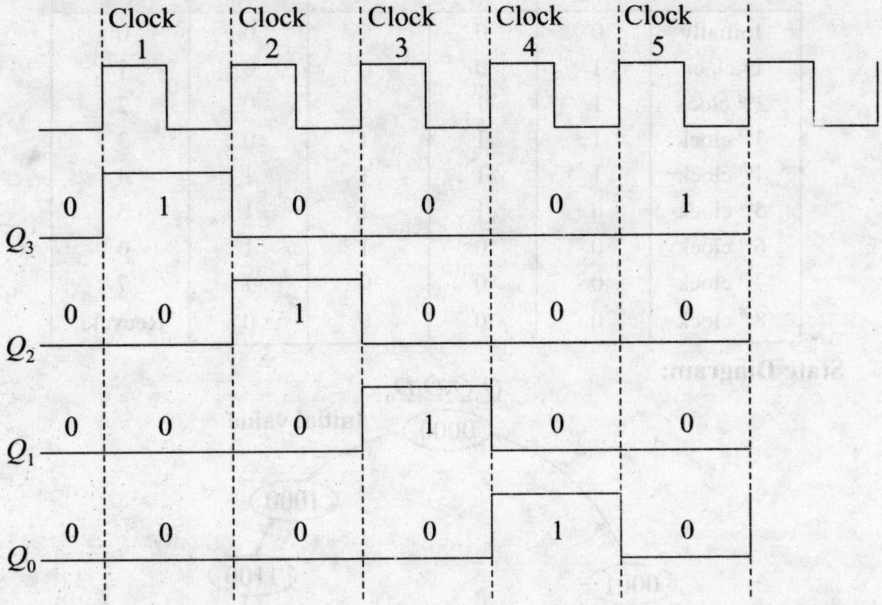

4.5.4 Johnson Counter

It is also known as **twisted Ring Counter** and another name as Moebius Counter and **self start counter.**

This is similar to Ring counter but only one difference is there that in this we feedback $\bar{Q}_0$ instead of Q_0 to serial input (D_3). It is shown in Fig. 4.23 and initially all the flip-flop are in reset state *i.e.* '0 0 0 0', after each clock pulse, the output of Q_3 is shifted to Q_2, the output of Q_2 is shifted to Q_1, Q_1 to Q_0 and in last complement output of Q_0 that is $\bar{Q}_0$ is shifted or send to Q_3 and so on. This sequence repeated after every eight clock pulse.

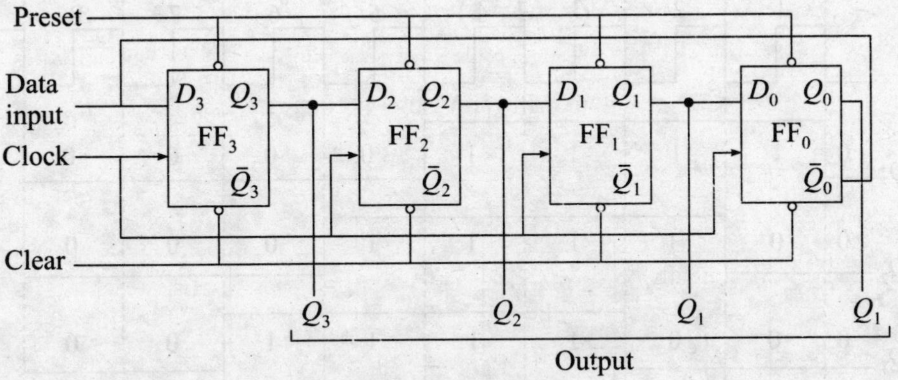

Fig. 4.23: Johnson Counter

Clock	Q_3	Q_2	Q_1	Q_0	*Counts*
Initially	0	0	0	0	0
Ist clock	1	0	0	0	1
2nd clock	1	1	0	0	2
3rd clock	1	1	1	0	3
4th clock	1	1	1	1	4
5th clock	0	1	1	1	5
6th clock	0	0	1	1	6
7th clock	0	0	0	1	7
8th clock	0	0	0	0	Recycle

State Diagram:

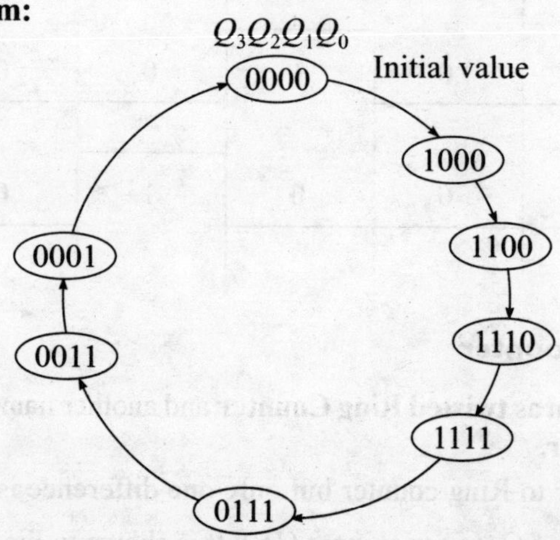

Twisted Ring Counter is capable of counting 8 states if 4 flip-flop is used. If, '*N*' flip-flop twisted ring counter is there, then it counts 2*N* states which is double of ring counter.

Waveform:

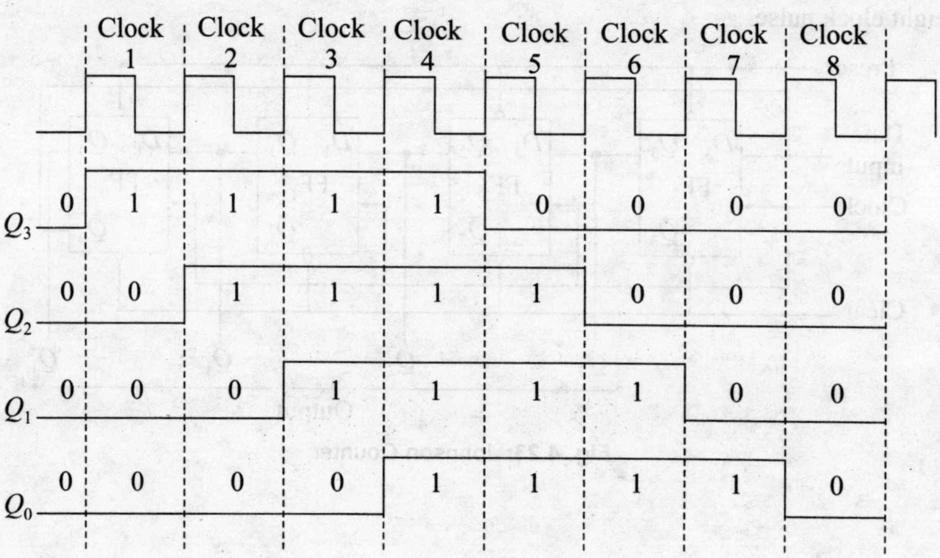

4.5.5 Divide By *N* Counter/Ripple Counter

"Divide by *N*" means it divides the input clock frequency by *N*. In this clock is not common to all the flip-flops. Now, we take two values of $N = 10, 12$ for examples.

4.5.5.1 Divide By 12 Counter: In this 12th state is reset means '1100' is reset. To make it reset we have to pass high value '11' of '1100' from NAND gate. The clear is used to clear or reset (0000) the flip-flop. After reset, counter starts counting from 0 to 11.

Clock Pulse	Q_4	Q_3	Q_2	Q_1	*Count*	
0	0	0	0	0	0	
1	0	0	0	1	1	
2	0	0	1	0	2	
3	0	0	1	1	3	
4	0	1	0	0	4	
5	0	1	0	1	5	Counter starts from o to 11 *i.e.* '12'
6	0	1	1	0	6	
7	0	1	1	1	7	
8	1	0	0	0	8	
9	1	0	0	1	9	
10	1	0	1	0	10	
11	1	0	1	1	11	
12	1	1	0	0	Reset after 12th clock	
Pass high value from NAND gate	0	0	0	0		

It is shown in Fig. 4.24.

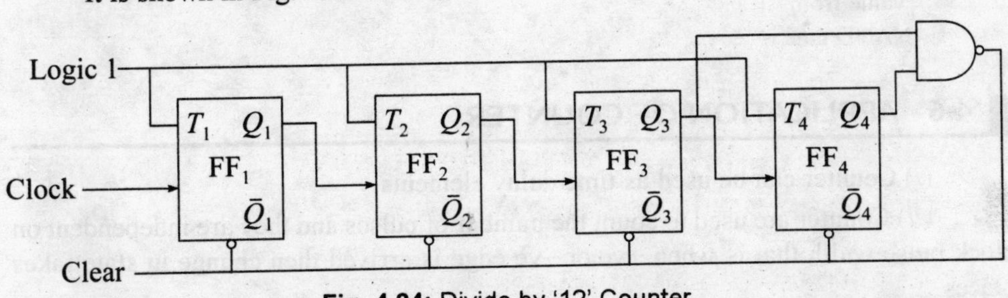

Fig. 4.24: Divide by '12' Counter

4.5.5.2 Decade Counter: It is also known as 'Divide by 10' counter. It counts values from 0 to 9 *i.e.* total ten pulses so also known as **BCD Counter** So, we have to reset the 10th state "1010" so, pass high value of 10th state by NAND gate like $\begin{matrix} Q_3 & Q_2 & Q_1 & Q_0 \\ 1 & 0 & 1 & 0 \end{matrix}$, then pass Q_3, Q_1 from NAND gate which clears all the flip-flop to 0 0 0 0 and again counters starts counting from 0 to 9. It is shown in Fig. 4.25.

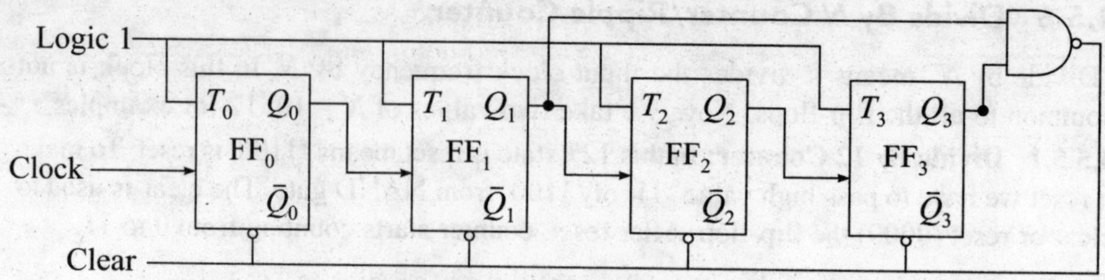

Fig. 4.25: Decade counter

So, finally after passing from NAND gate the reset value is sent to all flip-flop by clear value. Now, counting is starting from 0 to 9.

Clock Pulse	Q_3	Q_2	Q_1	Q_0	*Counting*
0	0	0	0	0	0
1	0	0	0	1	1
2	0	0	1	0	2
3	0	0	1	1	3
4	0	1	0	0	4
5	0	1	0	1	5
6	0	1	1	0	6
7	0	1	1	1	7
8	1	0	0	0	8
9	1	0	0	1	9
10	1	0	1	0	Recycle after 9th pulse
Pass High value from NAND gate	0	0	0	0	

4.6 APPLICATION OF COUNTER

(*i*) Counter can be used as time delay elements.

(*ii*) Counter are used to count the number of pulses and they are independent on clock pulse width that is when +ve or –ve edge is arrived then change in state takes places.

(*iii*) Counter converts the given number of pulses into predictable codes.

(*iv*) Counter are used for generating sequential codes.

(*v*) Counter are used in making of digital clock. Three types of counter are used, **first** one calculates the number of seconds and then give one increment to minute counter; **second** one calculates the number of minutes and then it give one increment in hour counter that is **third** one, when 60 counts are completed in minute counter.

(*vi*) Counter are used for frequency division as explained earlier single flip-flop can acts as divide by 2 and if we cascade them then we have multiple division of frequency.

4.7 SEQUENCE GENERATOR

A circuit which generates a prescribed sequence of bits, in synchronism with a clock, is referred to as a sequence generator. Such generator can be used as:

(*i*) Counters.

(*ii*) Random bit generator.

(*iii*) Prescribed period and sequence generators.

(*iv*) Code generators.

It is a combination of shift register and next state decoders that is shown in Fig. 4.26 in which output y of next state decoder is a function of $Q_{N-1}, Q_{N-2}, Q_{N-3}, \ldots Q_1, Q_0$. This system is similar to a ring counter ($y = Q_0$) or twisted-ring counter ($y = \bar{Q}_0$) which are special cases of sequence generation.

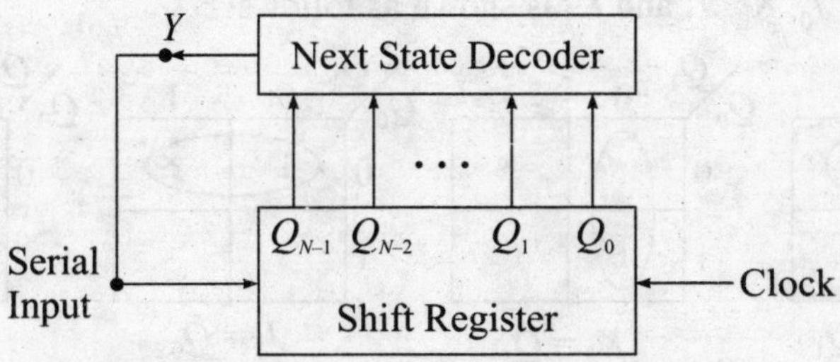

Fig. 4.26: Basic structure of sequence generator

4.8 DESIGNING OF SEQUENTIAL CIRCUITS (COUNTERS)

Counters can be classified into synchronous and Asynchronous. (*a*) **Synchronous Counter** designing has the following steps.

(*i*) Draw the state diagram which describes the operation of the counter.

(*ii*) From the state diagram write the count sequences in the form of table.

(*iii*) Find the number of flip-flops required.

(*iv*) Decide the type of flip-flop to be used for the design of counter. Then draw the present state and next state of counter and use the present and next state table in order to draw the excitation map.

(*v*) Prepare *k*-map for each flip-flop in terms of flip-flop output as the input variables. Simplified the *k*-map and obtain the minimized expressions.

(*vi*) Connect the circuit using flip-flop and other gates corresponding to minimized expressions.

4.8.1 Design of Modulus Counter

Example 4.7: Design MOD-3 counter using *J-K* flip-flop.

Solution: The excitation table for *JK* flip-flop is given as:

Q_n	Q_{n+1}	J	K
0	0	0	×
0	1	1	×
1	0	×	1
1	1	×	0

Table using present and next state is given as:

Present state		Next state		Flip-Flop			
Q_0	Q_1	Q_0^1	Q_1^1	J_0	K_0	J_1	K_1
0	0	0	1	0	×	1	×
0	1	1	0	1	×	×	1
1	0	0	0	×	1	0	×

K-map for J_0, K_0, J_1 and K_1 is shown as follows:

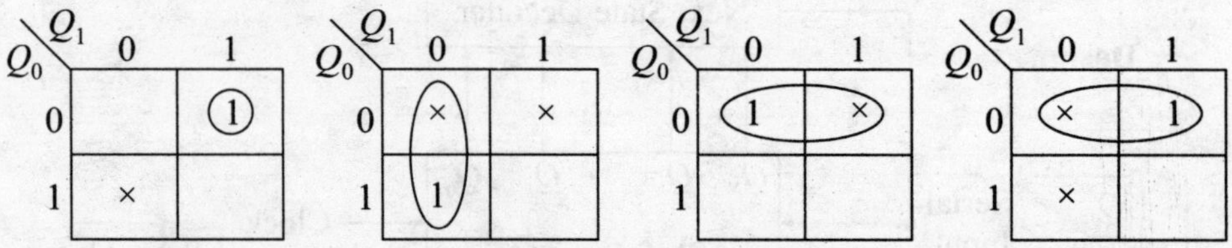

$J_0 = \bar{Q}_0 = Q_1$ $K_0 = \bar{Q}_1$ $J_1 = \bar{Q}_0$ $K_1 = \bar{Q}_0$

Designing using *J - K* Flip-Flop:

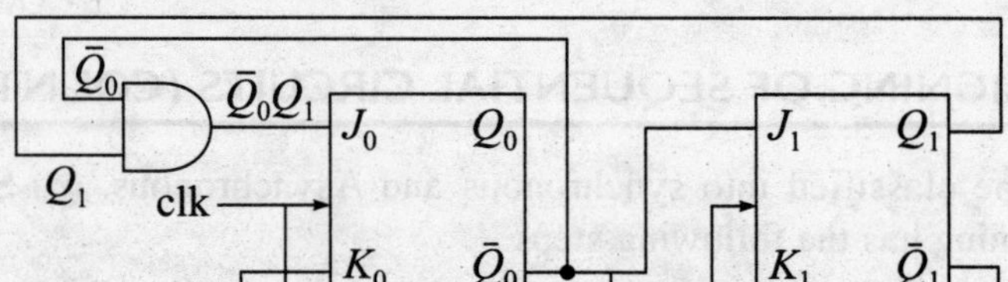

Example 4.8: *Design MOD - 5 counter using J - K flip-flop.*

Solution:

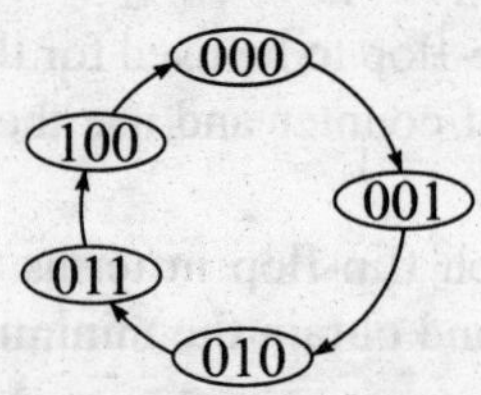

Present state			Next state			Flip-Flop					
Q_0	Q_1	Q_2	Q_0'	Q_1'	Q_2'	J_0	K_0	J_1	K_1	J_2	K_2
0	0	0	0	0	1	0	×	0	×	1	×
0	0	1	0	1	0	0	×	1	×	×	1
0	1	0	0	1	1	0	×	×	0	1	×
0	1	1	1	0	0	1	×	×	1	×	1
1	0	0	0	0	0	×	1	0	×	0	×

The K-map for $J_0, K_0, J_1, K_1, J_2, K_2$ is shown as follows:

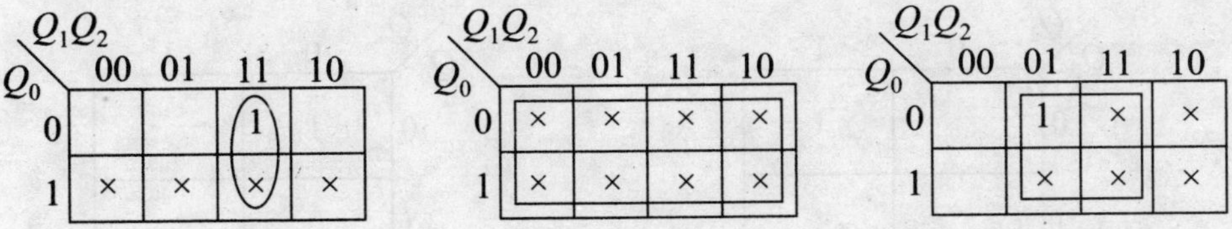

$J_0 = Q_1Q_2$ $\quad$ $K_0 = 1$ $\quad$ $J_1 = Q_2$

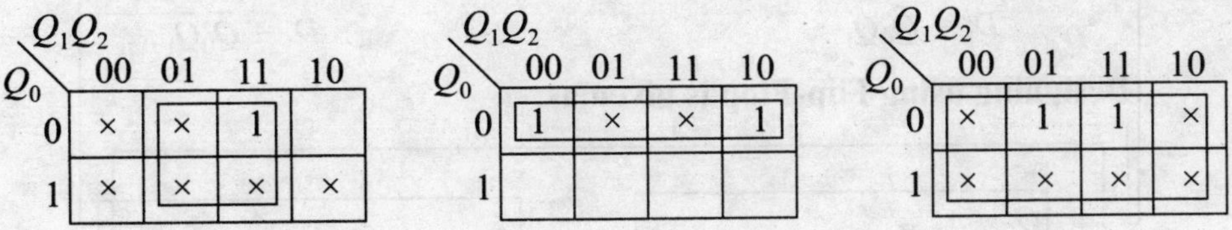

$K_1 = Q_2$ $\quad$ $J_2 = \bar{Q}_0$ $\quad$ $K_2 = 1$

Designing Using *J - K* Flip-Flop:

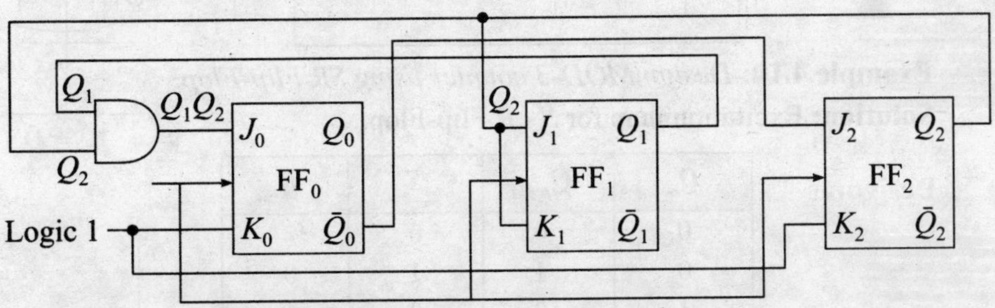

Example 4.9: *Design MOD-3 counter using D Flip-Flop.*

Solution: Excitation map for D Flip-Flop is given:

Q_n	Q_{n+1}	D
0	0	0
0	1	1
1	0	0
1	1	1

Table using present and next state is given as:

Present state		*Next state*		*Flip-Flop*	
Q_0	Q_1	Q_0^1	Q_1^1	D_0	D_1
0	0	0	1	0	1
0	1	1	0	1	0
1	0	0	0	0	0

K-map for D_0 and D_1 is given as:

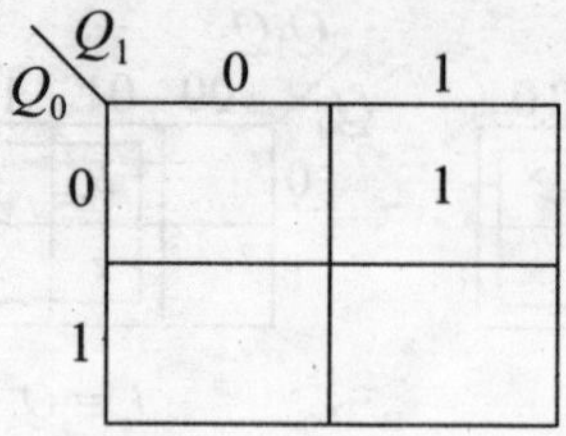

$D_0 = \bar{Q}_0 Q_1$

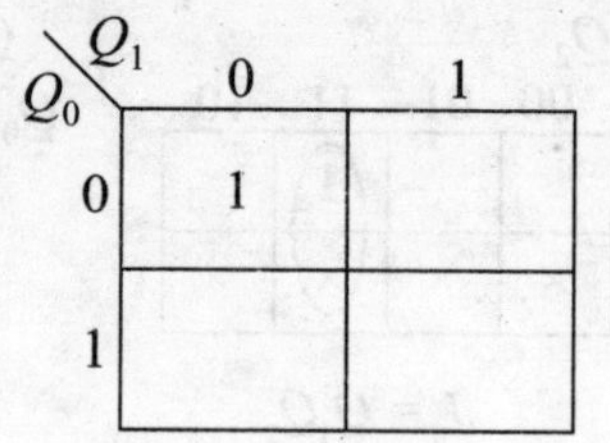

$D_1 = \bar{Q}_0 \bar{Q}_1$

Designing using Flip-Flop is given as:

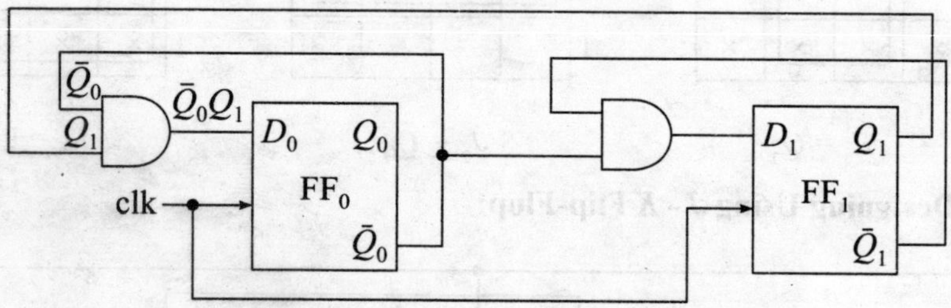

Example 4.10: *Design MOD-3 counter using SR Flip-Flop.*

Solution: Excitation map for S - R Flip-Flop:

Q_n	Q_{n+1}	S	R
0	0	0	×
0	1	1	0
1	0	0	1
1	1	×	0

The table for MOD-3 counter is given as follows:

Present state		*Next state*		*Flip Flop*			
Q_0	Q_1	Q_0^1	Q_1^1	S_0	R_0	S_1	R_1
0	0	0	1	0	×	1	0
0	1	1	0	1	0	0	1
1	0	0	0	0	1	0	×

K-map for S_0, R_0, S_1, R_1 is:

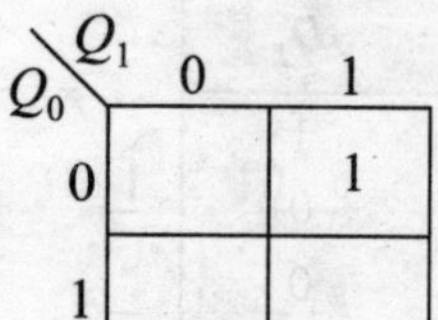

$S_0 = \bar{Q}_0 Q_1$

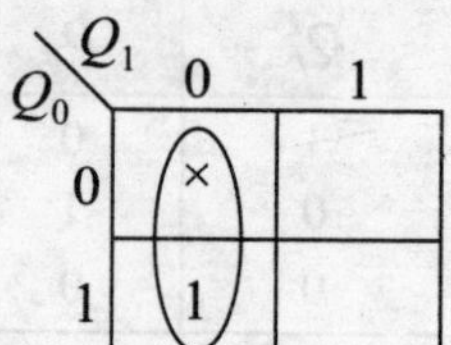

$R_0 = \bar{Q}_1$

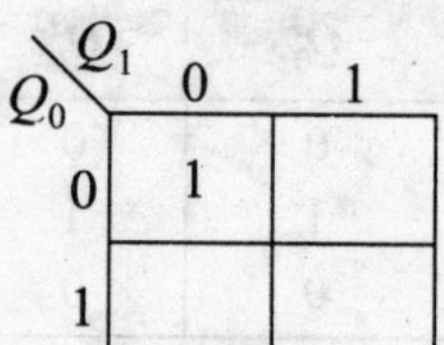

$S_1 = \bar{Q}_0 \bar{Q}_1$

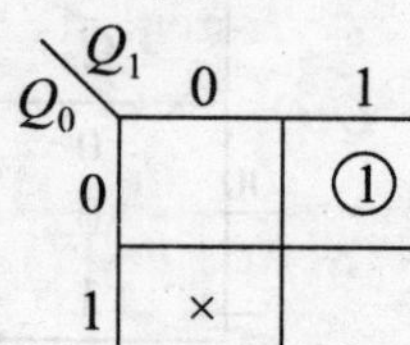

$R_1 = \bar{Q}_0 Q_1$

Designing using *S - R* Flip-Flop:

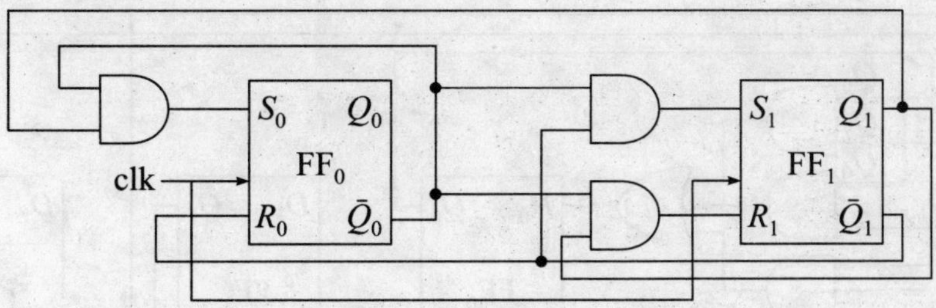

Example 4.11: *Design a MOD-8 Synchronous binary up counter using D Flip-Flop.*

Solution: The binary up counter counts from 0, 1, 2, 3, 4, 5, 6, and 7.

State diagram is given as:

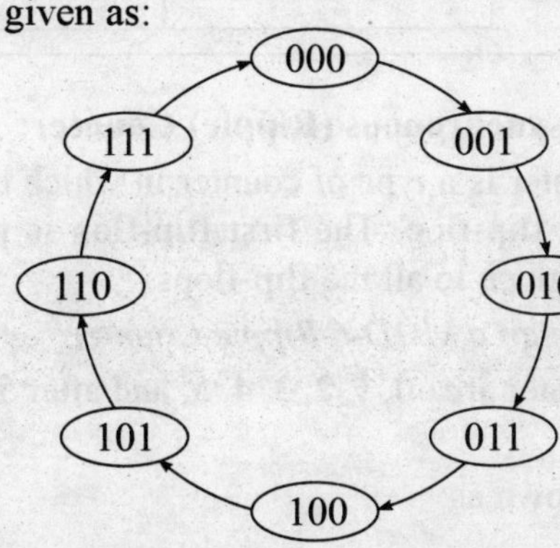

In order to design MOD-8 counter, minimum 3 flip-flops are required. The table representation is as shown as:

Present state			***Next state***			***Flip-Flops***		
Q_0	Q_1	Q_2	Q_0^1	Q_1^1	Q_2^1	D_0	D_1	D_2
0	0	0	0	0	1	0	0	1
0	0	1	0	1	0	0	1	0
0	1	0	0	1	1	0	1	1
0	1	1	1	0	0	1	0	0
1	0	0	1	0	1	1	0	1
1	0	1	1	1	0	1	1	0
1	1	0	1	1	1	1	1	1
1	1	1	0	0	0	0	0	0

***K*-map of D_0, D_1, D_2**

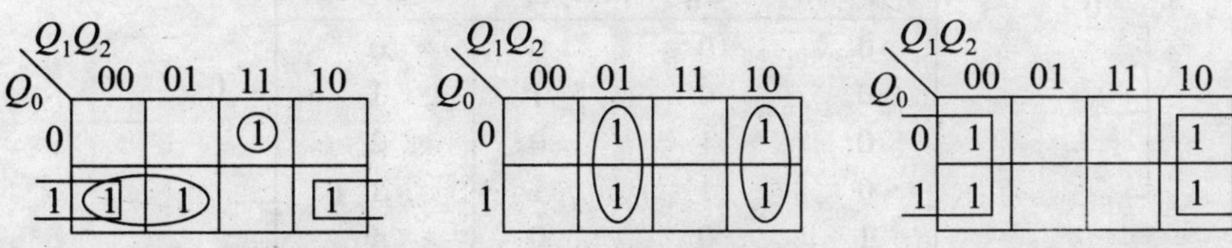

$D_0 = Q_0\bar{Q}_2 + Q_0\bar{Q}_1 + Q_1Q_2\bar{Q}_0$ $\quad$ $D_1 = \bar{Q}_1Q_2 + Q_1\bar{Q}_2$ $\quad$ $D_2 = \bar{Q}_2$

$\Rightarrow Q_1 \oplus Q_2$

Designing using *D* Flip-Flop:

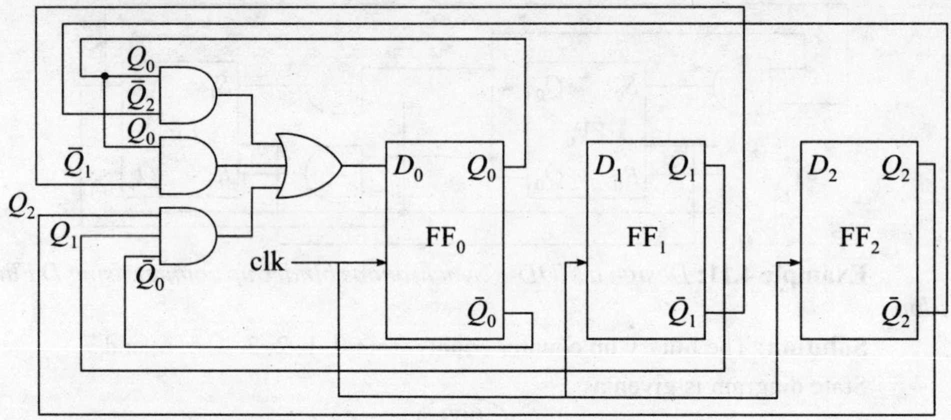

(*b*) **Designing of Asynchronous (Ripple) Counter:**

Asynchronous counter is a type of counter in which output of first flip-flop is acting as a clock to next flip-flop. The first flip-flop is provided with clock but simultaneous clock is not given to all the flip-flops.

Example 4.12: *Design a MOD-6 Ripple Counter.*

Solution: No. of stater are : 0, 1, 2, 3, 4, 5. and after 5 it reset and counts again from 0 to 5.

State diagram is shown as:

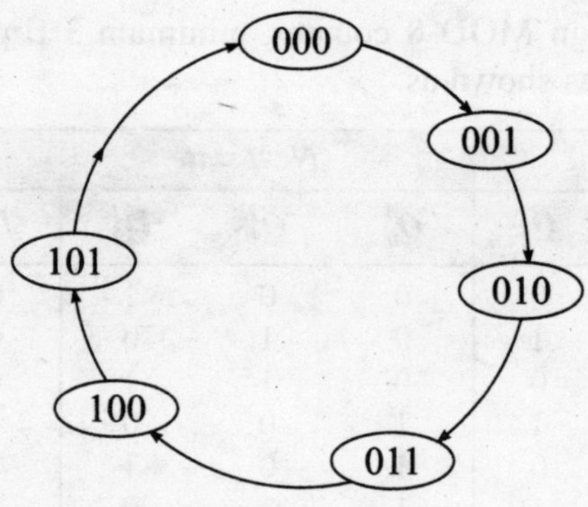

Truth Table for MOD-6 Counter is shown as:

Flip-Flop output			*Count*
Q_A	Q_B	Q_C	
0	0	0	0
0	0	1	1
0	1	0	2
0	1	1	3
1	0	0	4
1	0	1	5

In order to design this counter we need T Flip-Flop (or JK connected to logic 1). 3 flip flops are needed in order to design this counter. The NAND gate is used in order to reset when 6th pulse *i.e.* 6 count starts $\left(\underset{1}{Q_A}\ \underset{1}{Q_B}\ \underset{0}{Q_C}\right)$.

In this case the higher bits of 6 count *i.e.* $Q_A = 1, Q_B = 1$ is given to NAND gate and the output is low and hence the counter is reset and starts again counting from 0 to 5. It is shown is figure.

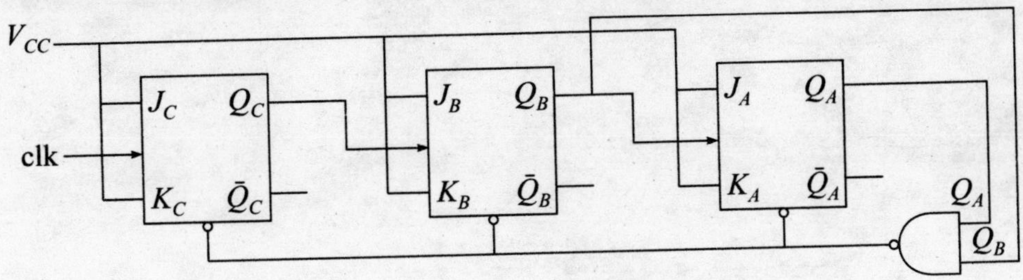

UNSOLVED QUESTIONS

1. What is counter?
2. What is difference between Latch and flip-flops.
3. Design 0, 1, 2, 3, 0 counter using D Flip-Flops.
4. Explain working of *SISO* shift Register.
5. Design the following Flip-Flop:
   (*i*) *SR* to *T* (*ii*) *JK* to *D*
   (*iii*) *D* to *T* (*iv*) *SR* to *JK*
6. Design and explain the working of MOD-5 ripple counter.
7. Design a Decade asynchronous counter.
8. Design a MOD-12 Ripple counter.
9. Design a 3-bit synchronous UP/DOWN counter with a Gray code sequence. The counter should UP when UP/DOWN control input $M = 1$ and count Down when control input $M = 0$.
10. Design a 3 bit Gray code counter using T Flip-Flops.

UNIT 5

DIGITAL LOGIC FAMILIES

Topics are:

- Switching Mode Operation of P-N Junction
- Bipolar and MOS Devices
- Bipolar Logic Family
- Characteristics of Digital ICs
- RTL (Resistor-Transistor Logic)
- DTL (Diode Transistor Logic)
- DCTL (Direct Coupled Transistor Logic)
- HTL (High Threshold Logic)
- TTL (Transistor-Transistor Logic)
- IIL (Integrator Injection Logic)
- ECL (Emitter Coupled Logic)
- MOS Family
- CMOS Logic Family
- Interfacing of CMOS and TTL Families
- Comparison of Logic Families

CHAPTER 5

Digital Logic Families

5.1 SWITCHING MODE OPERATION OF *p-n* JUNCTION

A *p-n* junction diode is formed by introducing *n*-type of impurity into one side and *p*-type impurity into other side of a single crystal of a semiconductor as shown in Fig. 5.1.

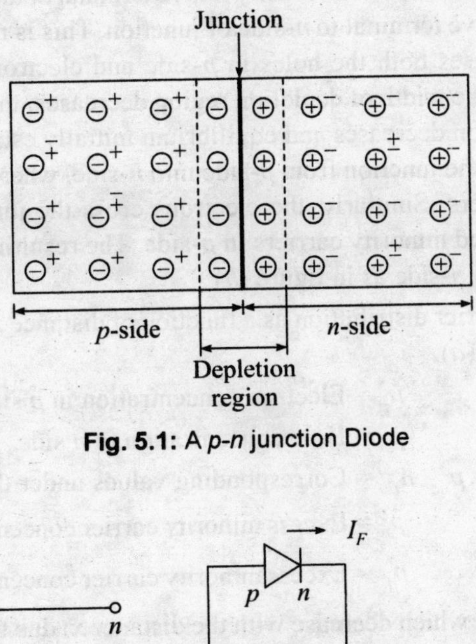

Fig. 5.1: A *p-n* junction Diode

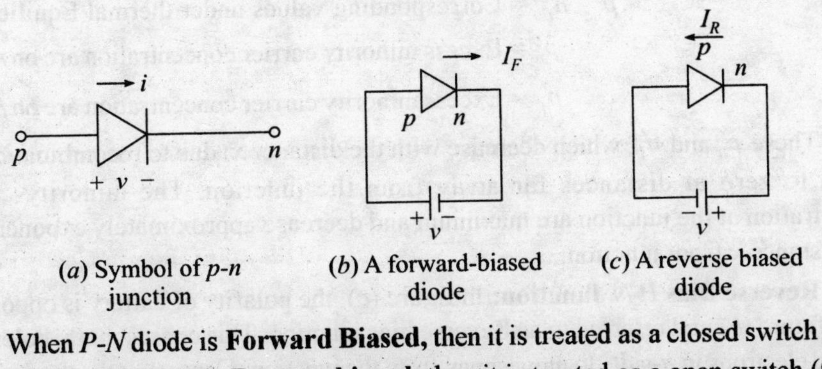

(*a*) Symbol of *p-n* junction

(*b*) A forward-biased diode

(*c*) A reverse biased diode

When *P-N* diode is **Forward Biased,** then it is treated as a closed switch (ON).

When *P-N* diode is **Reverse biased,** then it is treated as a open switch (OFF).

Recovery Time: The interval of time elapses before the diode recovers its steady state is called Recovery time.

Recovery Time (t_r)

Forward recovery time (t_{fr}) | Reverse recovery time (t_{rr})

(*i*) **Forward Recovery Time:** When the diode is switched from reverse-bias condition to forward-bias condition, then it takes certain time interval t_{fr} called as forward recovery time.

Reverse Bias → Forward Bias.

It is defined as time interval between instant of 10% diode voltage to instant this voltage attains 90% of diode voltage.

(*ii*) **Reverse Recovery Time:** When diode is switched from forward bias to Reverse bias, then time taken to achieve its steady state value is called as Reverse Recovery time.

Forward Bias → Reverse Bias.

Forward Bias *P-N* Junction: When positive terminal of the battery is connected to *p*-side and the negative terminal to *n*-side of junction. This is referred to as forward bias junction. This causes both the holes in *p*-side and electrons in *n*-side to move closer to junction. So, the width of depletion region decreases, the height of potential-energy barrier at junction decreases and equilibrium initially established is disturbed. Hence, the holes cross the junction from *p*-side into *n*-side, where they are referred as injected minority carriers. Similarly, the electrons cross the junction in the reverse direction and are injected minority carriers in *p*-side. The resulting current is I_F in the direction from *p*-side to *n*-side as in figure (*b*).

The minority carrier distribution as a function of distance *x* from the junction is illustrated in figure 5.2 (*a*).

n_p = Electron concentration in *p* side.

p_n = Hole concentration in *n* side.

p_{no}, n_{po} = Corresponding values under thermal Equilibrium.

p'_n = Excess minority carrier concentration are on *n*-side.

n'_p = Excess minority carrier concentration are on *p*-side.

These p'_n and n'_p, which decrease with the distance *x*, due to recombination and reduce to zero at distances far away from the junction. The minority-carrier concentration at the junction are maximum and decrease approximately exponentially with distance *x* from junction.

Reverse Bias *P-N* Junction: In figure (*c*), the polarity of battery is opposite to that in figure (*b*) and referred to as Reverse bias junction. This causes, both, holes in *p*-side and electrons in *n*-side, to move away from the junction. Consequently, the depletion layer width increase which prevents the holes from *p*-side to cross over to *n*-side and electrons from *n*-side to *p*-side. However, the minority charge carriers, the electron from *p*-side and holes from *n*-side, can easily cross the junction and constitute a reverse current I_R in the direction indicated in figure (*c*). This current is very small, of the order of a few microamperes for germanium diode.

The minority carrier near the junction cross over to the other side, recombine, and the concentration diminishes to zero at the junction. Far away from the junction, the minority carrier concentration are equal to their thermal equilibrium values shown in Fig. 5.2 (*b*).

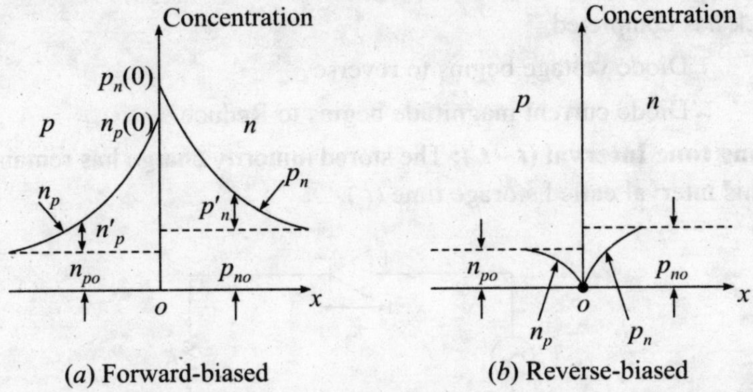

Fig. 5.2: Minority-carrier concentration in *p-n* junction Diode

5.1.1 Switching Characteristics of *P-N* Junction

In case of forward and reverse *P-N* Junction, when the voltage changes suddenly from forward to reverse bias, then minority carrier concentration requires a finite time, which is negligibly small when the bias voltage changes from reverse to forward condition, but is significant when voltage changes from forward to reverse bias condition. This is due to accumulation of excess minority-charge carriers on the two sides of junction under forward-bias condition, which must be removed before the diode comes to steady-state in reverse-bias condition. As soon as voltage changes sign to reverse bias the diode, a large current will flow in reverse direction due to excess charge carriers stored, till all the excess charge has been removed.

The time required for the removal of excess charge is referred as **storage time (t_s)**. Beyond that, the current changes exponentially and comes to a steady-state value.

The time required to reach the steady-state in reverse bias condition is referred as **Transition Time (t_r).**

The $\boxed{t_s + t_r = \text{Switching time}}$

Switching time determines the maximum frequency at which the diode can switch from ON to OFF and from OFF to ON. These are shown in Fig. 5.3.

At t_1: The input voltage (V_i) applied to diode resistance is reversed abruptly.

Upto time t_1 ($t = 0 - t_1$): Diode is conducting in forward direction and $V_i = V_F$ current flowing through R_L is given by I_F.

$$I_F = V_F/R_L$$

At instant of ($t = t_1$): The voltage v_i reverses abruptly to a value ($-V_R$). Now current does not drop to zero but instead reverses and remains at value ($-I_R = -V_R/R_L$) until ($t = t_2$).

At ($t = t_2$): The injected minority carrier density at $x = 0$ has reached its equilibrium state.

Diode Voltage (v):

At ($t = t_1$): The voltage falls slightly by $R_d\ (I_F + I_R)$ but does not reversed.

$$R_d = \text{Diode ohmic Resistance.}$$

At ($t = t_2$): The process of sweeping of excess minority carriers in vicinity of junction back has completed.

So, Diode voltage begins to reverse

Diode current magnitude begins to Reduce.

During time Interval ($t - t_2$): The stored minority charge has remained stored and hence this interval called storage time (t_s)

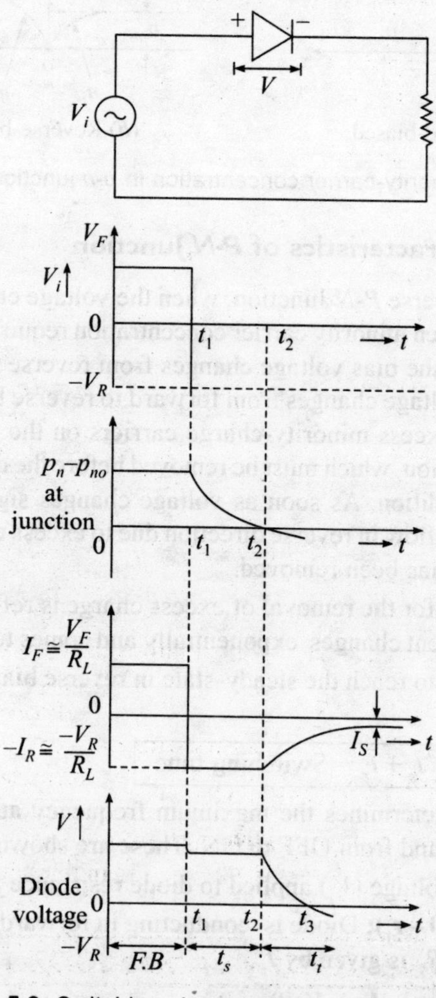

Fig. 5.3: Switching characteristics of *P–N* junction

5.2 BIPOLAR AND MOS DEVICES

5.2.1 Bipolar Devices

BJT (Bipolar Junction Transistor) is very popular bipolar device. It is a three terminals devices. In any transistor there are three terminals.

- Emitter (To emit charge carriers)
- Base (To bypass the charge carriers from Emitter to collector)
- Collector (To collect the charge carriers)

Two Configuration of Transistor:

(*i*) NPN Transistor 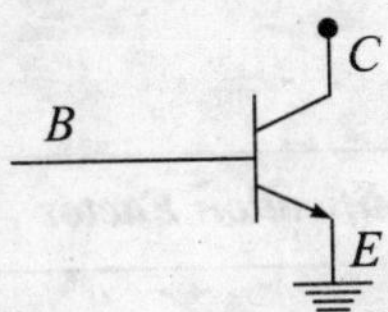	(*i*) PNP Transistor 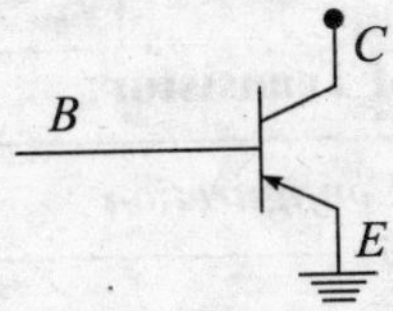
(*ii*) When layers of *P* type semiconductor material is sandwitched between two layers of *N* type, we get NPN transistor.	(*ii*) When layer of *N*-type type semiconductor material is sandwitched in between two layers of *P* type, we get PNP transistor.
(*iii*) The direction of current is outward.	(*iii*) The direction of current is inward.

NPN transistor as a switch:

Case 1: When V_{in} is +*ve* or we say V_{BE} = High, Q = Turns ON shown in Fig. 5.4. Collector to emitter junction becomes short. It place, GND, potential at V_{out}. So, V_{out} = 0 V.

Case 2: When V_{in} = Low (–*ve*) then V_{BE} = Low (–*ve*)/ zero /Ground.

Q = Turns OFF.

Collector to emitter junction becomes open as there is no current through R_C. So V_{out} = 5 *v*

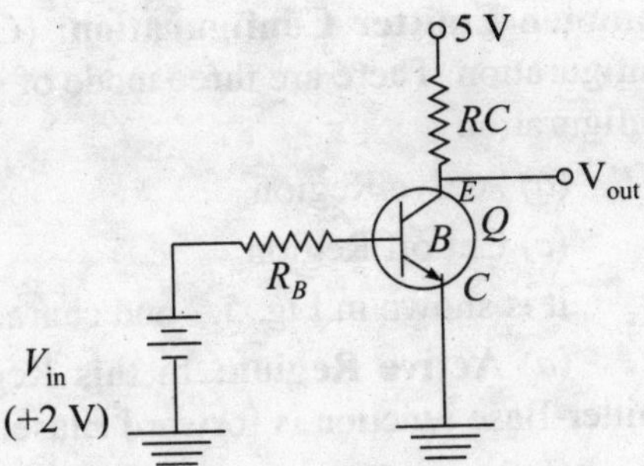

Fig. 5.4: Transistor As a switch

R_B and R_C = Used to limit base current and the collector current.

NPN Transistor: As *E-B* junction is always forward biased and *C-B* junction is always reverse biased. It means emitter terminal is connected to negative terminal of battery so that it can repel electrons. As base is thin and lightly doped, few of electrons recombine with hole in *P* region and remaining electron reach collector regions. As collector is connected to +*ve* terminal of battery, they are collected by +*ve* terminal of battery.

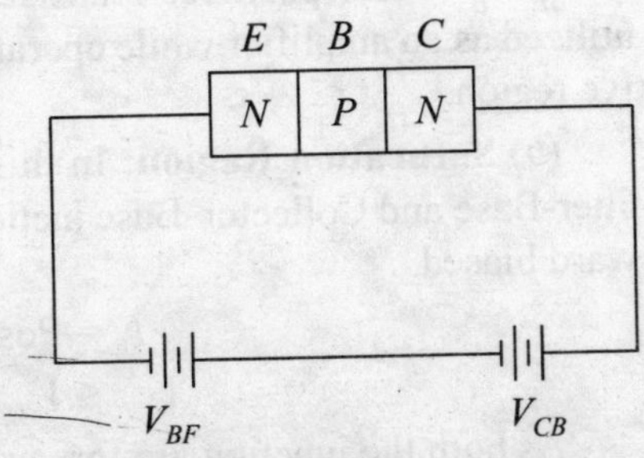

Fig. 5.5: NPN Transistor

In this way, continuous conduction of electrons takes place.

PNP Transistor: As *E-B* junction is always forward biased and *C-B* junction is always reverse biased. It means emitter terminal is connected to positive terminal of battery so that it can repel holes. As base is thin and lightly doped, few of holes coming from collector combine with electrons in base region and rest reach collector region. As collector is connected with *–ve* terminal of battery, they are absorbed by negative terminal of battery.

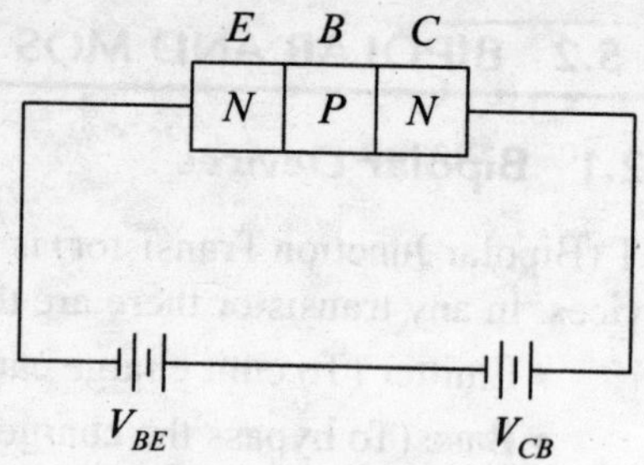

Fig. 5.6: PNP Transistor

$$\boxed{I_E = I_B + I_C}$$

Configuration of Transistor:

Configuration	*Current Amplification Factor*
1. Common Base (*CB*)	$\propto = \dfrac{I_C}{I_E}$ ($\propto = 0.95 - 0.99$)
2. Common Emitter (*CE*)	$\beta = \dfrac{I_C}{I_B}$ ($\beta = 10 - 1000$)
3. Common Collector (*CC*)	$\Upsilon = \dfrac{I_E}{I_B}$ ($\Upsilon = 1 + \beta \cong 10 - 1000$

Common Emitter Configuration: (*CE*) This is more important and widely used configuration. There are three mode of working of NPN transistor is common Emitter configuration.

(*a*) Active Region. (*b*) Saturation Region.

(*c*) Cut off Region.

It is shown in Fig. 5.7 and characteristics are shown in Fig. 5.8.

(*a*) **Active Region:** In this Region Emitter-Base junction is forward Biased.

Collector-Base junction is Reverse Biased.

In this region, V_{CE} is more positive than V_{BE}. I_B is also positive. Transistor can be utilized as an amplifier while operating in active region.

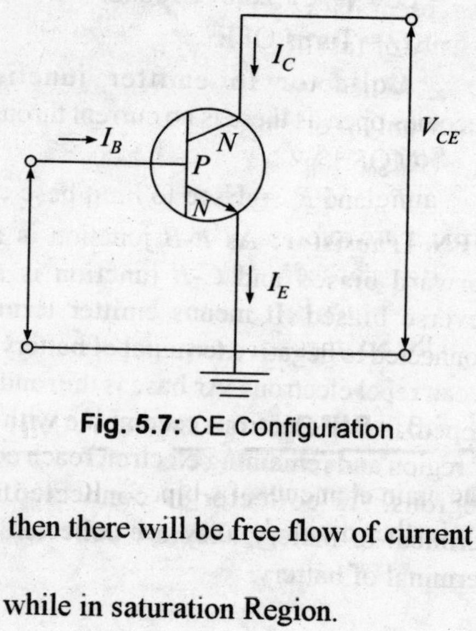

Fig. 5.7: CE Configuration

(*b*) **Saturation Region:** In this both Emitter-Base and Collector-Base juction are forward biased.

$$I_B = \text{Positive}$$

$$V_{CE} \leq V_{BE}$$

As both the junction are forward biased, then there will be free flow of current in both junction.

Transistor can be utilized as switch-ON while in saturation Region.

(*c*) **Cut-off Region:** In this both Emitter-Base and collector-Base junction are Reverse biased.

$$I_B = 0 \text{ or } -\text{ve}$$
$$V_{BE} = -\text{ve}$$
$$I_C = 0$$

Transistor can be utilized as switch off while operating in cut-off region.

Limitation of Transistor: Output impedance is Low. It means large amount of current flows through output.

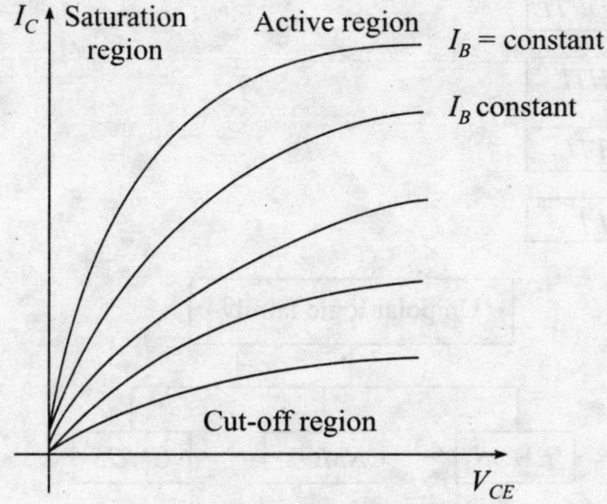

Fig. 5.8: Characteristics of CE

So, it leads to a great power dissipation.

So, transistors are replaced by FET (field effect Transistor).

5.2.2 MOS Devices

In present days, MOSFET have became very popular for logic circuits due to high density of fabrication and low power dissipation.

When MOS devices are used in logic circuit, there can be circuits in which either only *p* or only *n*-channel devices are used. Such circuits are referred as PMOS and NMOS logic. It is also possible to fabricate enhancement mode *p*-channel and *n*-channel MOS device on same chip. Such devices are referred as complementary MOSFET and logic based on these devices is known as **CMOS** logic. Power dissipation is very low in CMOS. So, these are very popular.

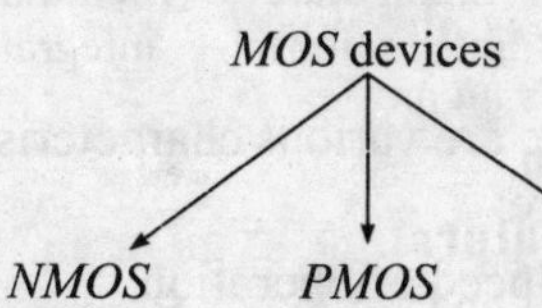

5.3 BIPOLAR LOGIC FAMILY

The main elements of a bipolar IC are diodes and transistor. In this classification they are further divided into two types based on the BJT operating mode:

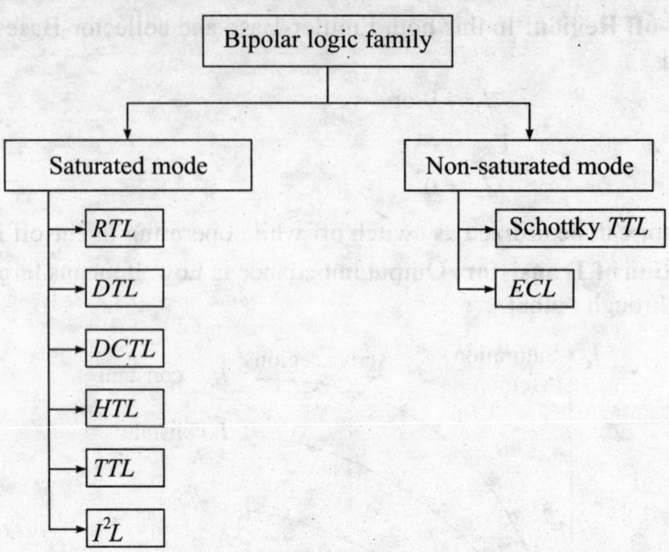

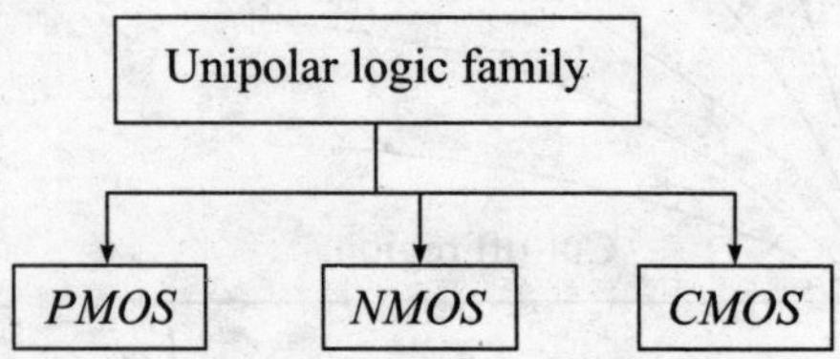

5.4 CHARACTERISTICS OF DIGITAL IC_S

With the widespread use of IC_s in digital systems and with the development of various technologies for the fabrication of IC_s, it has become necessary to be familier with characteristics of *IC* logic families and their advantage and disadvantages.

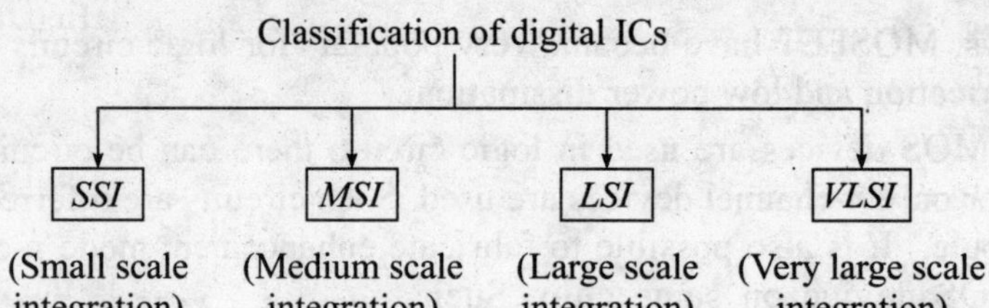

There are various characteristics of digital IC_s used to compare performance of digital *IC* are:

1. Speed of operation.
2. Power Dissipation.
3. Figure of merit.
4. Fan-In.
5. Fan Out
6. Current and voltage parameters.
7. Noise Immunity.
8. Operating Temperature Range.
9. Power supply requirements.
10. Flexibilities Available.

1. **Speed of operation:** (Propagation Delay) The speed of digital circuits are measured in terms of propagation delay time. The input and output waveform of logic gate are shown in Fig. 5.9.

It is defined as "Time taken for an output of a gate to change if there is a sudden change in applied input."

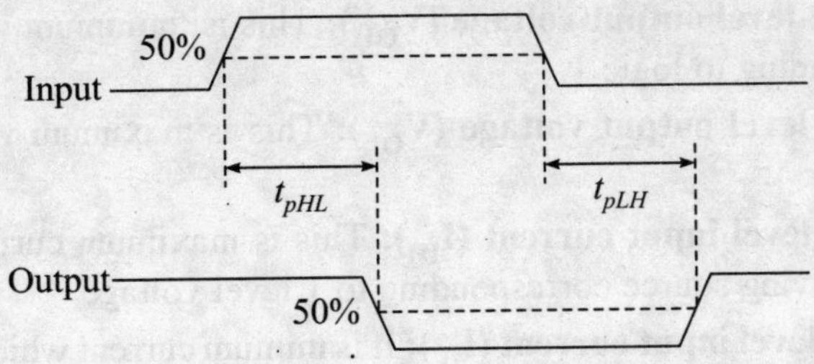

Fig. 5.9: Input and output waveform to define propagation delay times

$$\text{Average Propagation Delay} = \frac{t_{pLH} + t_{pHL}}{2}$$

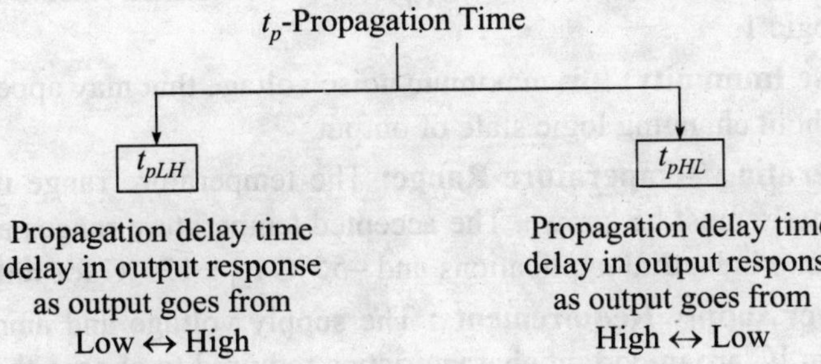

2. **Power Dissipation:** It is specified as power consumed by the logic gate when fully driven by all its inputs."

It is determined by current, I_{CC}, that it draws from the V_{CC} supply, and is given by $V_{CC} \times I_{CC}$. I_{CC} is average value of I_{CC} (0) and I_{CC} (1).

This power is specified in milliwatts.

3. **Figure of Merit:** The figure of merit of a digital IC is defined as the product of speed and power. The speed is specified in terms of propagation delay time expressed in nanoseconds.

Figure of merit = Propagation delay time (*ns*) × power (*mw*) It is specified in pico joules (*ns* × *mw* = *pJ*)

4. **Fan-In:** It is specified as number of inputs connected to the gate.

for example: If a gate has 8 inputs.

Fan – In = 8.

5. **Fan-Out:** This is the number of similar gates which can be driven by a gate. High fan-out is advantageous because it reduces the need for additional drivers to drive more gates.

6. **Current and Voltage Parameter:** The following currents and voltage are specified which are very useful in the design of digital systems.

(*i*) **High-level input voltage (V_{IH}):** This is maximum input voltage which is recognized by the gate as logic 1.

(*ii*) **Low-level input voltage (V_{IL}):** This is maximum input voltage which is recognized by the gate as logic 0.

(*iii*) **High-level output voltage (V_{OH}):** This is minimum voltage available at output corresponding to logic 1.

(*iv*) **Low-level output voltage (V_{OL}):** This is maximum voltage available at output at logic 0.

(*v*) **High-level input current (I_{IH}):** This is maximum current which must be supplied by a driving source corresponding to 1 level voltage.

(*vi*) **Low-level input current (I_{IL}):** It is minium current which must be supplied by a driving source corresponding to 0 level voltage.

(*vii*) **High-level output current (I_{OH}):** This is maximum current which the gate can sink in 1 level.

(*viii*) **Low-level output current (I_{OL}):** This is maximum current when the output of gate is at logic 1.

7. **Noise Immunity:** It is maximum noise voltage that may appear at input of a logic gate without changing logic state of output.

8. **Operating Temperature Range:** The temperature range in which an IC functions properly must be known. The accepted temperature ranges are: 0 to +70°C for consumer and industrial applications and –55°C to + 125°C for military purpose.

9. **Power supply Requirements:** The supply voltage and amount of power required by an IC are important characteristics required to choose the proper power supply.

10. **Flexibilities Available:** Various flexibilities are available in different IC logic families and these must be considered while selecting a logic family for particular job. Those are:

(*i*) **The Breadth of Series:** Type of different logic functions available in series.

(*ii*) **Popularity of Series:** The IC have low cost, that is popular.

(*iii*) **Wired-Logic Capability:** The output can be connected together to perform additional logic without any extra hardware.

5.5 RTL (RESISTOR-TRANSISTOR LOGIC)

It was the most popular form of logic in common use before the development of ICs. RTL circuits consists of resistors and transistors and was the earliest logic family to be integrated.

But RTL has become obsolete now because of

- Simplicity
- Historical Reasons

The basic RTL gate is a NOR gate as shown in Fig. 5.10. It can also be utilized to work as NAND gate. Transistors can be connected in series or parallel to perform logical operations.

RTL NOR Gate:

Truth table of NOR gate:

A	*B*	*C*(%)
0	0	1
0	1	0
1	0	0
1	1	0

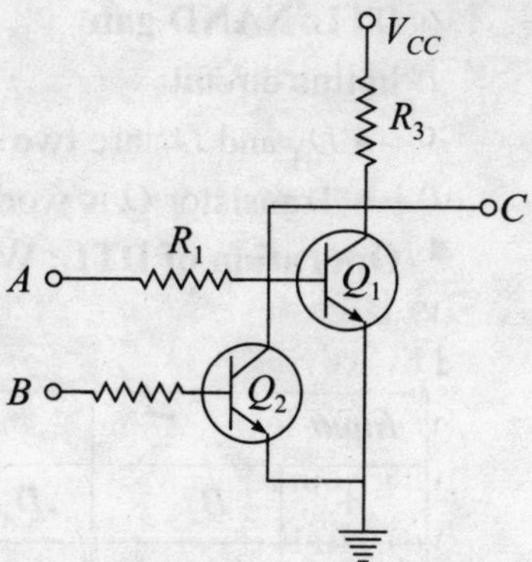

Fig. 5.10: RTL: NOR Gate

So, there are four different case of RTL NOR gate:

Case I: $A = 0$, $Q_1 = \text{OFF}$; $B = 0$, $Q_2 = \text{OFF}$ — $C = 1$. Point C will be at V_{CC}

Case II: $A = 0$, $Q_1 = \text{OFF}$; $B = 1$, $Q_2 = \text{ON}$ — $C = 0$. Point C will be at ground.

Case III: $A = 1$, $Q_1 = \text{ON}$; $B = 0$, $Q_2 = \text{OFF}$ — $C = 0$. Point C will be at ground

Circuit gives some value using NOR gate:

Propagation Delay	12 *n* seconds
Noise Margin	0.2 V
Power Dissipation	30 – 100 mw
Fan Out	4
Power Supply Voltage	5 V

Disadvantage of RTL Circuit: Low switching speed due to presence of R_1 and R_2.

5.6 DTL (DIODE TRANSISTOR LOGIC)

It eliminates the problem of decreasing output with increasing load.

In DTL input signals are applied to diodes in this logic family. The diode either conduct or cut-off and produce the desired output from transistor. DTL NAND gate is shown in Fig. 5.11.

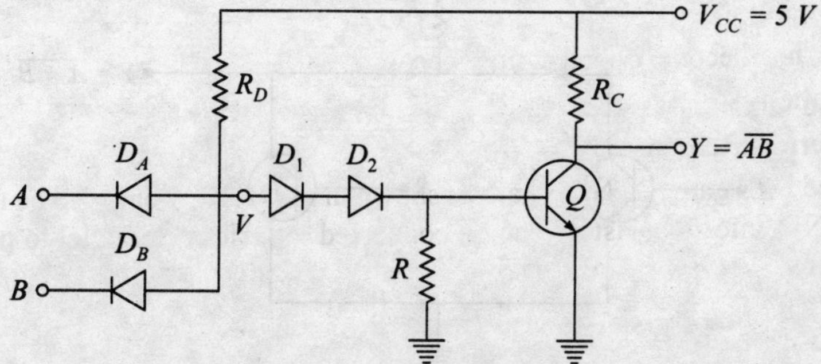

Fig. 5.11: DTL NAND gate

DTL NAND gate

In this circuit:

- D_A and D_B are two diodes which perform Logic AND operation.
- Transistor Q is working as an inverter.

Operation of DTL: Working Operation of DTL is shown by Table.

Table

Input		*Diode*				*Transistor*	*Output*
A	*B*	D_A	D_B	D_1	D_2	*Q*	$y = \bar{A} + \bar{B}$
0	0	*FB*	*FB*	OFF	OFF	OFF	1
0	1	*FB*	*RB*	OFF	OFF	OFF	1
1	0	*RB*	*FB*	OFF	OFF	OFF	1
1	1	*RB*	*RB*	ON	ON	ON	0

So, finally it is equal to NAND gate truth Table.

Characteristics of DTL:

Propagation Delay	25 ns
Fan-Out	8
Fan-In	8
Noise Immunity	0.8 V – 3.4 V

Comparison with RTL:

1. DTL have better Noise Margin/Immunity.
2. Higher fan out capability
3. Faster change.

5.7 DCTL (DIRECT COUPLED TRANSISTOR LOGIC)

It is same like RTL except that the circuit connections are given directly to base of transistor without use of resistors.

A DCTL NOR gate is shown in Fig. 5.12.

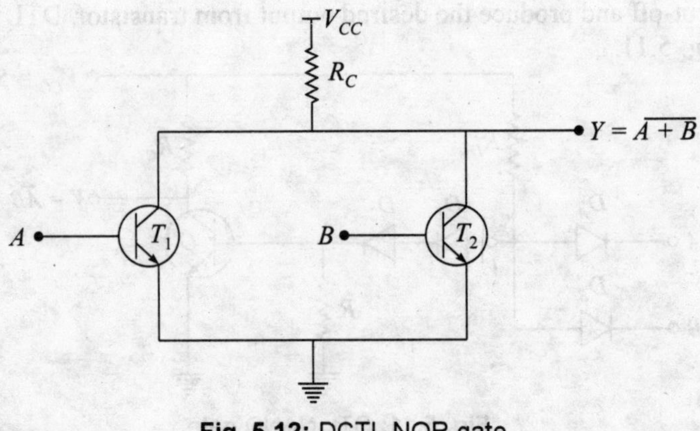

Fig. 5.12: DCTL NOR gate

Working Table:

Input		Transistor		output
A	B	T_1	T_2	$Y = \overline{A+B}$
0	0	OFF	OFF	$1(V_{CC})$
0	1	OFF	ON	0
1	0	ON	OFF	0
1	1	ON	ON	0

It gives the truth table of NOR gate.

Advantages:

1. Due to elimination of base resistor the DCTL is more simpler and economical to fabricate on IC.
2. More Flexible.
3. Package density is increased.
4. Noise Immunity is Increased.

In DCTL, the voltage corresponding to logic 1 and 0 levels are V_{BE}, sat (~ 0.8V) and V_{CE}, sat (~ 0.2 V).

The separation between the logic 1 and 0 level voltags, which is referred to as the logic swing, is very small.

Disadvantage:

1. Due to logic swing, Noise margin is poor.
2. It is simpler than RTL but it never become popular because of the problem of **Current Hogging:** As in DCTL mainly transistor are in parallel and shares the load current which bring them in saturation. The saturation is achieved by base voltage; but due to different V_{BE}, sat, the one transistor is in saturation and other not, so the saturated transistor carries bulk current and prevent other to saturate and result in error to the system.

5.8 HTL (HIGH THRESHOLD LOGIC)

HTL is shown in Fig. 5.13. This logic gate is similar to DTL. The supply voltage V_{CC} for this HTL gate is raised from 5V to 15V. This change in voltage causes increased noise immunity, because diode D_1 and D_2 must sustain a higher voltage and hence the two diodes are replaced by transistor to zener diodes in this DTL gate and the HTL gate resistance values are larger than other logic gates. This increased in resistance is necessary because of increased supply voltage or else it would result in large increase in current and therefore in power dissipation in HTL gate.

Zener diode in HTL gate has an operating voltage of about 6.9V.

Operation: HTL NAND gate performs the NAND operation for +ve logic.

Inputs (A, B) = Logic High (1)

Diodes (D_A, D_B) = Reverse Biased

Transistor (T_1) = Switch ON

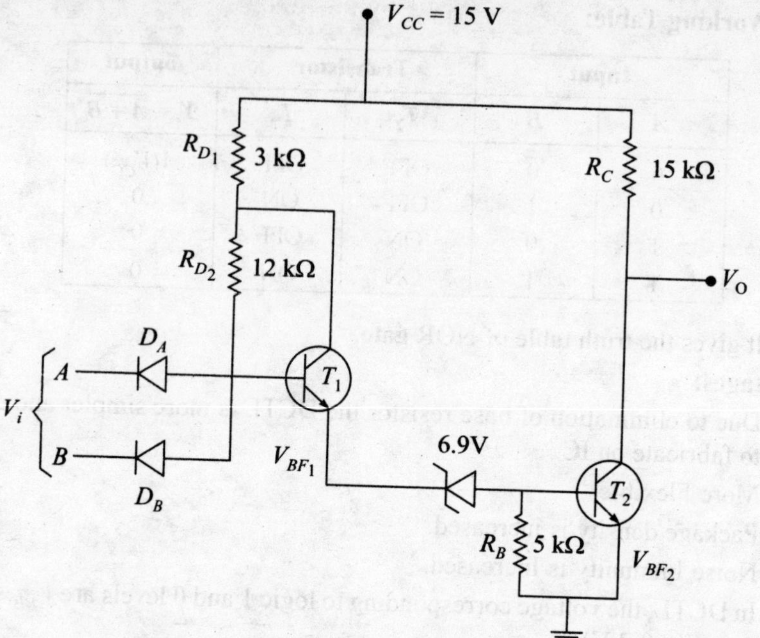

Fig. 5.13: HTL NAND Gate

Zener diode conducts when the input is above 6.9V.

V_0 = Low and required voltage to switch ON the transistor T_2 is 7.6 V.

$$V_0 = V_{BE1} + V_2 = 0.7 + 6.9 = 7.6 \text{ V}$$

If any one of input drops to ground potential (logic 0), the corresponding diode will conduct and flow through the R_B and R_{D2} resistor causes a voltage drop at input of transistor T_1.

These voltages are not enough to turn on the transistor T_2. The base voltage becomes Low and transistor T_2 remains cut-off and hence output is High.

Output = High.

Conclusion: **Noise signal must be higher than 7.5V to turn the transistor ON.**

Application:

HTL are used in:

- **Electrical Motors**
- **Relays**
- **Circuit Breakers.**

5.9 TTL (TRANSISTOR-TRANSISTOR LOGIC)

It refers to technology for designing and fabricating digital integrated circuits that employ logic gates consisting of primarily of bipolar transistor.

Because of speed limitation, DTL has become outdated. So it is completely replaced by other logic family TTL (Transistor Transistor Logic).

DTL speed limitation is overcome by the following modification in circuits, those are:

(*i*) The input diodes D_A and D_B are replaced by emitter-base junction of multiple-emitter transistor (T_1) which is easily and economically fabricated in IC.

(*ii*) The collector-base junction of T_1 acts as diode D_1.

(*iii*) The diode D_2 is replaced by emitter-base junction of another transistor (T_2).

The modified circuit is *TTL* NAND shown is Fig. 5.14.

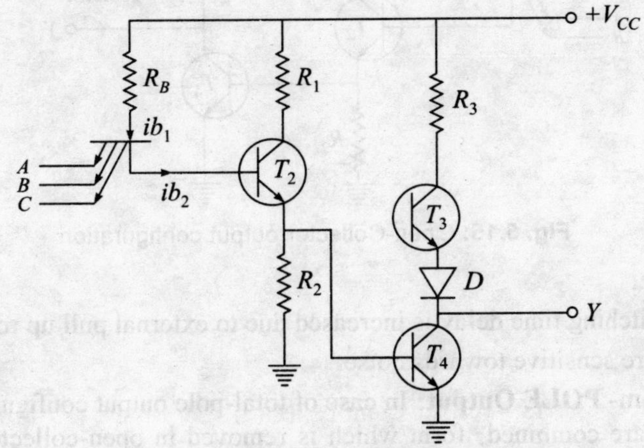

Fig. 5.14: TTL NAND gate

Operation: The working operation is shown in Fig. 5.14 and given by truth Table in that according to 3 input total cases take place.

Truth Table:

Inputs			*Transistors*				*Output*
A	**B**	**C**	T_1	T_2	T_3	T_4	**Y**
0	0	0	ON	OFF	ON	OFF	1
0	0	1	ON	OFF	ON	OFF	1
0	1	0	ON	OFF	ON	OFF	1
0	1	1	ON	OFF	ON	OFF	1
1	0	0	ON	OFF	ON	OFF	1
1	0	1	ON	OFF	ON	OFF	1
1	1	0	ON	OFF	ON	OFF	1
1	1	1	OFF	ON	OFF	ON	0

TTL Output circuit Configuration:

There are 3 types of circuit configuration of TTL:

1. Open-Collector output
2. Totem-pole output
3. Tristate output

1. **Open-Collector Output:** If we have output circuit configuration as open-collector output. In this circuit, if T_3, D and R_3 are removed from Fig. 5.14 and collector of T_4 becomes open. So, it is known as open-collector output configuration.

But this circuit will not work properly, so, to make it work we connect an external pull up resistor and output is still taken from collector of T_4. It is shown in Fig. 5.15.

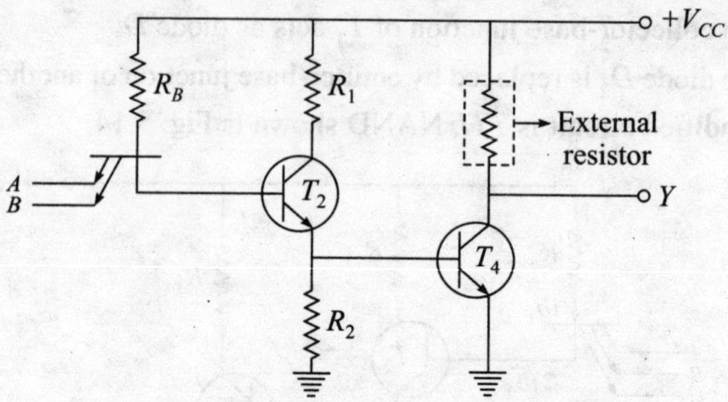

Fig. 5.15: Open-Collector output configuration

Disadvantage:

(*i*) Switching time delay is increased due to external pull up resistor.

(*ii*) More sensitive towards noise.

2. **Totem- POLE Output:** In case of total-pole output configuration there are T_4, T_3, R_3, D are combined form which is removed in open-collector output. This totem-pole arrangement is shown in Fig. 5.16 by dotted line.

Advantage:

(*i*) It can change states faster.

(*ii*) System can drive high value of capacitive loads.

(*iii*) Noise-sensitive becomes Low.

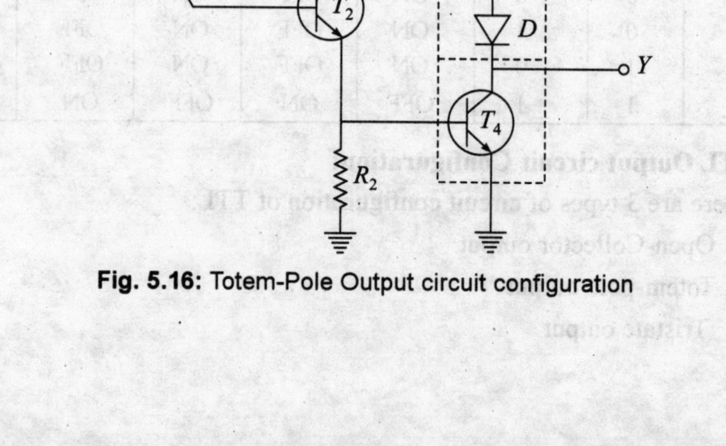

Fig. 5.16: Totem-Pole Output circuit configuration

Totem pole output is standard output of *TTL* gate. It is specially designed to reduce the propagation delay in the circuit. It is also designed is such a way so as to provide sufficient output power for a high fan out.

Disadvantage: This circuit cannot be connected to any other output, but is connected, there must be definitely loading problem.

3. **Tristate Output:** It means it has 3 output:

- Logic 0
- Logic 1
- High impedance state (open circuit)

Utility of open circuit of high impedance state is that circuit has disabled output means cut-off from other components.

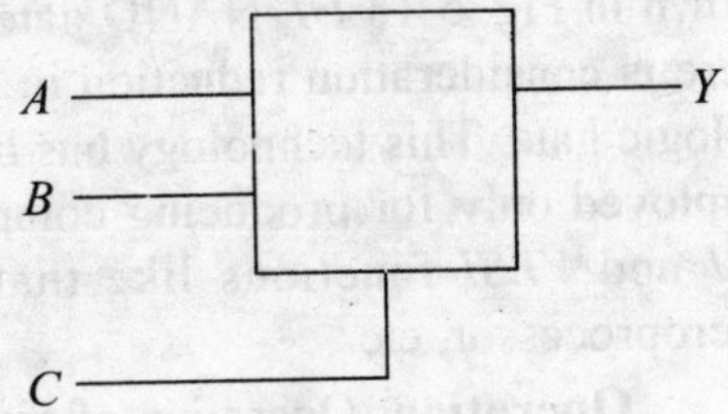

Tristate Gate

There are 3 inputs in Tristate output:

A, B = Normal input

C = Control/Select Input

when $C = 0$ Gate is disabled (High Impedance state)

$C = 1$ Gate is enabled (Normal function)

It is shown in Fig. 5.17

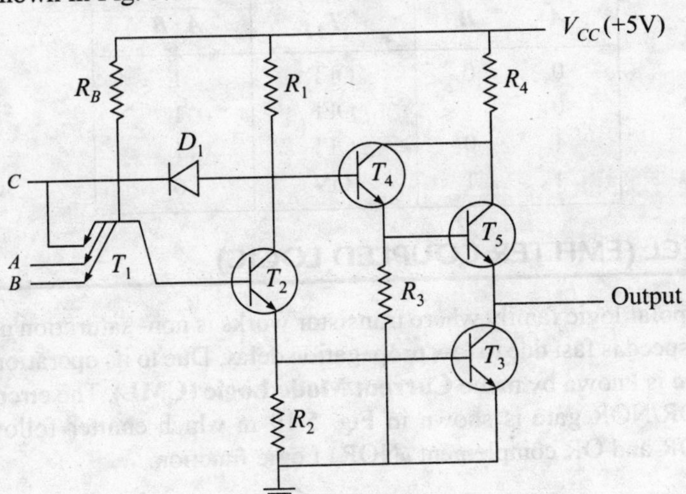

Fig. 5.17: Tristate Output

Operation of this tristate output is given in the truth table.

Truth Table:

When	*Inputs*		*Transistor*					*Output*
C = 1	*A*	*B*	T_1	T_2	T_3	T_4	T_5	*y (output)*
	0	0	ON	OFF	OFF	ON	ON	1
	0	1	ON	OFF	OFF	ON	ON	1
	1	0	ON	OFF	OFF	ON	ON	1
	1	1	OFF	ON	ON	OFF	OFF	0
C = 0	×	×	ON	OFF	OFF	OFF	OFF	High Impedance

5.10 IIL (INTEGRATOR INJECTION LOGIC)

For the *DCTL* problem, I^2L is designed, it is used very small silicon chip area, consumes little power, requires only four masks and two diffusions and hence, it is easy and cheaper to fabricate.

It does not contain resistors as shown in Fig. 5.18 I^2L NAND gate; so there is consideration reduction in size of logic gate. This technology has been employed only for producing complex *LSI* and *VLSI* functions like that of microprocessor, etc.

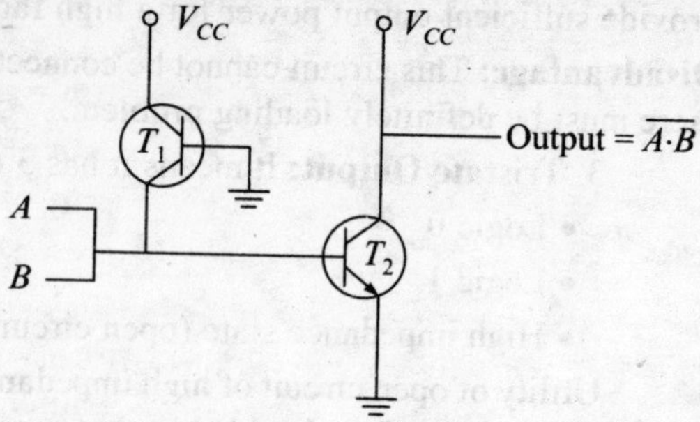

Fig. 5.18: I^2L NAND gate

Operation: Operation of given circuit is shown by the truth Table. Transistor T_1 is grounded.

Truth Table:

Input		*Transistor*	*Output*
A	*B*	T_2	$\overline{A \cdot B}$
0	0	OFF	1
0	1	OFF	1
1	0	OFF	1
1	1	ON	0

5.11 ECL (EMITTER COUPLED LOGIC)

It is part of bipolar logic family where transistor works is non- saturation mode and in the switching speed is fast due to low propagation delay. Due to its operation in current mode, the logic is known by name **Current Mode Logic (CML)**. The circuit diagram of an ECL OR/NOR gate is shown in Fig. 5.19 in which emitter follower output provides the OR and OR complement (NOR) Logic function.

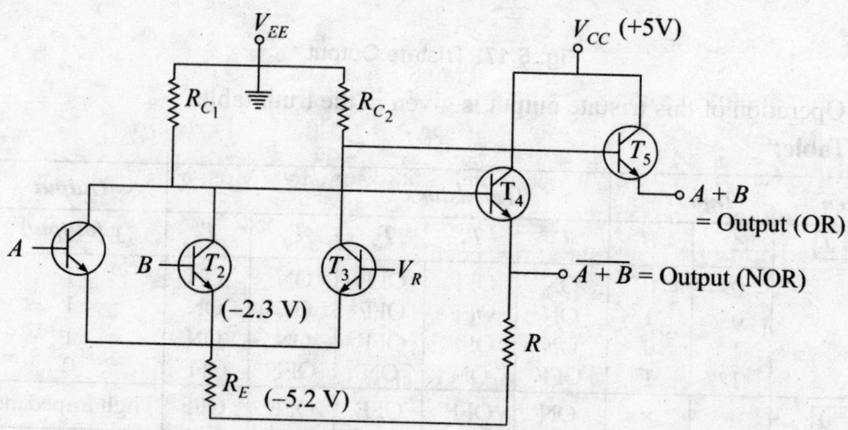

Fig. 5.19: ECL OR/NOR gate

Because of low output impedance of emitter follower and high input impedance differential amplifier input, High fan-out is possible.

V_{EE} pin is normal connected to ground, and V_{CC} pin is connected to –5.2V from power supply for best operation.

In +ve Logic: 1 = High = –0.9V

0 = Low = –1.7V

In –ve Logic 1 = High = –1.7V

0 = Low = –0.9V

Operation: When both the inputs are Low (–1.7V for +ve Logic). In this case T_1 and T_2 are at – 1.7V and base of T_3 is at –1.3V. Therefore, T_3 is more forward biased than T_1 and T_2. Once T_3 is ON, the emitter voltage of T_3 becomes –2.1V, make their base emitter junction reverse biased. As a result, T_1 and T_2 are OFF. When T_1 and T_2 are OFF, the current flowing through Rc_1 is only current through base of T_4. This makes emitter of T_4 as = (–0.9V) and see that NOR output is High. On the other hand, T_3 is ON, large current flowing through RC_2. The value of RC_2 is so adjusted that when T_3 ON, voltage drop across R_{C2} is –0.9V. Emitter of T_5 is –1.7V, OR output is Low and again Repeat when A = Low, B = High and A = High, B = Low and all.

Advantage:

(*a*) Large fan-out.

(*b*) Increasing overall switching speed.

(*c*) Low propagation delay.

Disadvantage:

(*a*) High power dissipation

(*b*) Large space requirement

(*c*) Low Noise Margins

5.12 MOS FAMILY

MOS = Metal Oxide Semiconductor

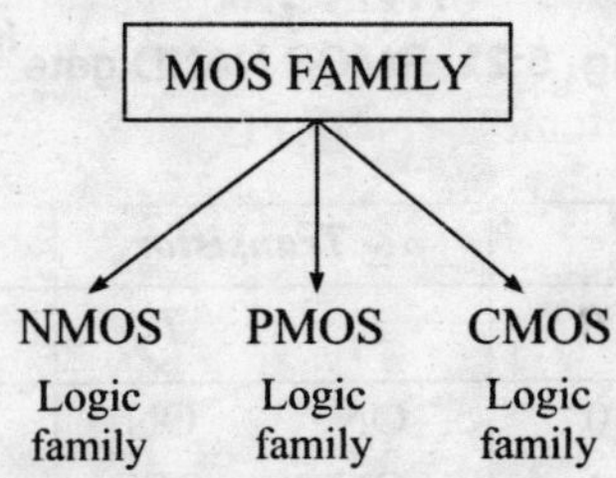

MOS family are used in Logic Gates and Registers/Memories.

Advantage: Good space Utilization.

Disadvantage: Low operating speed.

5.12.1 NMOS Logic Family

In this logic family *n*-channel is there and it is shown in Fig. 5.20. NMOS NAND gate.

In figure Two *n*-channel MOS FET. T_2 and T_3 are connected in series. T_1 is also *n*-channel MOSFET, acting as load which is ON permanently. Operation is shown in truth Table.

A	*B*	T_2	T_3	*Y*
0	0	OFF	OFF	1
0	1	OFF	ON	1
1	0	ON	OFF	1
1	1	ON	ON	0

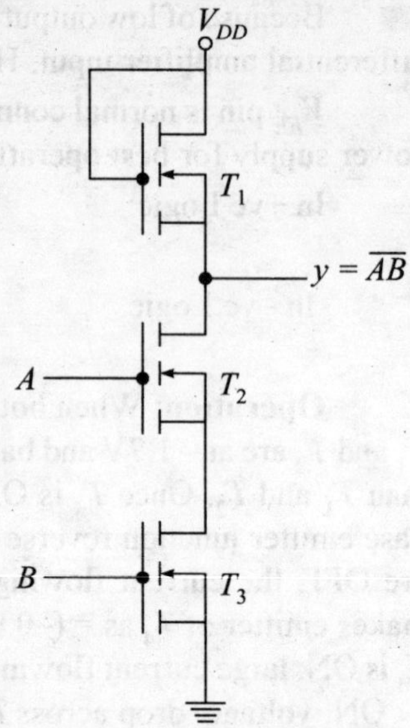

Fig. 5.20: NMOS NAND gate

5.12.2 PMOS Logic Family

PMOS logic family consists of *p*-channel Metal-Oxide semiconductor field effect transistor as element to build various gates in IC, PMOSFET is maily used as a resistor so as it is easier for manufacture to implement the logic using *P*-MOSFET.

It is shown in Fig. 5.21 PMOS NAND gate.

Working: Working of this PMOS logic family is shown by truth table.

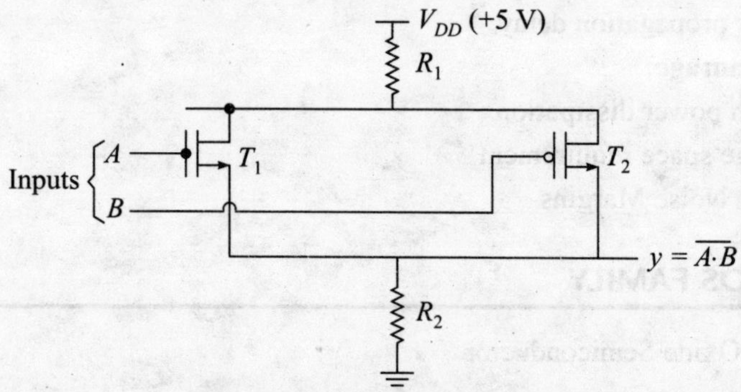

Fig. 5.21: PMOS NAND gate

Truth Table:

Inputs		*Transistor*		*Output*
A	*B*	T_1	T_2	*Y*
0	0	ON	ON	1
0	1	ON	OFF	1
1	0	OFF	ON	1
1	1	OFF	OFF	1

Advantage:

(*a*) Low Cost. (*b*) Easier to manufacture.

Disadvantage:

(*a*) Slow transition speed. (*b*) High power dissipation.

5.13 CMOS LOGIC FAMILY

CMOS = Complementary MOS.

It consists of one *N*-channel and one *P*-channel MOSFET. Gates of *N*-channel MOSFET and *P*-channel MOSFET are joined together and common terminal is connected to input voltage V_i.

Drains of *N*-channel MOSFET and *P*-channel MOSFET are joined together and common drain connection provides output.

Source of *P*-channel MOSFET is connected to a supply voltage V_{DD} and source of *N*-channel MOSFET is grounded.

N-channel MOSFET	*P*-channel MOSFET
V_{GS} = +ve ON	V_{GS} = –ve ON
V_{GS} = –ve OFF	V_{GS} = +ve OFF
if Gate = 1(+ve)	*N*-channel MOSFET-ON
	P-channel MOSFET-OFF
= 0(–ve)	*N*-channel MOSFET-OFF
	P-channel MOSFET-ON

CMOS NAND Gate: In case of CMOS NAND gate, the driver transistors, are connected in series while load transistors are connected in parallel. CMOS NAND gate is shown in Fig.5.22.

In figure T_1 and T_2 both *p*-channel MOSFET are connected in parallel, T_3 and T_4 both *n*-channel MOSFET are connected in series.

It is shown in truth table in which operation is given.

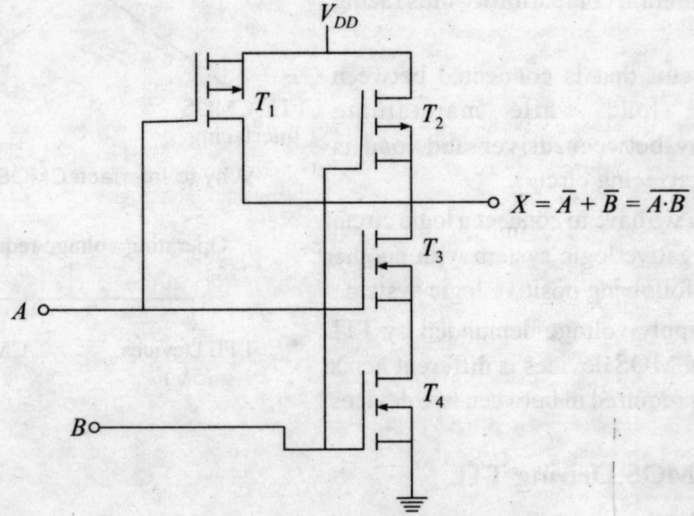

Fig. 5.22: CMOS NAND gate

Truth Table:

Inputs		*Transistor*				*Output*
A	*B*	T_1	T_2	T_3	T_4	*X*
0	0	ON	ON	OFF	OFF	1
0	1	ON	OFF	OFF	ON	1
1	0	OFF	ON	ON	OFF	1
1	1	OFF	OFF	ON	ON	0

Advantage:

(*i*) Low Power consumption (*ii*) Low Noise

Disadvantage:

Slow speed.

Application:

(*i*) Microprocessor and Interfacing (*ii*) Digital Calculator

(*iii*) Electronic watches.

5.14 INTERFACING OF CMOS AND TTL FAMILIES

To achieve optimum performance in a digital systems, devices from more than one logic family can be used, taking advantage of superior characteristics of each family for different parts of the system.

Interfacing: When gate from one family is feeding the gate of another family it is necessary to ensure that there are matching voltage and current levels. It is called Interfacing.

To maintain compatibility interfacing is required.

A circuit that is connected between driver and load while maintaining compatibility between driver and load is known as interfacing circuit.

When we have to connect a logic circuit following negative logic system with another logic circuit following positive logic system.

As supply voltage demanded by TTL devices and CMOS devices is different hence interfacing is required in between two devices.

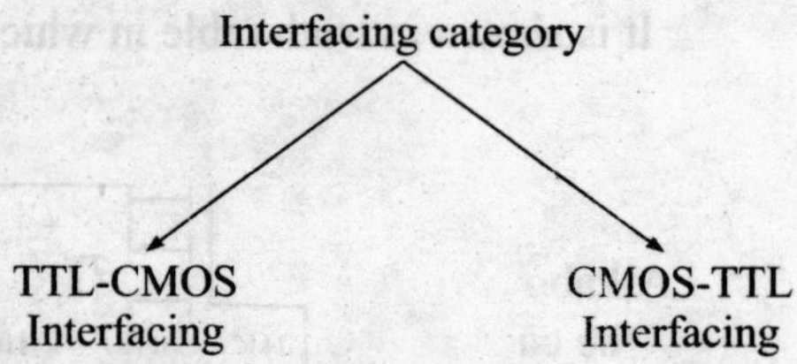

Why to Interface CMOS with TTL

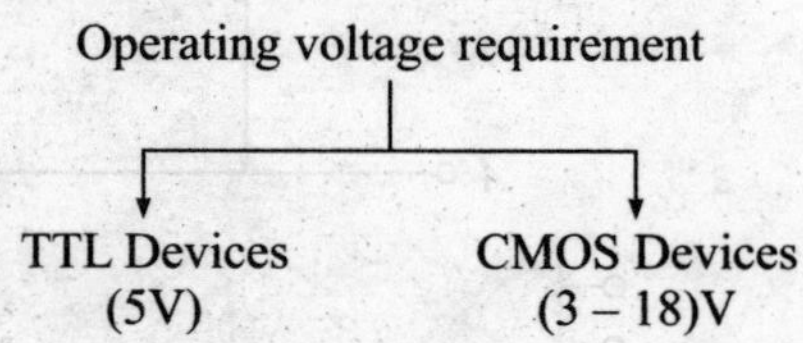

5.14.1 CMOS Driving TTL

Figure 5.23 shows a CMOS gate driving *N* TTL gates. For such an arrangement to operate properly the following conditions are required to satisfy.

$$V_{OH}(\text{CMOS}) \geq V_{IH}\,(TTL) \quad \text{...(1)}$$
$$V_{OL}(\text{CMOS}) \leq V_{IL}\,(TTL) \quad \text{...(2)}$$
$$-I_{OH}(\text{CMOS}) \geq NI_{IH}\,(TTL) \quad \text{...(3)}$$
$$I_{OL}(\text{CMOS}) \leq -NI_{IL}\,(TTL) \quad \text{...(4)}$$

Equation (3) is satisfied for small values of *N* but the equation (4) is not satisfied even for *N* = 1, except this difficulty can be overcome by using CMOS buffers having an adequate available output current.

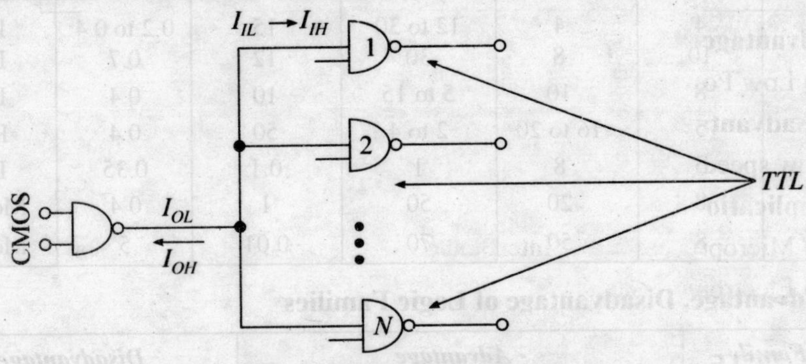

Fig. 5.23: CMOS Driving TTL

5.14.2 TTL DRIVING CMOS

Figure 5.24 shows a *TTL* gate driving N CMOS gate. For such an arrangement to operate properly the following conditions are required to be satisfied.

$$V_{OH}(TTL) \geq V_{IH}\,(\text{CMOS}) \quad \text{...(5)}$$
$$V_{OL}(TTL) \leq V_{IL}\,(\text{CMOS}) \quad \text{...(6)}$$
$$-I_{OH}(TTL) \geq NI_{IH}\,(\text{CMOS}) \quad \text{...(7)}$$
$$I_{OL}(TTL) \geq -NI_{IL}\,(\text{CMOS}) \quad \text{...(8)}$$

All above equations are satisfied for high value of *N*, for 74 HC/74 HCT series, whereas the case of 74C. Series the equation (5) not satisfied.

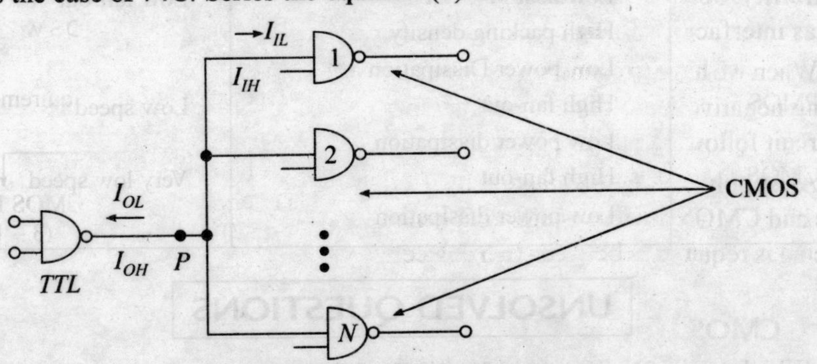

Fig. 5.24: TTL driving CMOS

A circuit modification is used to overcome value of V_{OH} (*TTL*) above 3.5V is obtained by connecting a resistance (≈ 2kΩ) between points *P* and V_{CC}. This acts as

passive pull-up, which pulls up the voltage at P to higher value after the transistor T_4 of TTL becomes non-conducting.

5.15 COMPARISON OF LOGIC FAMILIES

Logic Family	*Fan In*	*Fan Out*	*Propagation Time (nsec)*	*Power Dissipation (mW)*	*Noise Margin (volts)*	*Cost*
RTL	4	4	12 to 30	15	0.2 to 0.4	Low
DTL	10	8	30	12	0.7	Low
TTL	8	10	5 to 15	10	0.4	Low
ECL	5	16 to 20	2 to 4	50	0.4	High
I^2L	5	8	1	0.1	0.35	Low
PMOS	8	20	50	1	0.4	Medium
CMOS	8	50	70	0.01	5	Medium

Advantage, Disadvantage of Logic Families

Logic Family	*Advantage*	*Disadvantage*
RTL	Low power dissipation margin	Low speed Low Noise Low Fan-out
DTL	Low power dissipation	Low speed
TTL	Low power dissipation High speed High fan-out Low cost	Low V_{CC} tolerance Susceptible to transients
ECL	High speed High fan-out Low Noise	High cost Interfacing problem
I^2L	Low cost High packing density Low power Dissipation	Low fan-out
PMOS	High fan-out Low power dissipation	Low speed
CMOS	High fan-out Low power dissipation	Very low speed

UNSOLVED QUESTIONS

1. Which type of logic gate is faster? Explain.
2. Draw the TTL circuit with totem pole output and explain its working.
3. Explain the operation of ECL Family. How it is different from other logic families?

4. Write a short note on the characteristic of digital IC.
5. Define the following terms:
    (*a*) Fan in
    (*b*) Fan out
    (*c*) Figure of Merit
    (*d*) Speed of operation
    (*e*) Wired Logic
    (*f*) Noise Immunity.
6. Why ECL Logic is faster than *TTL*?
7. What is tri-state Logic?
8. Write brief note on interfacing *TTL* with CMOS.
9. Compare Bipolar and CMOS devices.
10. Explain working of *N*-MOSFET.
11. Draw the circuit of *DTL*.
12. Draw the circuit of *DCTL*.
13. Explain the working of *ECL*.
14. Explain the switching mode operation of *PN* junction.

UNIT 6

A/D CONVERTER AND D/A CONVERTER

Topics are:

- Introduction
- Types of Converter
- Sample and Hold
- D/A Converter (Digital to Analog Converter)
- Specification of D/A Converter
- A/D Converter (Analog to Digital Converter)
- Specifications of A/D Converter

CHAPTER 6

A/D Converter and D/A Converter

6.1 INTRODUCTION

The process of converting an analog signal to a digital signal form involves a sequence of four processes. These are:

1. Sampling.
2. Holding.
3. Quantizing.
4. Encoding.

These processes are not referred as separate operations. **Sampling and Holding** operations are done simultaneously using a circuit known as **Sample and Hold** (S/H) circuit. These operations are required to be performed for the conversion of time varying analog signals and not for *d.c* signals.

Quantizing and Encoding processes are done simultaneously using a circuit referred to as **Analog-to-Digital converter** (A/D converter or ADC). The process of conversion of an analog signal to digital signal is referred to as an Analog-to-Digital converter.

But the output of system may be required to be in analog form and therefore, the digital output has to be converted back to analog form. The process is referred to as a Digital-to-Analog conversion, and system used for this is **Digital-to-Analog converter.**

Analog data are used to drive motor/other analog devices. Digital data are used for easy processing.

6.2 TYPES OF CONVERTER

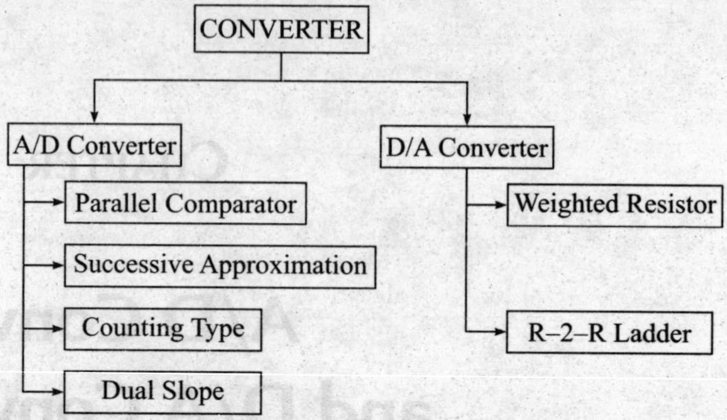

6.3 SAMPLE AND HOLD

A time varying analog signal can have any value in a given range. For example: $v(t) = V\sin \omega t$. Shown in Fig. 6.1 will have different value of voltage between range $+v$ and $-v$ at different instants of time.

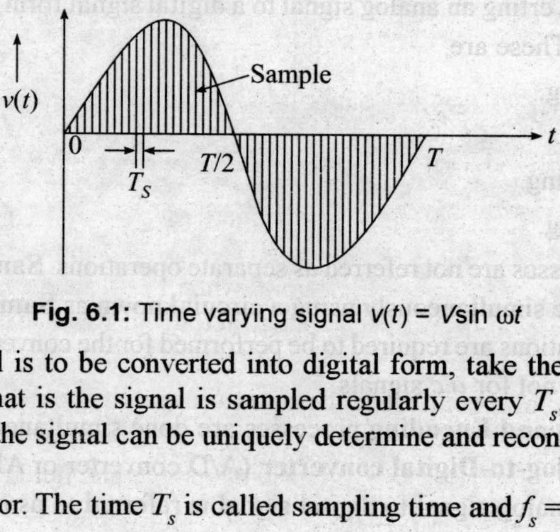

Fig. 6.1: Time varying signal $v(t) = V\sin \omega t$

If this signal is to be converted into digital form, take the samples of $v(t)$ at regular intervals, that is the signal is sampled regularly every T_s, then according to sampling theorem, the signal can be uniquely determine and reconstructed from these samples with no error. The time T_s is called sampling time and $f_s = \dfrac{1}{T_s}$ sampling rate.

According to sampling theorem:

$$\boxed{f_s \geq 2f}$$

Where f is frequency of analog signal. Above relation conclude that the sampling rate be rapid enough for at least two samples to be taken during course of time period of analog signal.

The constant voltage corresponding of each sample is obtained using Sample and Hold circuit.

6.3.1 Sample and Hold Circuit

Basic sample and Hold circuit is shown in Fig. 6.2. In this circuit the voltage across the capacitor follows the circuit's' input signal voltage (V_i) during the time switch S is closed. The capacitor holds the instantaneous value of signal voltage attained just before the switch is opened. Thus for every T_s, the switch is closed for a short duration and then opened.

The *dc* voltage across capacitor gives the value of the signal at the instant the switch is opened. Then *d.c* voltage represents a sample of signal and is converted to digital signal using A/D converter circuit during Hold period.

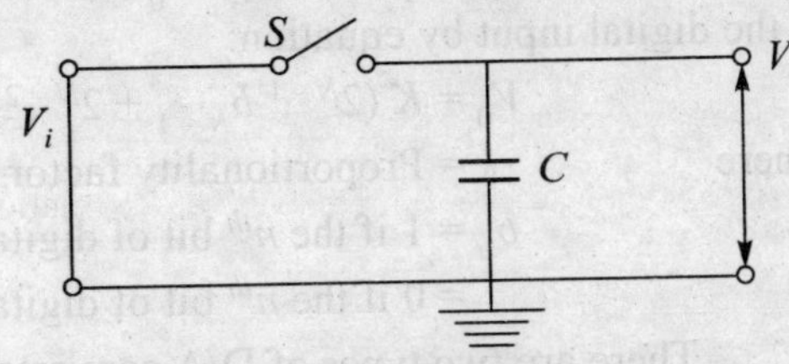

Fig. 6.2: Sample and Hold circuit

Now, in Fig. 6.3 **practical S/H circuit** is shown. An enhancement mode MOSFET is used as a switch which is controlled by a control voltage V_s.

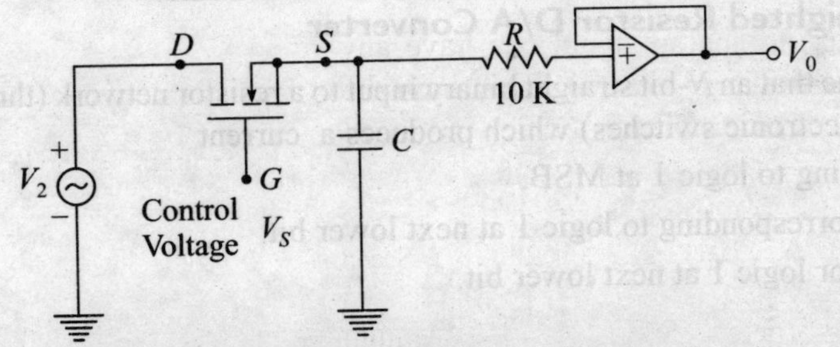

Fig. 6.3: Practical sample and Hold circuit.

An OP-AMP is used to avoid discharge of capacitor due to loading effect of A/D converter. During the +ve values of V_s ($V_s > V_t$) the MOSFET conducts, input voltage appears across *c* and at the output, when V_s is less then V_t (threshold voltage), the MOSFET is cut-off (*s* open) and voltage across *c* is retained because input resistance of OP-AMP is very high.

Accuracy of circuit depends upon the holding of charge in the capacitor, therefore, a capacitor with a very low leakage must be used.

Now, in Fig. 6.4 output waveform of sample and hold circuit is shown.

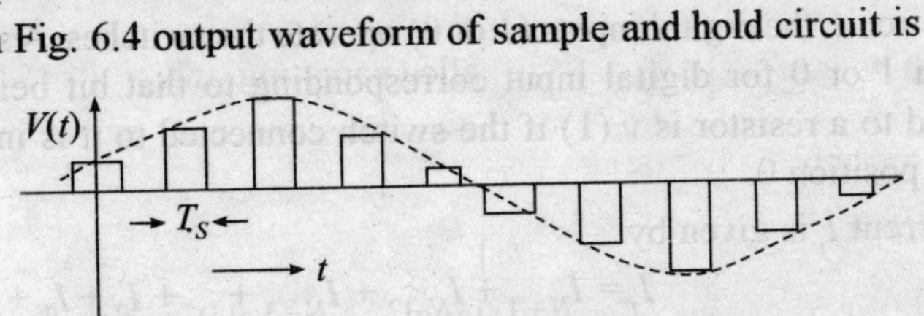

Fig. 6.4: Output waveform of sample and Hold Circuit

6.4 D/A CONVERTER (DIGITAL TO ANALOG CONVERTER)

The input to a D/A converter is an N-bit binary signal, available in parallel form. Normally, digital signals are available at output of the latches or registers and the voltages, corresponding to logic 0 and logic 1, available to drive the converter are in general not precisely fixed voltages. Therefore, these voltages are not applied directly to converter but are used to operate digitally controlled switches. The switch is thrown to one of the two positions depending on digital signal (1 or 0) which connects precisely fixed voltages $V(1)$ or $V(0)$ to converter input, corresponding to 1 or 0, respectively.

The analog output voltage V_0 of an N-bit straight binary D/A converter is related to the digital input by equation:

$$V_0 = K(2^{N-1} b_{N-1} + 2^{N-2} b_N - 2 + ... + 2^2 b_2 + 2b_1 + b_0) \quad ...(1)$$

where k = Proportionality factor.

$b_n = 1$ if the nth bit of digital input is 1.

$= 0$ if the nth bit of digital input is 0.

There are two types of D/A converter:

1. Weighted Resistor D/A converter.
2. R-2-R ladder D/A converter.

6.4.1 Weighted Resistor D/A Converter

Let us assume that an N-bit straight binary input to a resistor network (through digitally controlled electronic switches) which produces a current

I corresponding to logic 1 at MSB,

$I/2$ Corresponding to logic 1 at next lower bit.

$I/2^2$ for logic 1 at next lower bit.

⋮

$I/2^{N-1}$ for logic 1 at LSB.

The total current thus produced will be proportional to digital input. This current can be converted into a corresponding voltage, by using an OP-AMP, which will be proportional to digital input.

The circuit of Fig. 6.5 can be used for converting the digital input to analog output which operates according to above principle. This circuit is referred to as **weighted-resistor D/A converter.** Since the resistance values are weighted in accordance with binary weights.

In the circuit, the digital inputs (1 or 0) operate the switches. A switch is thrown to the position 1 or 0 for digital input corresponding to that bit being 1 or 0, The voltage applied to a resistor is $v(1)$ if the switch connected to it is in position 1 and $v(0)$ if it is in position 0.

The current I_i is given by:

$$I_i = I_{N-1} + I_{N-2} + I_{N-3} + ... + I_2 + I_1 + I_0 \quad ...(2)$$

where

$$I_{N-1} = V_{N-1}/R$$

$$I_{N-2} = V_{N-2}/2R \qquad \text{where } V_n = V(1) \text{ if } b_n = 1$$

$$= V(0) \text{ if } b_n = 0$$

$$I_{N-3} = V_{N-3}/2^2R$$

⋮

$$I_0 = V_0/2^{N-1}R$$

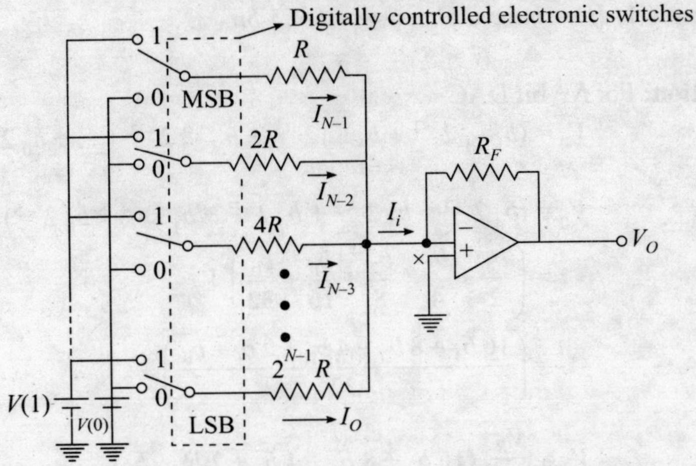

Fig. 6.5: Weighted Resistor D/A converter

For straight binary input, $V(0) = 0$ and $V(1) = -V_R$, and output voltage V_0 is given by:

$$V_0 = -(-V_R)\left(\frac{R_F}{R} b_{n-1} + \frac{R_F}{2R} b_{n-2} + \frac{R_F}{2^2 R} b_{n-3} + ... + \frac{R_F}{2^{N-1} R} bo\right) \quad ...(3)$$

It is same form as equation 1.

with $\quad k = \frac{R_F}{2^{N-1} R} V_R$

The output swings in only one direction and therefore is unipolar.

With $V(1)$ and $V(0)$ as the voltage applied to resistor network for 1 and 0, then output voltage of Fig. 6.5 is given by:

$$V_0 = \frac{R_F}{2^{N-1} R}(2^{N-1} V_{N-1} + 2^{N-2} V_{N-2} + ... + 2^1 V_1 + 2^0 V_0) \quad ...(4)$$

$b_n = b_0, b_1, ... b_{N-1} = 1$ [If n^{th} *bit is* –1]

else

$b_n = 0$ [If n^{th} bit is 0]

where $\quad V_R$ = Reference voltage.

MSB of binary word correspond to $b_{N-1} = \frac{\text{weight}}{V_R/2}$

LSB of binary word correspond to $b_0 = \frac{V_R}{2^n}$

Advantage:

1. It is simple to construct.
2. Cost is low.

Example 1: *If we have a binary word of 5-bit*

$$N = 5$$

Solution: For N^{th} bit DAC

$$V_0 = (b_{n-1}\, 2^{-1} + b_{N-2}\, 2^{-2} + b_{N-3}\, 2^{-3} + ... + b_0\, 2^{-N})\, V_R$$

$$N = 5$$

$$V_0 = (b_4\, 2^{-1} + b_3\, 2^{-2} + b_2\, 2^{-3} + b_1\, 2^{-4} + b_0\, 2^{-5})\, V_R$$

$$= \left(\frac{b_4}{2} + \frac{b_3}{4} + \frac{b_2}{8} + \frac{b_1}{16} + \frac{b_0}{32}\right) V_R$$

$$= \left(\frac{16\, b_4 + 8\, b_3 + 4\, b_2 + 2\, b_1 + b_0}{32}\right) V_R$$

$$V_0 = \frac{V_R}{32}\,(16\, b_4 + 8\, b_3 + 4\, b_2 + 2\, b_1 + b_0)$$

Now assume $V_R = 32.$

$$V_0 = 16\, b_4 + 8\, b_3 + 4 b_2 + 2 b_1 + b_0$$

Take simplest example:

MSB				LSB
a_4	a_3	a_2	a_1	a_0
↓	↓	↓	↓	↓
0	0	1	0	1

$$V_0 = 4 a_2 + a_0 \;\Rightarrow\; 4 + 1 \;\Rightarrow\; 5$$

V_0 Analog voltage proportional to digital input.

6.4.2 R-2R Ladder Type DAC

Limitation of Weighted Resistor Type DAC:

- Concept was very easy to understand but requires too many values of resistances. *N*-bit converter requires *N-bit* resistors.

An R-2R Ladder D/A converter is shown in Fig. 6.6. In uses resistors of only two values *R* and 2*R*. The inputs to the resistor network are applied through digitally controlled switches. A switch is in position 0 or 1 corresponding to digital input for that bit position being 0 or 1, To analyse this circuit, for simplicity we consider a 3-bit R-2R ladder D/A converter network shown in Fig. 6.7. In this circuit we assume the digital input as 001.

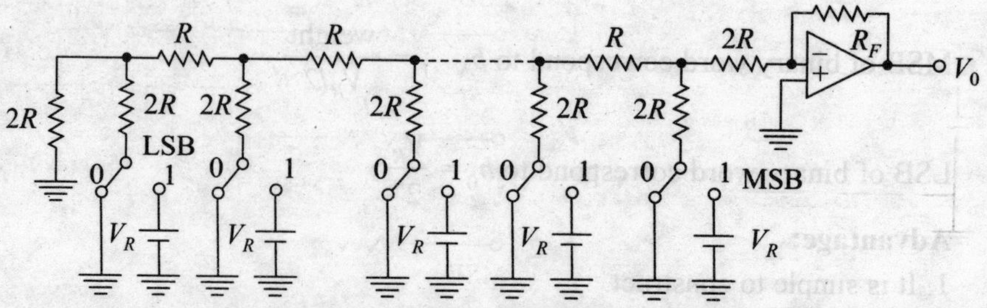

Fig. 6.6: R-2R Ladder D/A converter

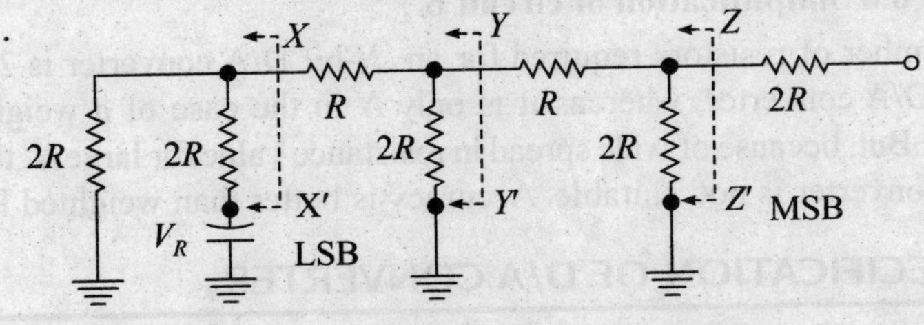

Fig. 6.7: 3-bit R-2R Ladder D/A network

The circuit is simplified using thevenin's theorem. Applying thevenin's theorem at *XX'*, we get the circuit of Fig. 6.8 (*a*). Simplify at *YY'* and *ZZ'*, we obtain the circuit of Fig. 6.8(*b*) and 6.8 (*c*) respectively.

Here LSB has assumed as 1 and equivalent voltage obtained is $V_R/2^3$.

Similarly, for digital input of 010 and 100, the equivalent voltages are $V_R/2^2$ and $V_R/2^1$. The value of equivalent resistance is 3*R* in each case. So, we get the equivalent circuit of Fig. 6.7, which is shown in Fig. 6.8 (*d*), for this output voltage V_0 is:

$$V_0 = -\left(\frac{R_F}{3R}\frac{V_R}{2^3}b_0 + \frac{R_F}{3R}\frac{V_R}{2^2}b_1 + \frac{R_F}{3R}\frac{V_R}{2^1}b_2\right)$$

$$= -\left(\frac{R_F}{3R}\right)\left(\frac{V_R}{2^3}\right)[4b_2 + 2b_1 + 1b_0] \quad \text{...(5)}$$

Equation (5) shows that analog output voltage is proportional to digital input. In general, for an *N*-bit *D*/*A* converter, the output voltage can be similarly determined and is given by:

$$V_0 = (2^{N-1}b_{N-1} + 2^{N-2}b_{N-2} + ... + 2^2 b_2 + 2^1 b_1 + 2^0 b_0)$$

where $R_F = 3R,\ V_R = -2^N\ V$ ...(6)

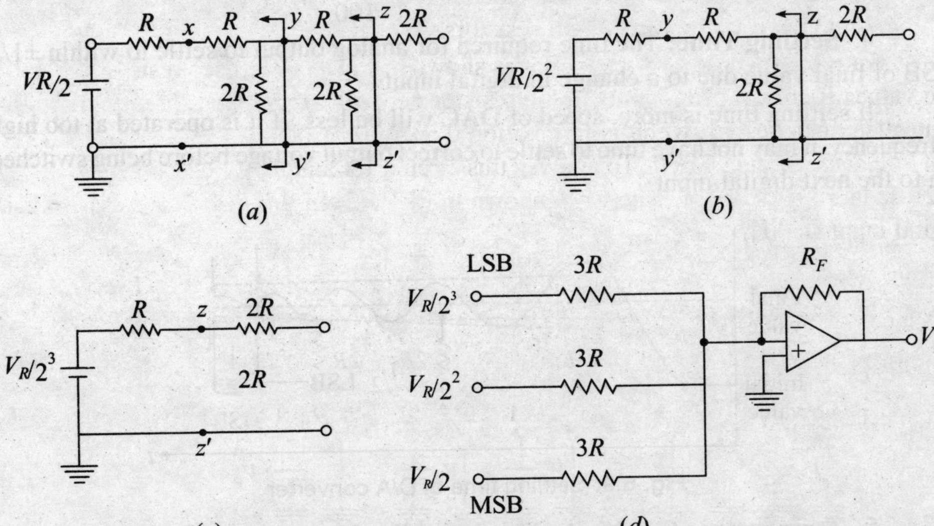

Fig. 6.8: Equivalent circuit of figure 6.7

Figure 6.8 Simplification of circuit 6.7

The number of resistors required for an N-bit D/A converter is $2N$ in case of R-2R ladder D/A converter, whereas it is only N in the case of a weighted-resistor D/A converter. But, because of wide spread in resistance value for large N, the weighted-resistor D/A converter is not suitable. Accuracy is better than weighted Resistor.

6.5 SPECIFICATION OF D/A CONVERTER

The characteristics of a D/A converter, which are generally by manufactures are:

1. Resolution
2. Linearity
3. Accuracy
4. Settling time.
5. Temperature sensitivity.

Explanation:

1. **Resolution:** This is possible smallest change in output voltage as a fraction or percentage of full-scale output range. It is very important to note that change in output voltage is as small as possible.

$$\text{Formula for the resolution for an } N\text{-bit DAC} = \frac{1}{2^N - 1} \times 100$$

2. **Linearity:** In D/A converter, equal increments in the numerical significance of digital input should result in equal increments in analog output voltage.

In an actual circuit, the I/O relationship is not linear. It is because of two reasons:

(*a*) Errors in resistors. (*b*) Losses in switches connected.

So, it is not able to maintain the linear relationship between input and output.

3. **Accuracy:** We can measure accuracy at a DAC by measuring the difference between actual output voltage obtained practically and expected output voltage. It is specified as a percentage of full scale or maximum output voltage.

$$\text{Maximum Error} = \frac{\text{Accuracy} \times \text{fullscale output voltage}}{100}$$

4. **Settling Time:** The time required for analog output to settle to within ±1/2 LSB of final value due to a change in digital input.

If settling time is more, speed of DAC will be less. If it is operated at too high a frequency, it may not have time to settle to correct output voltage before being switched on to the next digital input.

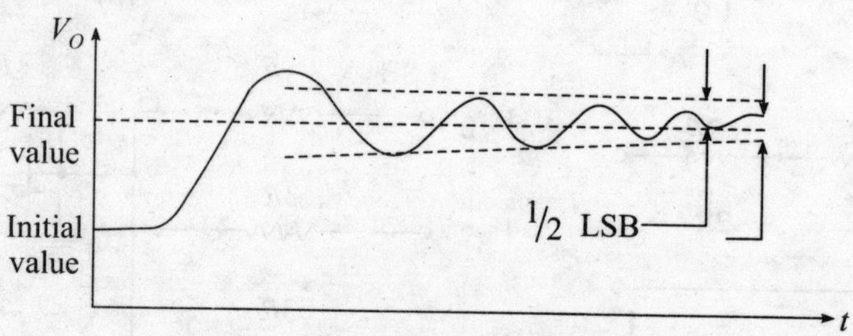

Fig. 6.9: Settling time of D/A converter

5. **Temperature Sensitivity:** It is basically change in output voltage from its expected value due to change in temperature.

The analog output voltage for any fixed digital input varies with temperature. This is due to temperature sensitivities of reference voltage source, resistors, OP-AMP etc. It is specified as ± ppm/°C.

6.6 A/D CONVERTER (ANALOG TO DIGITAL CONVERTER)

The block diagram of ADC is shown in Fig. 6.10. It is opposite to that of a DAC. ADC (Analog to Digital converter) accepts an analog input voltage (V_0) and produces an output binary word $D_1, D_2, D_3, ..., D_n$ of functional value D, so that:

$$D = d_1 2^{-1} + d_2 2^{-2} + ... + d_n 2^{-n}$$

where d_1 = MSB

d_n = LSB.

An ADC has two additional control lines:

The **start input** to tell the ADC when to start the conversion and **EOC** (End of conversion) output to announce when conversion is complete.

This is given in Fig. 6.10.

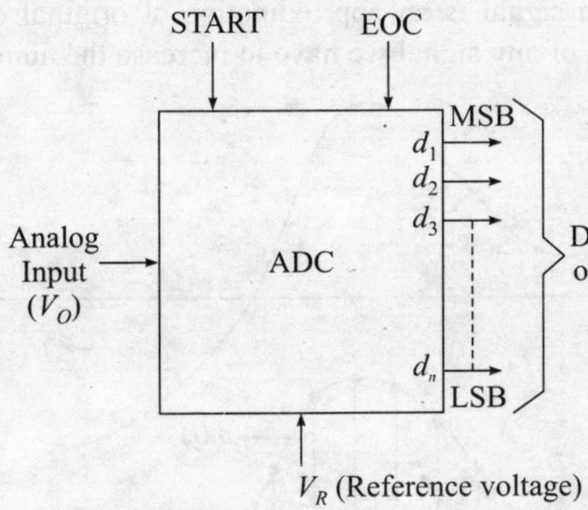

Fig. 6.10: Block Diagram of ADC

6.6.1 Quantization

In a digital-to-analog converter, the possible number of digital inputs is fixed. For example. in a 3-bit D/A converter, there are 8 possible inputs. In contrast, in an analog -to-digital converter, the input analog voltage can have any value in a range, but the digital output can have only 2^N discrete value for an N-bit A/D converter. Therefore, the whole range of analog voltage is required to be represented suitably in 2^N intervals. This process is known as **Quantization.**

Each interval is then assigned a unique N-bit binary code, which is referred to as **Encoding.** Simple diagram for coding of quantization and Encoding is in Fig. 6.11.

Quantization is defined as approximating the signal value. It refers as the use of a finite set of amplitude levels and the selection of a level nearest to a particular sample value of message signal as representation for it. This operation combined with sampling permits the use of coded pulses for representing the message signal.

Analog voltage	Equivalent digital value
V	
7/8 V	111
6/8 V	110
5/8 V	101
4/8 V	100
3/8 V	011
2/8 V	010
1/8 V	001
0	000

Fig. 6.11: Quantization and Encoding

Quantization is used in pulse code modulation. It is shown in Fig. 6.12 (*c*) the amplitudes of signal $m(+)$ lie in range $(-m_p$ to $m_p)$ which is divided into L intervals each of magnitude $\Delta V = \dfrac{2m_p}{L}$.

The quantization signal is an approximation of original one. If we want to increased the resolution of any signal we have to increase the number of levels L.

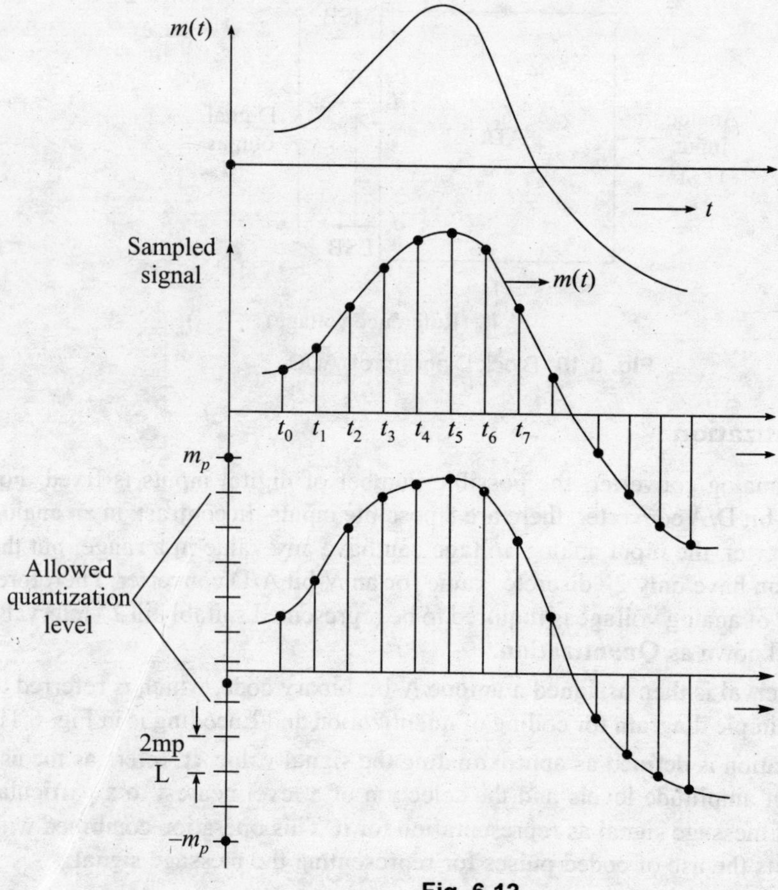

Fig. 6.12(*a*) Analog signal

Fig. 6.12(*b*) Sampled of Analog signal

Fig. 6.12(*c*) Quantization

Fig. 6.12

6.6.2 Types of A/D Converter

1. Parallel comparator (Simultaneous Type)
2. Successive Approximation
3. Counting Type
4. Dual Slope.

6.6.3 Parallel Comparator (Simultaneous Type)

It is also a flash type analog to digital converter. The simplest parallel comparator ADC is shown in Fig. 6.13(*a*) that is fastest and most expensive technique. The circuit consists of a resistive divider network, 7 op-Amp comparator and an 8 line to 3 line priority encoder. To construct a parallel comparator we need $2^n - 1$ comparators. The comparator and its truth table is shown in Fig. 6.13(*b*). The analog voltage (V_a) is applied in (+) terminal of comparator. The comparator compares the analog voltage with corresponding node voltage and produces the logical output as shown in Fig. 6.13(*b*).

From Fig. 6.13(*a*) at each node of resistive divider, comparison voltage is applied. Since, all the resistor are of equal value, the voltage level of each node is shown in Fig. 6.13(*a*).

The comparator output drives an 8 line to 3 line priority encoder. The encoder logic outputs a binary code that represents the analog input.

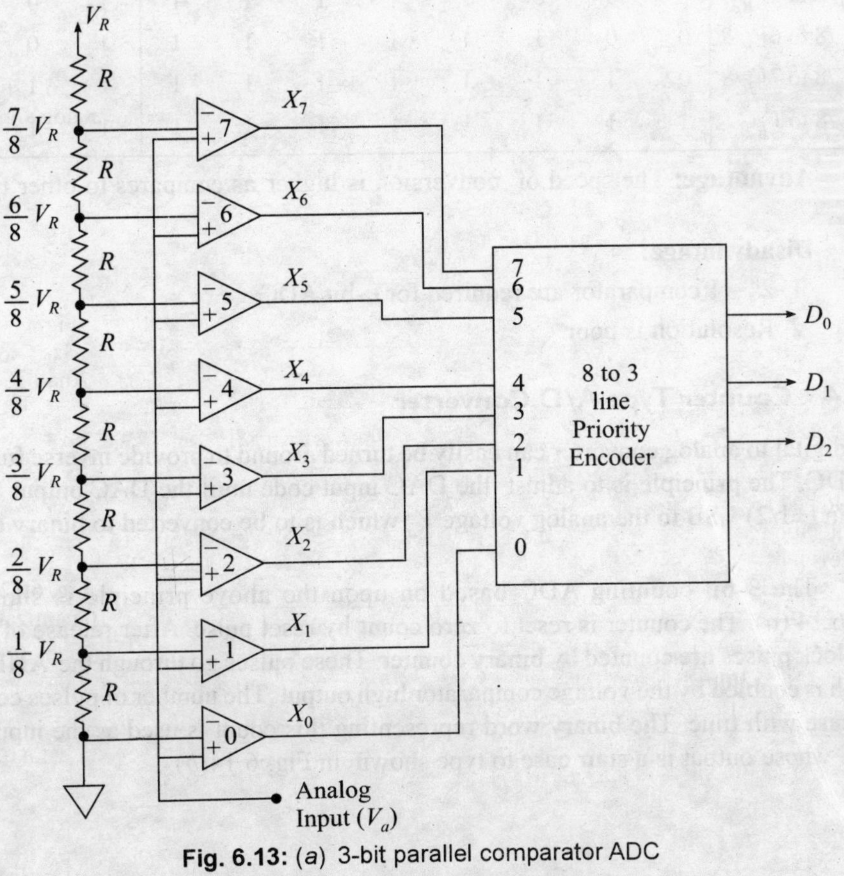

Fig. 6.13: (*a*) 3-bit parallel comparator ADC

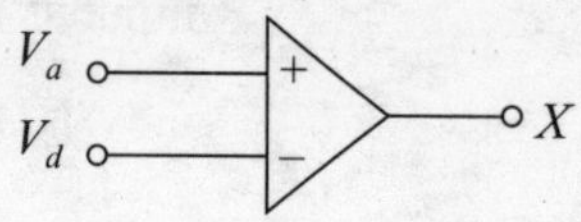

Voltage Input	*Logic Output X*
$V_a > V_d$	$X = 1$
$V_a < V_d$	$X = 0$
$V_a = V_d$	Previous value

Fig. 6.13 (*b*) Comparator and Truth Table

For example: Suppose that analog input ($V_a < 6/8\ V_R$), the output of comparator 1 through 5 would go high and 6 through 8 would go low. The priority encoder gives the output of 101. The 7th comparator has higher priority and Ist comparator has lowest priority. Table shows the truth table of 3-bit parallel comparator.

Truth Table of 3-bit Parallel Comparator

Input Voltage (V_a)	X_7	X_6	X_5	X_4	X_3	X_2	X_1	X_0	D_2	D_1	D_0
0 to $V_R/8$	0	0	0	0	0	0	1	0	0	0	0
$V_R/8$ to $2V_R/8$	0	0	0	0	0	0	1	1	0	0	1
$2V_R/8$ to $3V_R/8$	0	0	0	0	0	1	1	1	0	1	0
$3V_R/8$ to $4V_R/8$	0	0	0	0	1	1	1	1	0	1	1
$4V_R/8$ to $5V_R/8$	0	0	0	1	1	1	1	1	1	0	0
$5V_R/8$ to $6V_R/8$	0	0	1	1	1	1	1	1	1	0	1
$6V_R/8$ to $7V_R/8$	0	1	1	1	1	1	1	1	1	1	0
$7V_R/8$ to V_R	1	1	1	1	1	1	1	1	1	1	1

Advantage: The speed of conversion is higher as compares to other type of ADC.

Disadvantage:

1. $2^n - 1$ comparator are required for *n*-bit ADC.
2. Resolution is poor.

6.6.4 Counter Type A/D Converter

The digital to analog converter can easily be turned around to provide inverse function of ADC. The principle is to adjust the DAC input code until the DAC output comes within (±1/2) LSB to the analog voltage V_a which is to be converted to binary digital form.

The 3-bit counting ADC based on upon the above principle is shown in Fig. 6.14 (*a*). The counter is reset to zero count by reset pulse. After release of reset, the clock pulses are counted by binary counter. These pulses go through the AND gate which is enabled by the voltage comparator high output. The number of pulses counted increase with time. The binary word representing this count is used as the input of a DAC whose output is a stair case to type shown in Fig. 6.14(*b*).

The analog output V_d of DAC is compared to an analog input V_a by the comparator . If $\boxed{V_a > V_d}$, the output of **comparator becomes high** and AND gate is enabled to allow the transmission of clock pulses to the counter. When $\boxed{V_a < V_d}$ the output of **comparator becomes low** and the AND gate is disabled. This stops the counting at time $V_a \leq V_d$ and digital output of counter represents the analog input voltage V_a.

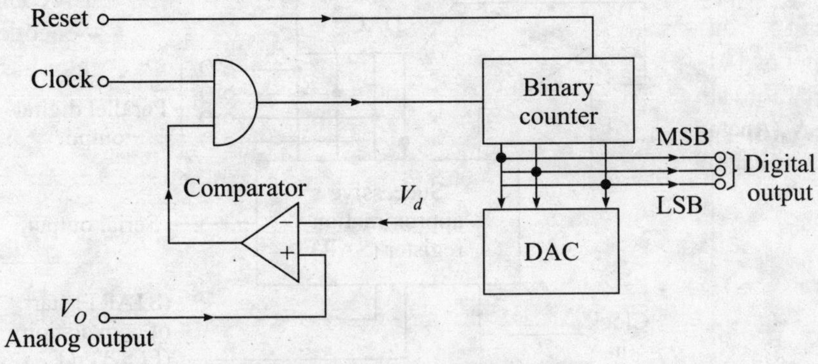

Fig. 6.14: (a) Counter Type ADC

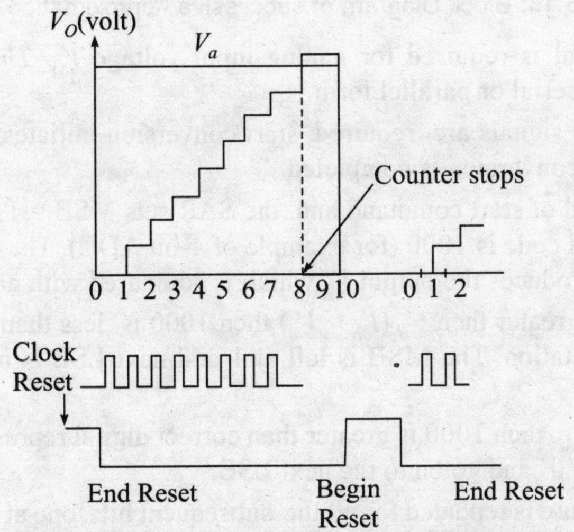

Fig. 6.15: (*b*) DAC Staircase Output Waveform

For a new value of analog input V_a, a second reset pulse is applied to clear the counter. The counter frequency must be low enough to give sufficient time for DAC to settle and for comparator to respond.

Advantage: It offers more and better resolution.

Disadvantage:

1. Slow speed
2. Conversion time will be long.

6.6.5 Successive Approximation Type A/D Converter

Basic Principle: This method is based on the principle that: "The unknown analog input voltage has been approximated with an n-bit digital value by trying one bit at a time, beginning with MSB."

Block Diagram: It is shown in Fig. 6.15. It consists of DAC, a comparator and successive approximation register (SAR).

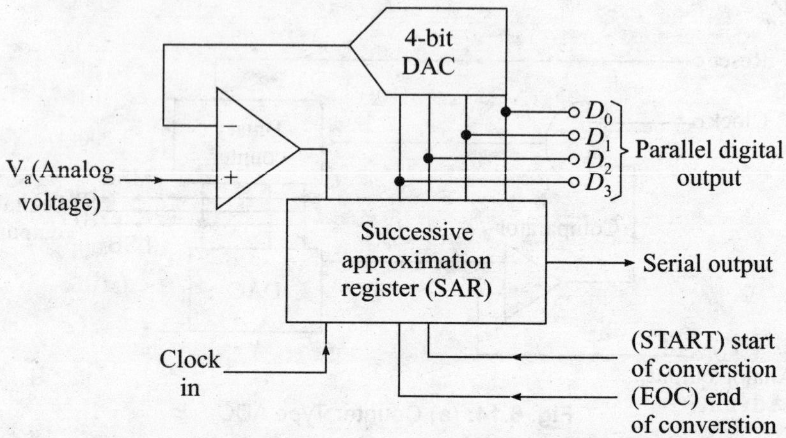

Fig. 6.15: Block Diagram of successive Approximation ADC

One terminal is required for analog input voltage V_a. *The* digital output is available in either serial or parallel form.

Two control signals are required, start conversion initiates an ADC sequence and EOC tells the conversion is completed.

With arrival of start command and, the SAR sets MSB = 1, with all other bits to zero so that trial code is 1000 (for example of 4-bit ADC). The DAC receives this digital input and produces the output V_d which is compared with analog input V_a.

1. If V_a is greater then V_d ($V_a > V_d$) then 1000 is less than the correct digital representation. The MSB is left at 1 and next LSB is made 1 and further tested.
2. If $V_a < V_d$, then 1000 is greater than correct digital representation. So, reset MSB is '0' and go on to the next LSB.

This procedure is repeated for all the subsequent bits, one at a time until all bits positions have been tested. Figure 6.16 shows two principle of operation of SAR (successive approximation register).

Whenever the DAC output crosses V_a, the comparator changes state and this can be taken as the end of conversion (EOC) command.

From Fig. 6.16 one clock pulse is required for the SAR to compare each bit. However, an additional clock pulse is usually required to reset the SAR to performing a conversion. This time for one analog to digital conversion depends on both clock's period T and number of bits.

The relationship is:

Conversion time $(T_c) = T(n + 1)$ ...(7)

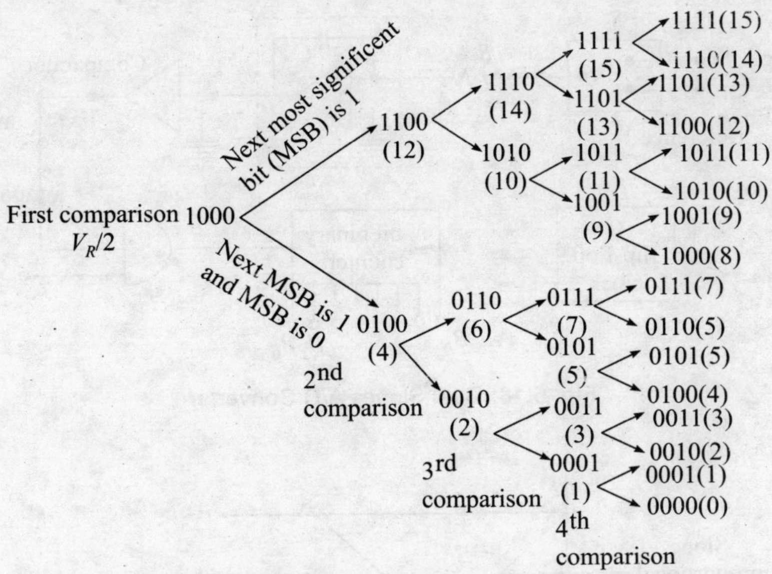

Fig. 6.16: Principle of Successive Approximation

Advantage:

(*i*) Conversion time is constant.

(*ii*) Conversion time α number of bits in digital output.

6.6.6 Dual Lope A/D Converter

It is one of widely used A/D converter because of its low cost. It named dual slope because of two slope *i.e.*, one from input analog signal (voltage) and another from reference signal (voltage).

It consists of following parts:

(*i*) Integrator

(*ii*) comparator

(*iii*) A binary counter

(*iv*) control circuit (switch)

The block diagram for dual slope A/D converter is shown in Fig. 6.17 and operation is understood by Fig. 6.18.

Initially, the switch S_1 is connected to analog input signal (voltage) V_a and values stored with counter is all 0's *i.e.*, counter is cleared.

As soon as analog input is applied to an OP-Amp integrator having time constant $\boxed{\tau = RC}$, the output voltage V_i is given as:

$$\left.\begin{aligned} V_i &= -\frac{1}{\tau}\int_0^t V_a\, dt \\ V_i &= -\left(\frac{V_a}{\tau}\right)t \end{aligned}\right] \qquad \ldots(8)$$

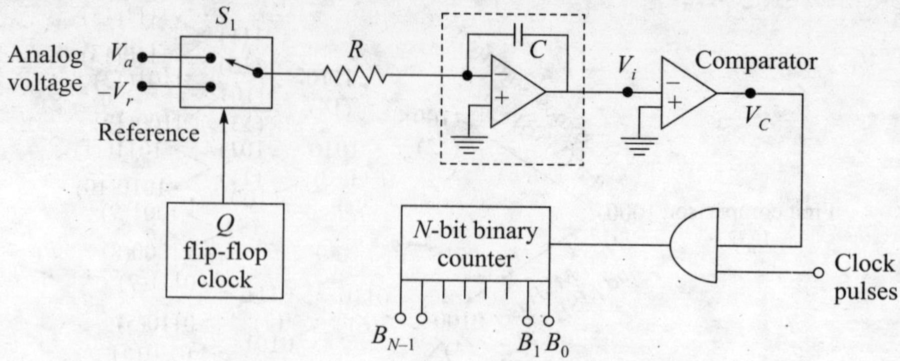

Fig. 6.18: Dual Slopes A/D Converter

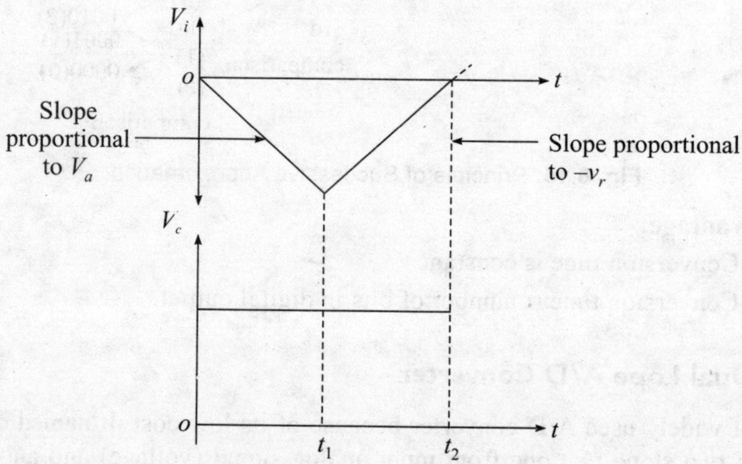

Fig. 6.19: Dual Slopes

As shown in the equation (8) V_i is negative, which is feeded to comparator circuit having output V_C is high, which enables the clock pulses of counter from 00...00 to 11...11 when 2^{N-1} pulses are arrived.

As the 2^N th clock pulses are arrived at time t_1, counter output is all 0's and yield to flip-flop output Q to 1. Now switch S_1 will be connected to $-V_r \cdot -V_r$ is feeded to integrator and output V_i starts to increase in +ve direction at time t_2 it crosses zero voltage. The comparator output V_c becomes low at time t_2 and disable the clock pulses of counter and counter yield to digital value of analog voltage applied.

The same procedure is repeated again and again.

Advantage:

(*i*) Low cost.

(*ii*) Very Good Conversion Accuracy.

Disadvantage: Slow speed.

Uses: It is widely used in digital voltmeter.

6.7 SPECIFICATIONS OF A/D CONVERTER

Different characteristics are usually specified by manufactures of A/D converter:

1. Range of Input voltage.
2. Input Impedance.
3. Accuracy
4. Conversion time.
5. Format of digital output

Explanation:

1. **Range of Input voltage:** It is difference if maximum and minimum analog input voltage (V_a) which is applied to ADC.
2. **Input Impedance:** It can be set in such a way that limits the maximum input current to ADC without effecting the performance of system or any leading damage.
3. **Accuracy:** It is measure of effectiveness involved in conversion from analog to equivalent digital signal.

   Error is also measure of accuracy. If error is more than accuracy will be less and vice-versa. It is represented in terms of %.
4. **Conversion Time:** When analog signal are converted into digital signal then time required during this conversion is known as conversion time.
5. **Format of Digital Output:**
   1. Serial output format
   2. Parallel Output format.

Format is decided according to need of output which is further interface with any other network.

SOLVED EXAMPLES

Example 1: *A 9-bit DAC has full scale analog voltage of 5V. Determine step size, percentage resolution, analog output voltage to input is 11010000001.*

Solution: For 9-bit DAC, $N = 9$

Percentage Resolution

$$= \frac{1}{2^N - 1} \times 100$$

$$= \frac{1}{2^9 - 1} \times 100$$

$$= \frac{1}{512 - 1} \times \Rightarrow \frac{100}{511} \quad \Rightarrow \quad 0.19\%$$

Step Size $= \dfrac{\text{Full scale deflection}}{\text{Number of steps}} \Rightarrow \dfrac{5V}{2^N - 1}$

$$= \frac{5V}{511} \Rightarrow 0.0097 \Rightarrow 9.7\text{mv}$$

Analog output for the input $(1101000001)_2$

$$= 9.7 \times 1665$$

$$\Rightarrow 16150.5 \text{ mv}$$

$$\text{output voltage} = 16.1505 \text{ V}$$

UNSOLVED QUESTIONS

1. Write short note on following terms:
   (*i*) Resolution
   (*ii*) Accuracy
   (*iii*) Encoding.
2. Determine the analog output voltage of 6-bit R-2R ladder network. DAC with $V_{ref.}$ as 5V when digital input is 000111.
3. Specification of analog to Digital converter.
4. Working of counting type A/D converter.
5. Explain working of R-2R D/A converter.
6. What are the specification of Digital ICs.
7. Explain the term quantization error.

Unit 7

Programmable Logic Devices

Topics are:

- PLD Introduction
- ROM (Read Only Memory)
- PLA (Programmable Logic Array)
- PAL (Programmable Array Logic)
- Comparison between ROM, PLA, PAL
- FPGA (Field Programmable Gate Array)
- CPLD (Complex Programmable Logic Device)

CHAPTER 7

Programmable Logic Devices

7.1 PLD INTRODUCTION

It is a device which can be used to generate desired logic function. We have studied various ICs for performing basic digital operations and other functions, such as multiplexer, demultiplexer, adders, comparator, code converters, shift register and counter, etc. have been studied. These ICs are referred to as fixed-function ICs *i.e.*, each one of them performs a specific, fixed function.

It is very difficult to manufacture different ICs to satisfy all the designers. There are many situations, for which there are no LSI or VLSI chips to achieve necessary logic function. For those types of situations led to development of PLD (programmable Logic devices).

PLD can be configured by individual user for his specific application. PLD is an IC that contain large number of gates, flip-flops and registers that are interconnected on the chip. This IC is said to be programmable because the specific function IC for a given application is determined by selected value of some of interconnection. This process is called programming because and product is desired circuit pattern of gates, flip-flops, register, etc.

Advantages of PLD:

1. High density.
2. Low space requirement.
3. Lesser power requirement.
4. Easy Programming.
5. Low cost of production.
6. Short design cycle.
7. Higher switching speed.
8. Low development cost.
9. Design Security.
10. Compact circuitry.

7.1.1 Types of PLD

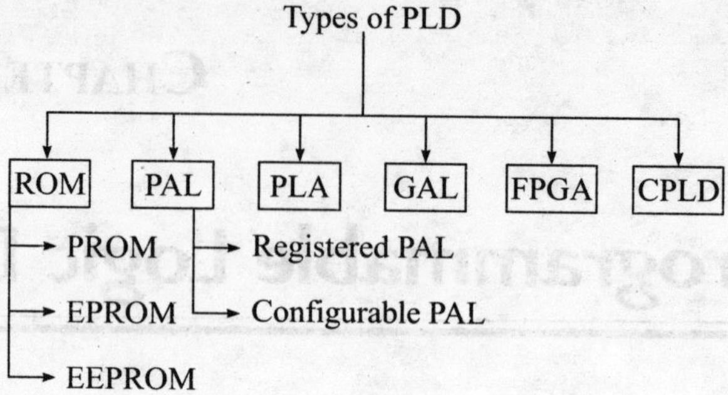

7.1.2 Structure of PLD

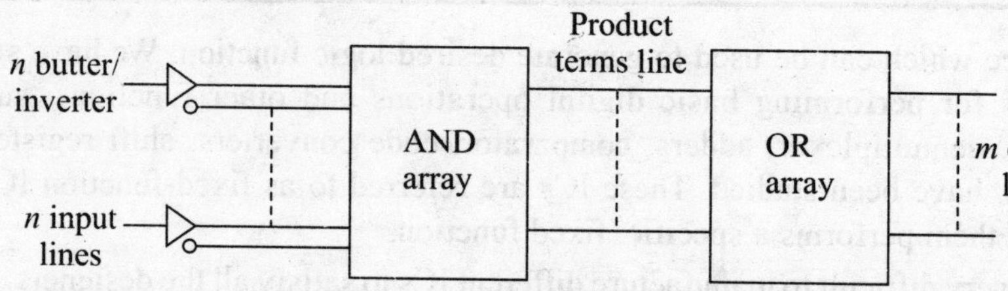

Fig. 7.1

PLD consists of 3 important things:

1. **Buffers/Inverters:** These are used to provide true value of inputs as well as complemented values of inputs. These also provide necessary device for the AND array which consists of large number of AND gates that is next block to buffer/inverters.
2. **AND Array:** It produces '*P*' numbers of product terms from '*n*' numbers of input variables and their complements. These are given to next block.
3. **OR Array:** It consists of several numbers of OR gates and realize a set of '*m*' numbers of outputs at SOP form.

PLD can be programmed according to Boolean function. The following table. Show the programmable array:

Device Type	*AND Array*	*OR Array*
ROM	Fixed	Programmable
PAL	Programmable	Fixed
PLA	Programmable	Programmable

In programmable array, the connections of gate can be selected. The approach for fabricating the programmable gate is to employ fuse links at each of inputs of gate as shown in Fig. 7.1.

Some of fuses are programmed to blow out to achieve desired output from gate.

7.2 ROM (READ ONLY MEMORY)

It is permanent or semiconductor device used to store the information which is permanent in nature and these information cannot be changed. It can only be read by us from memory. There is no write operation because specified data is programmed into device by manufactures.

It is non-volatile memory. It have variety of applications in digital system such as implementation combinational logic up program storage, etc.

ROM are well suited for LSI manufacturing process and are available in many forms. Two major semiconductor technologies are used for manufacturing of ROM.

(*a*) Bipolar (*b*) MOS

ROM is a device that includes both the decoder and OR gate within a single IC package.

Block Diagram:

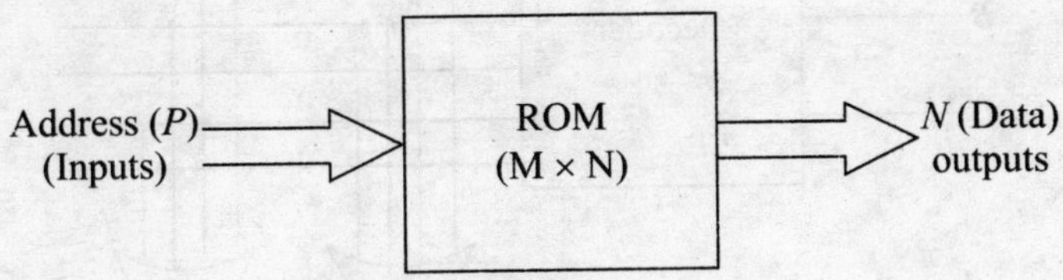

Fig. 7.2

A ROM of Size M × N has *M* number of locations and *N* number of bits can be stored at each location. The number of address inputs is *P*, where $\boxed{2^P = M}$ and number of data output line is *N*.

These are shown in Fig. 7.2

Types of ROM:

1. PROM (Programmable Read only Memory)
2. EPROM (Erasable PROM)
3. EEPROM (Electrically Erasable PROM)

Application of ROM:

(*i*) For implementing any complex combinational circuit from their truth table.

(*ii*) It is memory unit with fixed word pattern that can be read out only.

(*iii*) These are useful for converting one binary code to another (such as ASCII to EBCDIC and vice-versa).

(*iv*) Used to perform arithmetic function such as multipliers.

(*v*) Used to design of control unit of digital systems.

Advantages of ROM:

(*i*) Ease of design since no simplification or minimized is required.

(*ii*) Design can be changed, modified rapidly.

(*iii*) It is usually faster than discrete SSI/MSI circuit.

(*iv*) Cost is reduced.

Disadvantages of ROM: ROM will design the function of minterm or maxterm only. It will not give designing for the don't care conditions. It is implementing the all minterms whether they are needed or not.

7.2.1 Designing of Circuit Using ROM

Example 1: *Design following function using ROM:*

$$F_1 (A, B, C) = \Sigma m\ (0, 2, 4, 5)$$
$$F_2 (A, B, C) = \Sigma m\ (1, 3, 4, 6)$$

Solution: Here 3 variables A, B, C are given and output is from (0–7). So, we will use 3 to 8 line decoder.

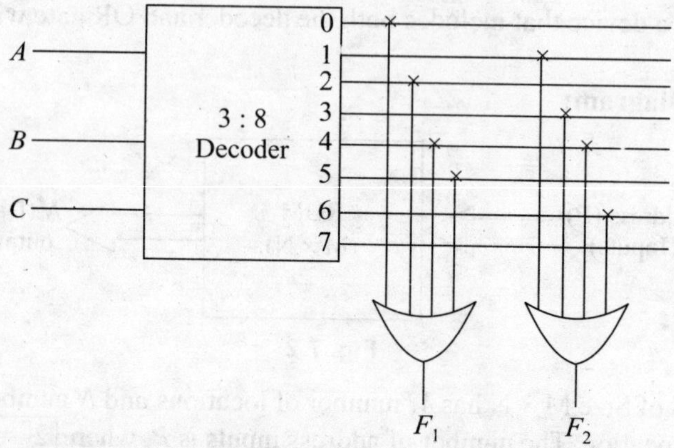

Example 2: *Design circuit of Half adder using ROM*

Solution: Truth Table of Half adder.

A	*B*	*S*	*C*
0	0	0	0
0	1	1	0
1	0	1	0
1	1	0	1

$S = \Sigma m(1, 2)$

$C = \Sigma m(3)$

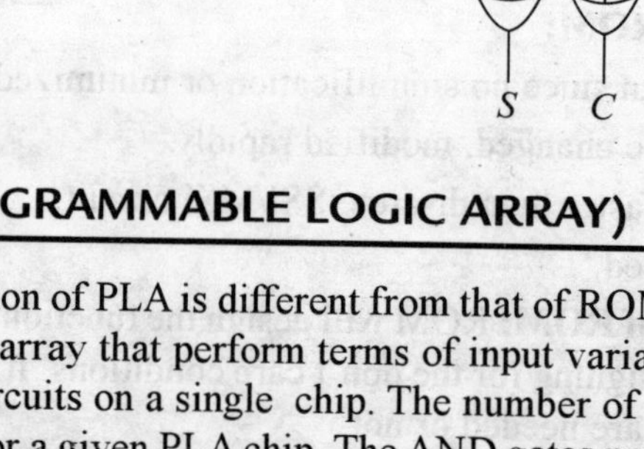

7.3 PLA (PROGRAMMABLE LOGIC ARRAY)

The internal organization of PLA is different from that of ROM. The decoder of ROM is replaced with AND array that perform terms of input variables. A PLA consists of two-level AND-OR circuits on a single chip. The number of AND and OR gates and their inputs are fixed for a given PLA chip. The AND gates provides the product terms and OR gates logically sum these product terms and thereby generate a SOP expression.

It has M inputs, n product terms, and N output with $n < 2^M$, and can be used to implement a logic function of M variables with N outputs. Since all of the possible 2^M minterms are not available.

Architecture of PLA:

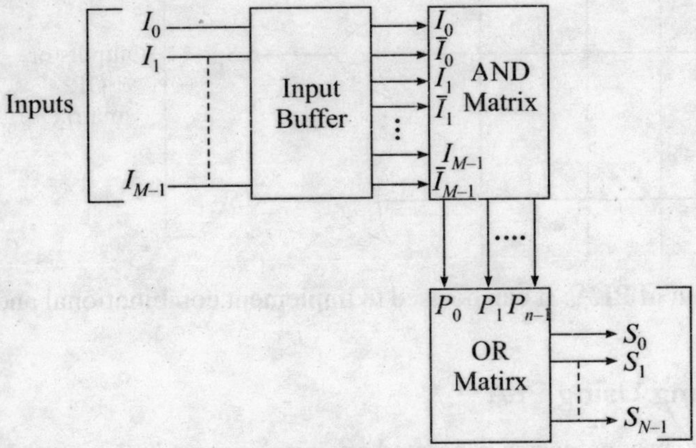

Fig. 7.3: Block Diagram of PLA

(*a*) **Input Buffer:** The buffer circuit at input are required to limit loading of sources that drive the inputs.

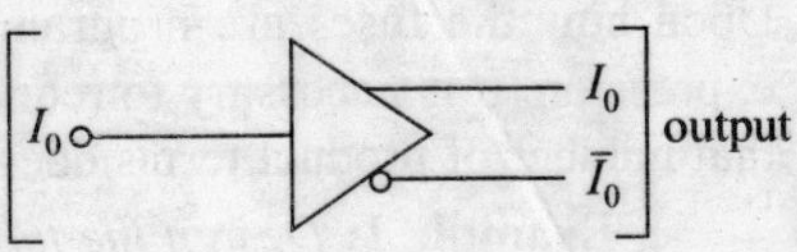

It produces inverted as well as non-inverted inputs at output shown in Fig. 7.3.

(*b*) **AND Matrix:** It is used to form product terms. It has n AND gate with outputs P_0 through P_{n-1} for each AND gate.

$$P = I_0\overline{I}_0 \cdot I_1\overline{I}_1, \ldots I_{M-1}\overline{I}_{M-1}$$

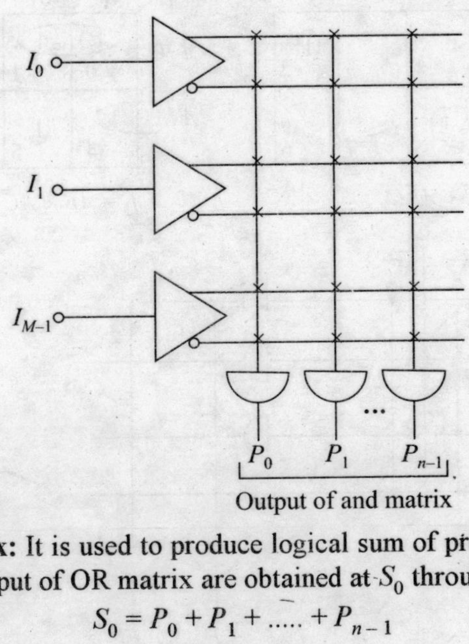

Output of and matrix

(*c*) **OR Matrix:** It is used to produce logical sum of product terms outputs of AND matrix. The output of OR matrix are obtained at S_0 through S_{N-1}.

$$S_0 = P_0 + P_1 + \ldots + P_{n-1}$$

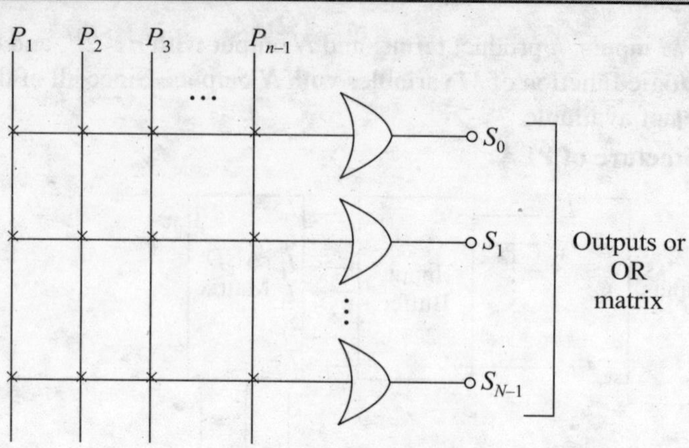

Application of PLA: It can be used to implement combinational and sequential circuits.

7.3.1 Designing Using PLA

In case of PLA, the product terms generated are not necessarily the minterms, as these product terms generated are not necessarily the minterms, as these product terms depend. Upon how the fuses are programmed. The realization using PLA is based on SOP expression. It is necessary to reduce or minimize the boolean functions is such a way that number of product terms does not exceed the number of AND gates in PLA.

Example 1: *Design the following function using PLA:*

$$F_1(A, B, C) = \Sigma m\ (2, 5)$$

$$F_2(A, B, C) = \Sigma m\ (0, 1, 4, 6)$$

Solution: Now to solve these two function F_1, F_2 use *k*-map.

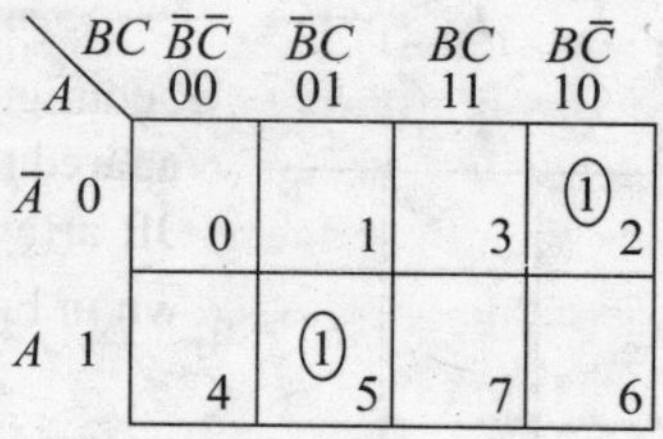

$f_1 = \bar{A}B\bar{C} + A\bar{B}C$

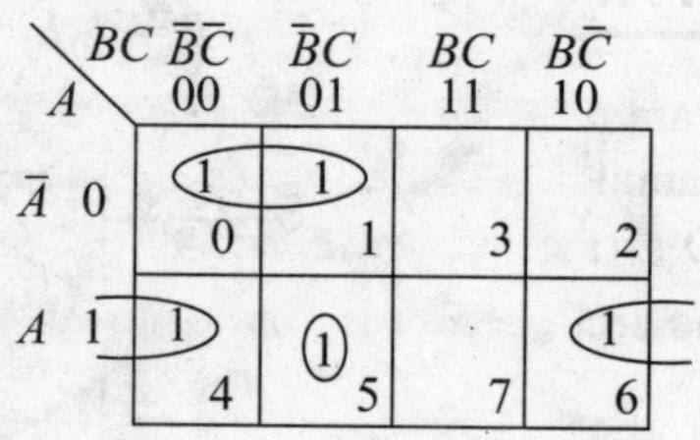

$f_2 = \bar{A}\bar{B} + A\bar{C}$

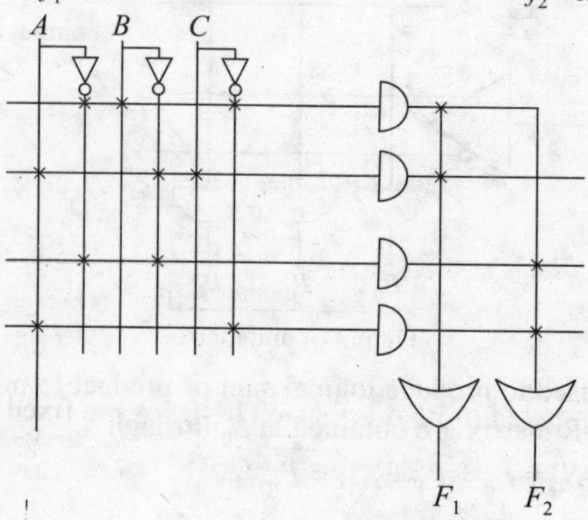

Example 2: *Design the circuit of half adder using PLA:*

Solution:

Truth Table:

A	*B*	*S*	*C*
0	0	0	0
0	1	1	0
1	0	1	0
1	1	0	1

$$S = \bar{A}B + A\bar{B}$$

$$C = AB$$

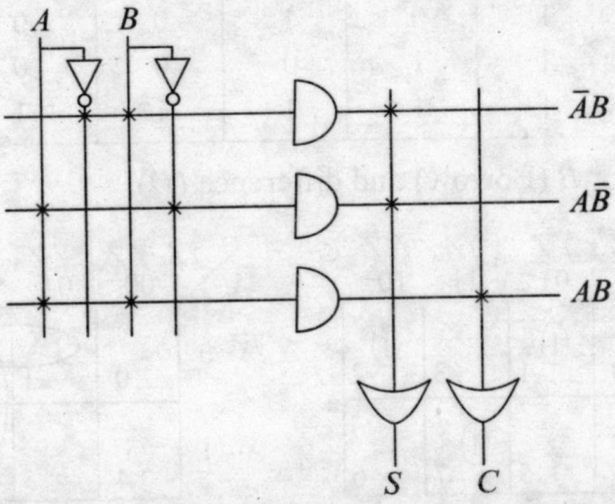

7.4 PAL (PROGRAMMABLE ARRAY LOGIC)

It is programmable array of logic gates on a single chip in AND-OR configuration. It has programmable AND array and a fixed OR array. It is cheap compared to PLA as only AND array is programmed. It is less flexible than PLA due to OR array is fixed.

The architecture is similar to PLA but simple diagram is shown in Fig. 7.4.

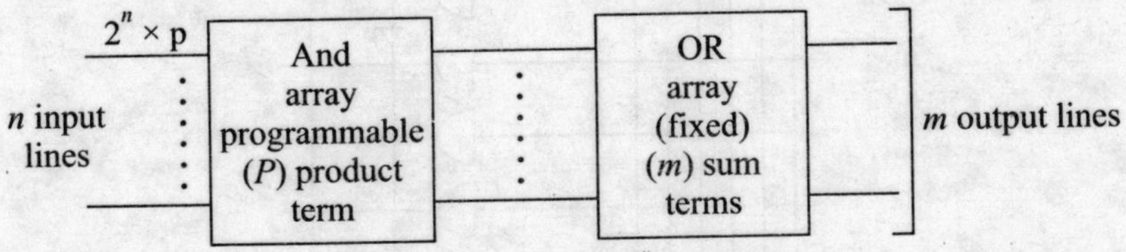

Fig. 7.4: Block diagram of PAL

Advantage:

(*i*) Easy to program than PLA.

(*ii*) Less cost than PLA.

Disadvantage:

It is less flexible as compared to PLA because OR gates are fixed only, AND gates are programmable.

In this number of AND gates are equal to 2^n (where n = number of input)

If $n = 3$

$2^3 = 8$ (8 AND gates are used).

Example 3: *Implement circuit of full subtractor using PAL.*

Solution: Truth Table:

X	*Y*	*Z*	*D*	*B*
0	0	0	0	0
0	0	1	1	1
0	1	0	1	1
0	1	1	0	1
1	0	0	1	0
1	0	1	0	0
1	1	0	0	0
1	1	1	1	1

Now *k*-map for *B* (Borrow) and difference (*D*):

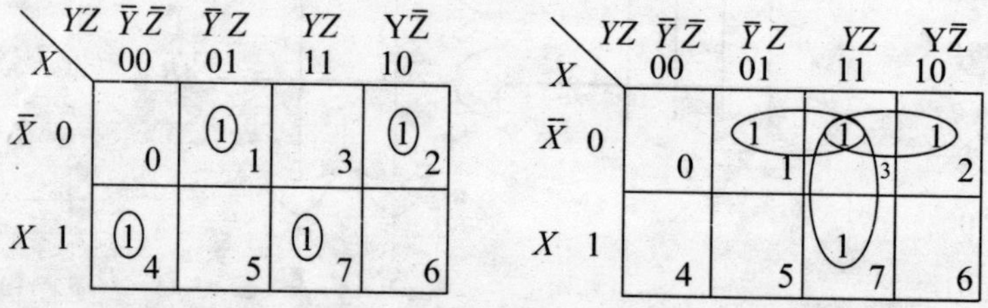

$D = \bar{X}\bar{Y}Z + \bar{X}Y\bar{Z} + X\bar{Y}\bar{Z} + XYZ$ $B = \bar{X}Z + \bar{X}Y + YZ$

Number of gates $= 2^3 \Rightarrow 8$ (AND gate)

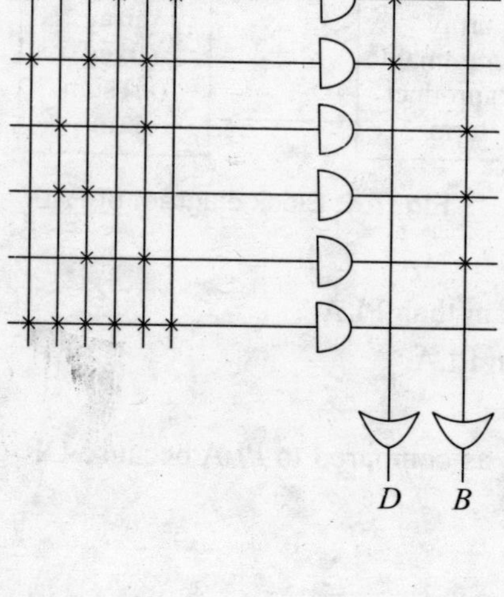

Example 4: *Solve BCD to Excess-3 code using PAL and PLA.*

Solution: First of all draw the truth table of BCD to excess-3 code.

Truth Table:

Decimal	*BCD Code*				*Excess-3 Code*			
Number	*A*	*B*	*C*	*D*	*W*	*X*	*Y*	*Z*
0	0	0	0	0	0	0	1	1
1	0	0	0	1	0	1	0	0
2	0	0	1	0	0	1	0	1
3	0	0	1	1	0	1	1	0
4	0	1	0	0	0	1	1	1
5	0	1	0	1	1	0	0	0
6	0	1	1	0	1	0	0	1
7	0	1	1	1	1	0	1	0
8	1	0	0	0	1	0	1	1
9	1	0	0	1	1	1	0	0

***K*-map for *W, X, Y, Z*:** 10–15 are don't care condition.

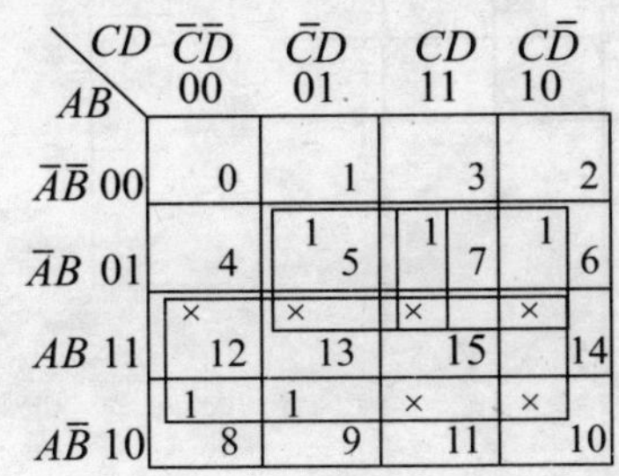

$W = A + BD + BC$

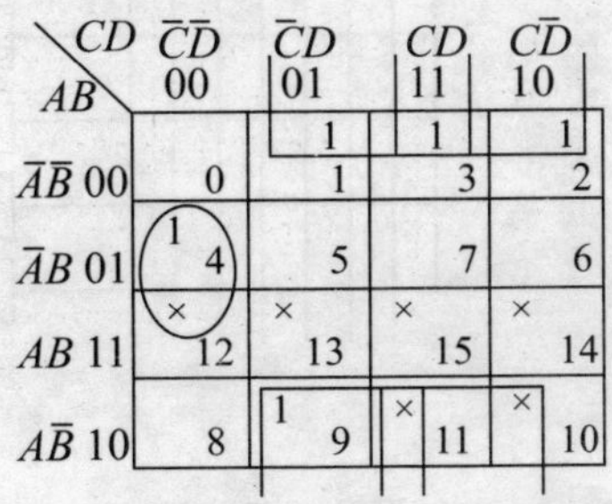

$X = B\bar{C}\bar{D} + D\bar{B} + C\bar{B}$

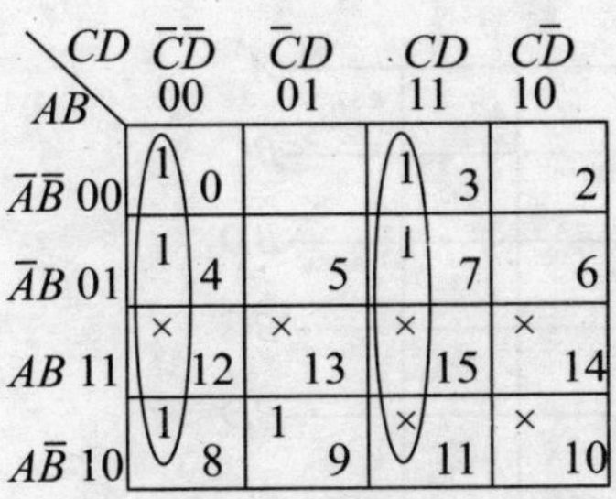

$Y = \bar{C}\bar{D} + CD$

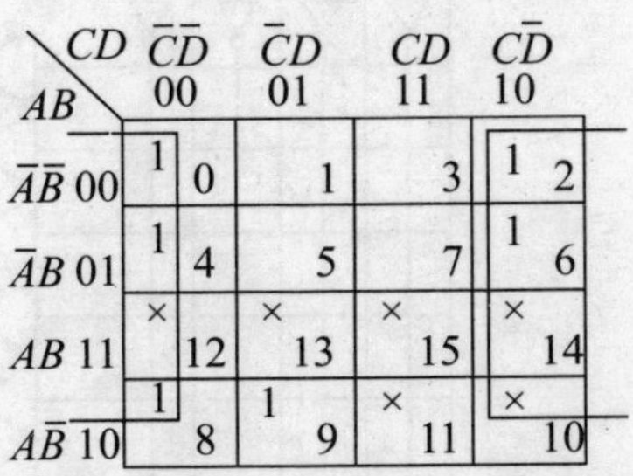

$Z = \bar{D}$

Designing using PAL:

No. of AND gate $= 2^n \Rightarrow 2^4 \Rightarrow 16$

n = number of Input Variable.

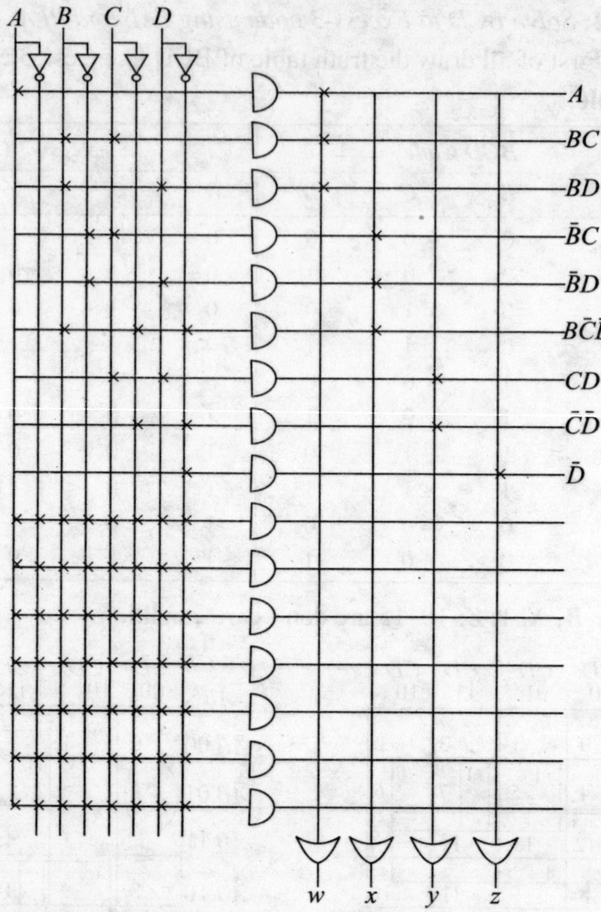

Designing Using PLA:

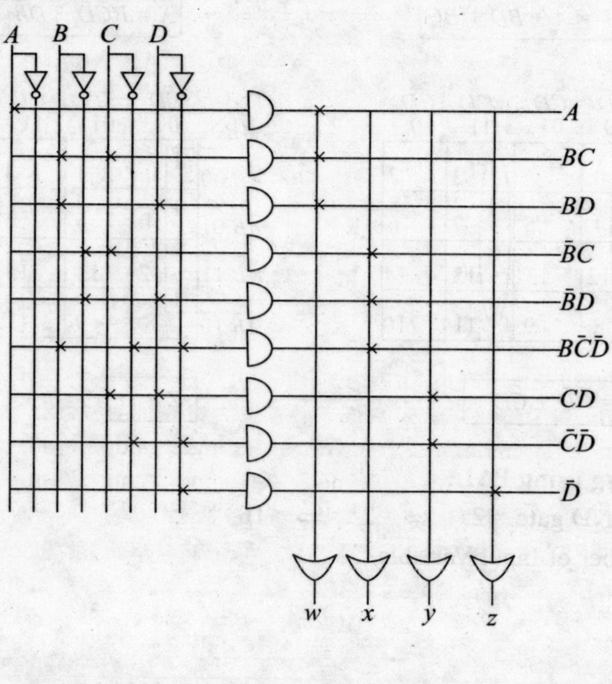

7.5 COMPARISON BETWEEN ROM, PLA, PAL

Comparison Table

S.No.	*ROM*	*PLA*	*PAL*
1.	AND array is fixed and OR array is programmable.	Both AND and OR array are programmable.	AND array is programmable and OR array is fixed.
2.	Flexible to the designer.	More flexible to the designer	Flexible to the designer.
3.	Cheap and simple to use.	Costly and complex as compared to ROM and PAL.	Cheaper and simple
4.	All minterms are decoded.	And array can be programmed to get the desired minterm.	AND array can be programmed to get desired minterm.
5.	Only Boolean function in sum of minterm form can be implemented.	Any boolean function is sop form can be implemented.	Any Boolean function in sop for can be implemented
6.	It is useful when the combination circuit uses more minterms.	It is useful when the circuit has more don't cares	It is useful when the circuit has 4 functions and atleast any one function is part of other.

7.6 FPGA (FIELD PROGRAMMABLE GATE ARRAY)

It is type of logic chip that can be programmed. It is similar to PLD, but PLD are generally limited to hundred of gates, FPGA supports thousands of gates. These are used for integrated circuit designs. Once design is set, hardwired chips are produced for faster performance.

FPGA is a semiconductor device containing programmable logic components called "Logic blocks" and programmable interconnects. FPGA do not contain AND, OR planes.

FPGA is composed of a number of relatively independent configurable logic modules, configurable *I*/0 cells and programmable interconnection path known as Routing channels. Logic blocks can be programmed to perform the function of basic logic gates such as AND, XOR or more complex combinational functions such as decoders or mathematical functions. In most FPGA, logic blocks also include memory element, which may be simple flip-flop or complete blocks of memory.

FPGA differ primarily in size and configuration of their logic modules and interconnection requirements. FPGA with larger logic modules may not be sufficiently utilized, therefore wasting the logic modules for performing simple logic functions. Use of smaller logic modules leads to much larger number of interconnect paths in device causing significant propagation delay as well as consuming large percentage of FPGA area.

FPGA contain 3 main types of resources:

1. Logic Blocks.
2. Input/output–Input/output blocks for connecting to pins of the package.
3. Interconnection wires and switches.

Logic blocks are arranged in a two-dimensional array, and the interconnection wires are organized as horizontal and vertical routing channels. The routing channels contain wires and programmable switches that allows the logic blocks to be interconnected in many ways. FPGA is used to implement logic circuits of more than a few hundred thousand equivalent gates in size.

Figure 7.5 shows two locations for programmable switches, the brown boxes adjacent to logic blocks holds switches that connect the logic block input and output terminal to the interconnection wires and brown boxes that are diagonally between logic blocks connects one interconnection wires to another. Programmable connections also exist between *I*/0 blocks and interconnection wires.

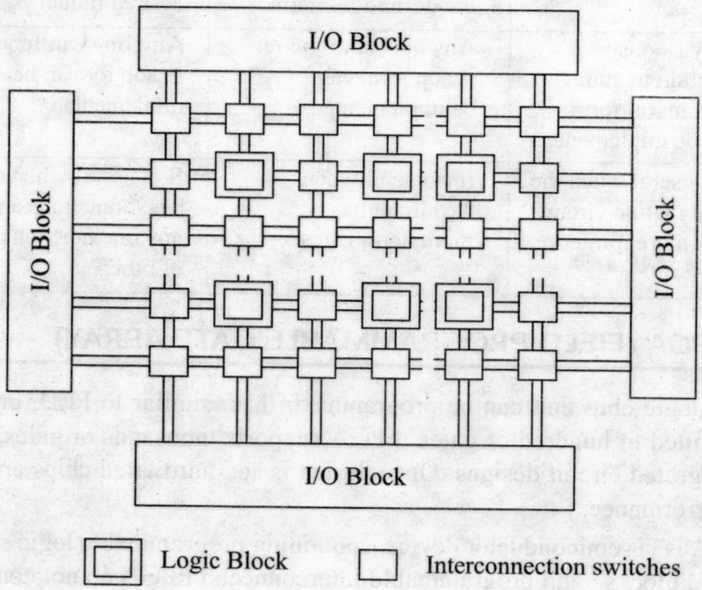

Fig. 7.5: FPGA

Types of FPGA:

1. Logic cell Array
2. Actel ACT
3. Antifuse
4. Plessey Era

The most commonly used Logic block is **Look-up-Table (LUT).** It has 2 inputs x_1, x_2 and one output function *f*. It is capable of implementing any logic function of two variables. Because a two-variable truth table have 4 rows, this LUT has 4 storage cells. The input variables x_1 and x_2 are used as select inputs of 3 multiplexers. Which, depending on the valuation of x_1 and x_2, select two content of one of the four storage cells as the output of LUT.

To realized a logic function by using the two-input LUT, consider the truth table and Fig. 7.6 is circuit for two-input LUT.

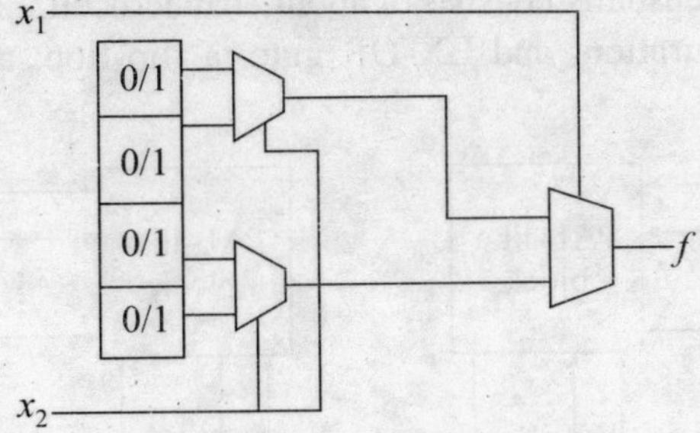

Fig. 7.5: Circuit for Two-Input LUT

Truth Table:

x_1	x_2	y_1
0	0	1
0	1	0
1	0	0
1	1	1

$$f_1 = \overline{x}_1\,\overline{x}_2 + x_1\,x_2.$$

Advantage of FPGA:

(*i*) It is used to design Large Logic circuits.

(*ii*) It is used to implement logic circuits of more than a few hundred thousand equivalent gates in size.

Disadvantage of FPGA:

The storage cells in LUT in an FPGA are volatile. Due to this FPGA has to be programmed every time power is applied. Often a small memory chip that holds its data permanently called PROM.

7.7 CPLD (COMPLEX PROGRAMMABLE LOGIC DEVICE)

The PLD such as PLA and PAL have limited number of inputs, product terms, and outputs. These devices can support upto about 32 total number of inputs and outputs only.

For implementation of circuits that require more inputs and outputs than that are available in a single PLD chip or more sophisticated type of chip, referred to as **CPLD.** A typical PAL has 8 macrocell, if each macrocell represents about 20 equivalent gates, then the PAL can accommodate a circuit that needs upto about 160 gates. For circuit requiring a very large number of gates, CPLD having large number of macrocell can implement circuits of upto about 10000 equivalent gates. There are a number of manufacturers of CPLD. Manufacturing a wide range of products with different features.

7.7.1 Block Diagram of CPLD

Figure 7.6 gives block diagram of CPLD. It consists of a number of PAL-like blocks, *I*/0 blocks, and set of interconnection wires. The PAL-like blocks are connected to set of interconnection wires and each block is also connected to an I/0 block to which a number of chip's input and outputs pins are attached.

A PAL-like block usually consists of about 16 macrocell. Each macrocell consists of an AND-OR configuration, and EX-OR gate, a flip-flop; an Multiplexer, and a tristate buffer.

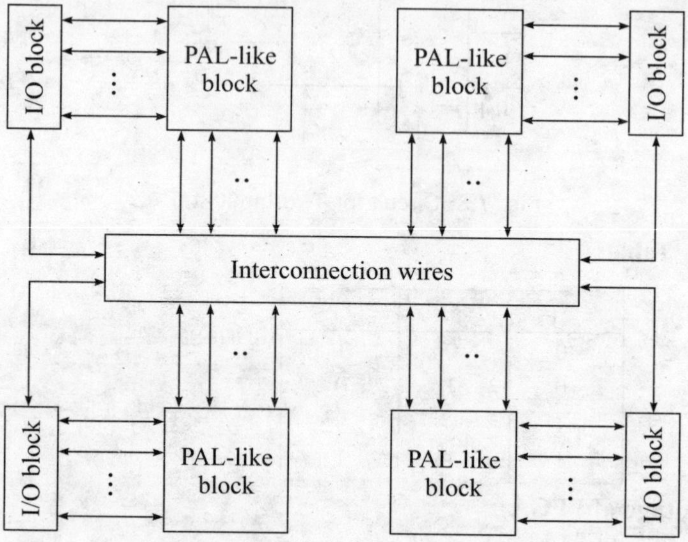

Fig. 7.6: Block diagram of CPLD

Now after this a typical macrocell is shown in Fig. 7.7. Each AND-OR configuration usually consists of 5 to 20 AND gates and an OR gate with 5 to 20 inputs. An **EX-OR** gate is used to obtain the output of OR gate is inverted or non-inverted form depending upon its other input being 1 or 0 respectively. A ***D*-FF** stores the output of the EX-OR gate, a multiplexer selects either the output of ***D*-FF** or the output of Ex-OR gate, depending upon its select input (1 or 0).

The **Tristate buffer** acts as a switch which enables the chip's pin to be used either as an output (tri-state enabled)/as an input (tri-state disabled). In case the chip's pin is used as an input pin, an external source can drive a signal onto the pin, which can be connected to other macrocell using interconnection wiring. When used as an input pin, the macrocell becomes redundant and it is wasted.

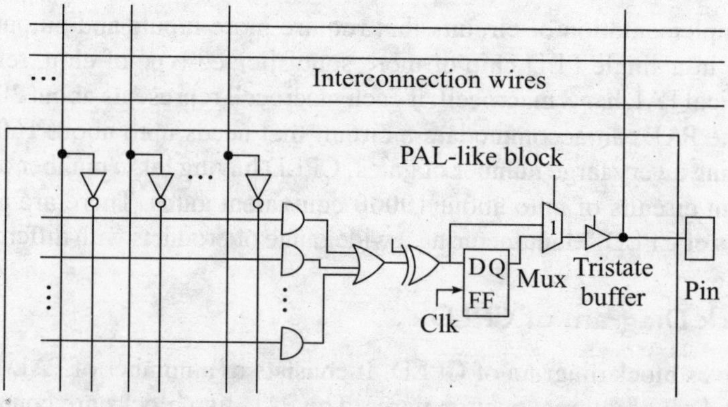

Fig. 7.7: Circuit Diagram of CPLD used Macrocell

Advantage:

1. More flexible.
2. Less time require to start working.
3. Less power consumption.

Disadvantage:

More complex.

UNSOLVED QUESTIONS

1. Differentiate PAL and PLA.

2. Design the circuit of Half adder using PAL and PLA.

3. Compare FPGA and CPLD.

4. Design the following function using PAL:

$y = \Sigma m\ (0, 2, 5, 7)$

5. Design the circuit of full adder using ROM, PAL and PLA.

6. Design the circuit of full subtractor using PLA and PAL.

7. Write short note on:

(*i*) ROM as PLD.

(*ii*) CPLD advantage over FPGA.

UNIT 8

SEQUENTIAL LOGIC DESIGN

Topics are:

- Sequential Logic Based Design Circuit
- Design at RTL (Register Transfer Level)
- ASM (Algorithmic State Machines)
- Asynchronous Sequential Logic: (ASLC)
- Analysis Procedure
- Design Procedure
- Reduction of State and Flow Table
- Race Free State Assignment
- Hazards
- State Reduction and Assignment

CHAPTER 8

Sequential Logic Design

8.1 SEQUENTIAL LOGIC BASED DESIGN CIRCUIT

Circuit is which output at any instant of time, not only depends on present input but also on previous outputs. These circuits consists of storage elements *i.e.,* flip-flops, counters, Register.

The beaviour of every sequential logic based digital circuit is described by state table.

State table consists of 4 sections:

(*i*) present state (*ii*) Next state

(*iii*) Input (*iv*) output.

For practical circuit implementation, we need to design digital circuits based system and system designed is so large and complex that it is difficult to describe that by using state table. It is because of the reason that, number of state would be large.

To handle such large digital system, we use modular approach of designing.

In modular approach, large system is divided into various sub-systems and interconnections between them is provided via data bus.

control signals are also provided to maintain synchronization of various activities.

8.2 DESIGN AT RTL (REGISTER TRANSFER LEVEL)

Register is a group of flip-flop which store digital information in form of bits.

Digital information stored in register can be processed by certain operation like:

- LOAD
- CLEAR
- SHIFT
- COUNT

These operations are called Register Transfer operation.

Digital System Specifications:

- How many number of registers and what is the capacity of each register?
- What are the various register transfer operations?
- What are central signals needed to maintain synchronization?

RTL Operations Include:

- Loading certain information in a register.
- Shifting available information in register etc.

Registers in Digital System:

Every digital system has storage elements and it requires registers. Registers can be represented by different way in different types of digital systems *e.g.*

1. *AX*

   Take a rectangular box and put name of register inside.

2. | A_{15} | A_{14} | A_{13} | A_{12} | A_{11} | A_{10} | A_9 | A_8 | A_7 | A_6 | A_5 | A_4 | A_3 | A_2 | A_1 | A_0 |
   |---|---|---|---|---|---|---|---|---|---|---|---|---|---|---|---|

   AX is a 16 bit register consists of 16 cells. One cell represent one bit. Each cell is assigned as a letter with a subscript No.

3. 15 8 7 0

AH	AL

   AX is a 16 bit register which is segmented into two bytes

   i.e., Higher byte and Lower byte.

   ↓ AH ↓ AL

4. 15 0

Current Address Register

   Current address register is a 16 bit register. It is not partitioned into 16 cells.

Example of RTL Operation:

Let take two 16 bit register *AX* and *BX*. Suppose we want to put data of *AX* to *BX*, then it can shown like this–

$$BX \leftarrow AX$$

This statement includes 16 bit register *AX*, 16 bit register *BX* and a replacement operation in between them (shown by arrow). Arrow indicates transfer of information with direction.

AX = Source Register

BX = Destination Register

Contents of source register *AX* are transferred to destination register *BX*. Contents of *BX* register are modified, yet contents of *AX* remains unaltered. It signifies just copy operation from *AX* to *BX*.

We can control this operation by applying control function prior to register transfer operation. Control function and register transfer operation are separated by colons.

Example: $xT_1 : BX \leftarrow AX$

Control functions with statement just signifies particular interval of time at which register transfer operations occurs.

It indicates:

At $x = 1$ and at the arrival of clock cycle T_1,

when $xT_1 = 1$, then only register transfer operations takes place.

Its hardware is shown in figure 8.1

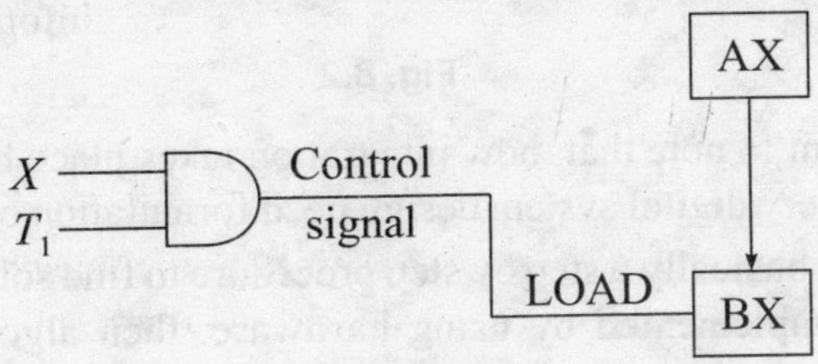

Fig. 8.1

8.3 ASM (ALGORITHMIC STATE MACHINES)

All Digital systems are based on sequential logic and hence consists of storage elements *e.g.*, Registers. Registers are group of flip-flops which stores binary information.

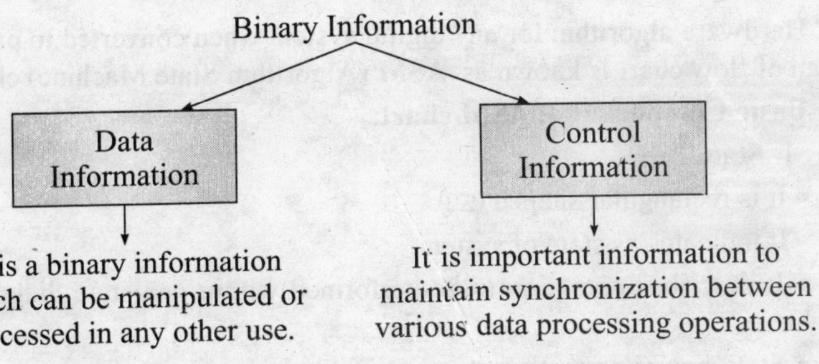

Data processing Task are:

→ Left shift

→ Right Shift

→ Arithmetic Operation

→ Logical operation.

Designing of digital system requires two units:

Data path Unit   Control Unit

Data Path unit: It accepts the input data and accomplish data processing as per requirements of the system.

Control Unit: It decides about sequence of various data processing operations executing in data unit and maintain synchronization by sending the control signal to data path unit. Also called FSM (Finite State Machine). Figure 8.2 shown the block diagram of digital system showing data path unit and control unit.

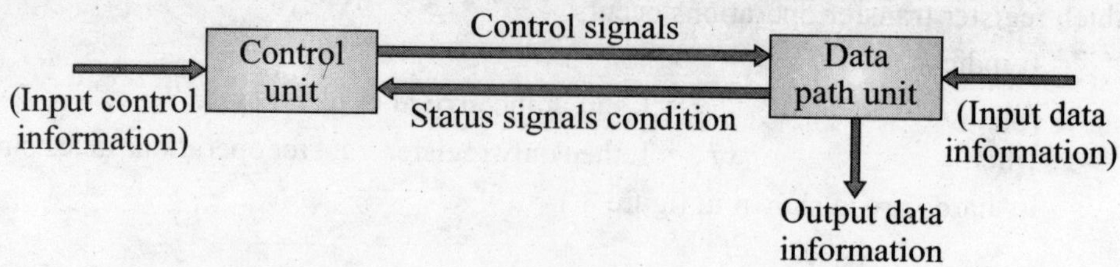

Fig. 8.2

It is important to note that, how integration takes place between data path unit and control unit. Every digital system design need formulation of hardware algorithm.

Algorithm is basically a step by step procedure to find solution to a big problem and procedure is implemented by using hardware, then algorithm is specified as hardware algorithm.

It is more easy to solve the problem, if algorithm is converted into a flowchart.

Flowchart is basically a pictorial representation of solution to a typical problem. Step by step procedure can be converted to pictorial format by using:

→ State Box

→ conditional Box

→ Decision Box

Hardware algorithm for any digital system when converted to pictorial format by mean of flow chart is known as **ASM** (Algorithm State Machine) chart.

Basic Component of ASM chart:

1. **State Box:**

• It is rectangular shaped (▭)

• It indicates as state of system.

• It describes operation to be performed within one state diagram or row in state table.

• State box is equivalent to a node/bubble in a state diagram or row in state table.

In this output is always asserted.

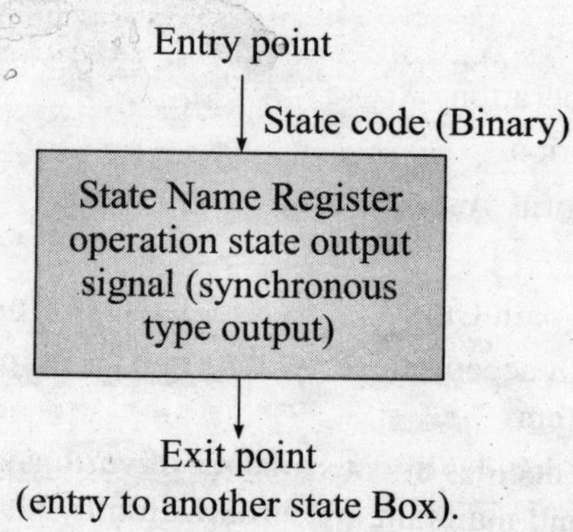

2. **Decision Box:**

• It is diamond shape (◇)

• It indicated that a given condition is tested and exit path is chosen accordingly. Test condition is basically a boolean expression composed of input signals and plays a similar role to logic expression in transition of state diagram.

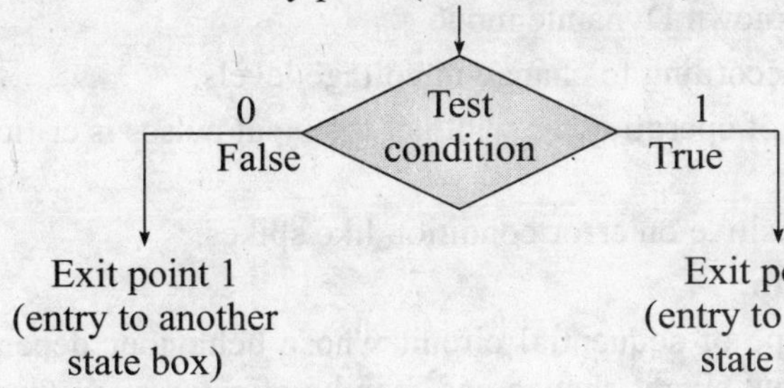

3. **conditional Output Box:**

• Rectangular box with Rounded corner (▢)

• This describes asserted output signal. However, it can only be placed after an exit path of a decision box. It implies that these output signals can be asserted only, if condition of previous decision box is met.

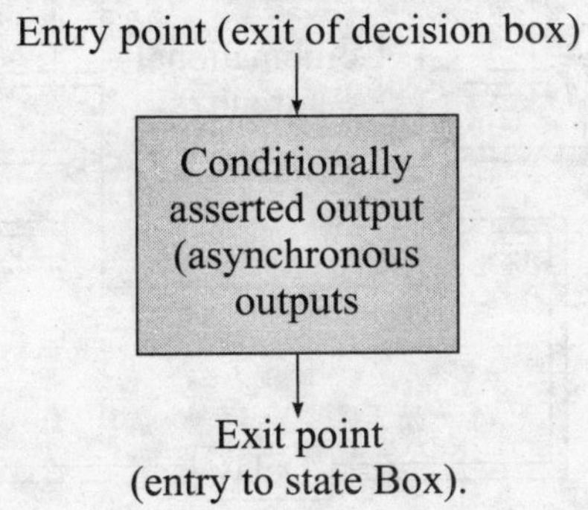

8.4 ASYNCHRONOUS SEQUENTIAL LOGIC (ASLC)

In asynchronous sequential circuit the operation is independent of clock signals and is depends only on the propagation delays of various gates used in the construction. It means that transition in ASLC does not depend on any external clock frequency signals rather they are controlled by the propagation delays of various gates in the circuit for producing the transition of the states.

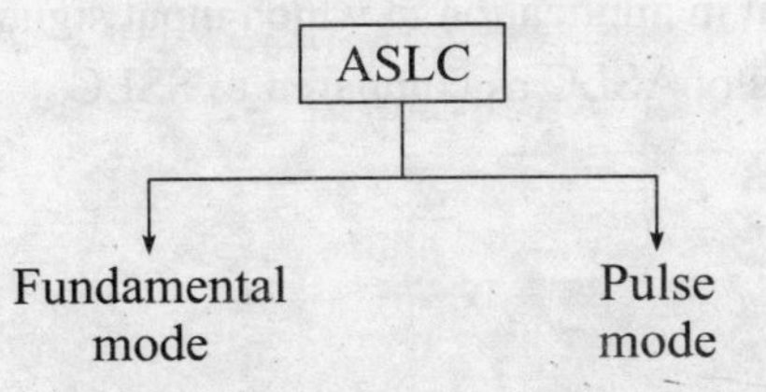

(*i*) **Fundamental Mode:**

- It is also known as static Mode.
- It operates according to voltage level of inputs.
- In this operation assumes that the input signals will be changed only when the circuit is in a stable state and that only one variable can change at a given time.

(*ii*) **Pulse Mode:**

- It is is also known Dynamic mode.
- It operates according to change in voltage levels.
- In this mode of operation the width of the input pulses is critical to the circuit operation
- They are sensitive on error condition like spikes.

Block Diagram:

ASLC is a type of sequential circuit whose behaviour depends only on the order in which its input signal change and can be affected at any instant of time. It consists of combinational circuit with feedback which are connected through delayed lines as shown in Fig. 8.3.

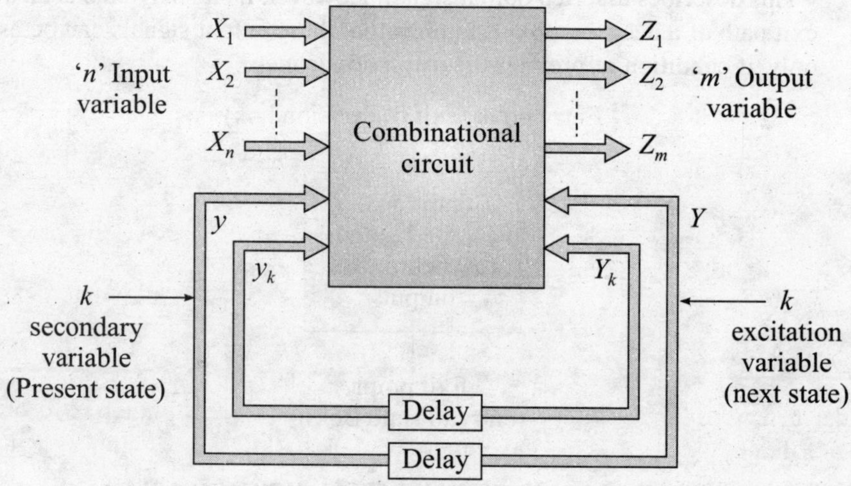

Fig. 8.3

Here, state transition occurs when there is an input change and no clock pulse is required. Asynchronous sequential logic circuit (ASLC) are faster as compared to synchronous sequential logic circuit (SSLC). Since ASLC respond immediately whenever there is a change in any input variable without having to wait for a clock pulse.

These are useful in application in which input signals may change at any time. Cost is also less of ASLC as compared to SSLC.

8.4.1 Comparison of ASLC and SSLC

ASLC	SSLC
1. Asynchronous sequential logic circuit.	1. Synchronous sequential logic circuit.
2. It des not require clock pulse.	2. It require clock pulse.
3. The storage elements used in ASLC are time-delay devices.	3. The storage elements used in SSLC are flip-flops.
4. It is more difficult to design and analyze.	4. It is easy to design and analyze.
5. It is used in cases where speed is important.	5. It has less speed as compared to ASLC.
6. Behaviour depends upon the input signals at any instant of time and the order in which the input change.	6. Behaviour can be defined from the knowledge of its signals at discrete instant of time.
7. It is economical to use.	7. It is costly.
8. Timing is major problem because of delays.	8. No timing problem because circuit is synchronise with clock pulse
9. 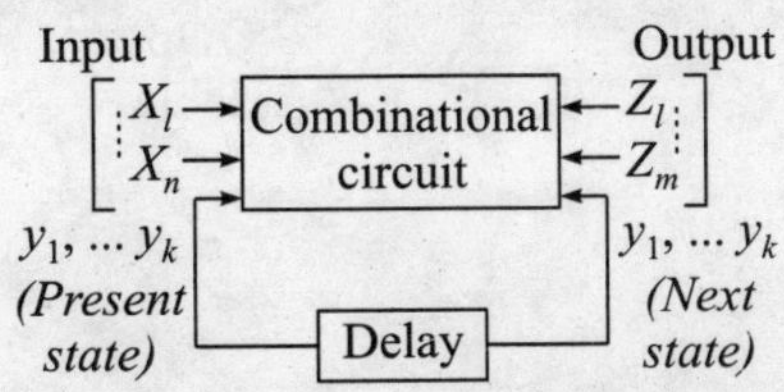	9. 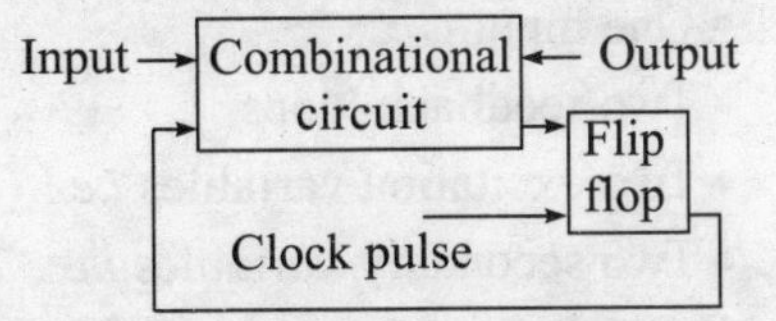

8.5 ANALYSIS PROCEDURE

In order to analyses the asynchronous sequential circuit the first step is to obtain a table or a diagram that describes the sequence of internal states and output as a function of changes in the input variables.

Transition Table:

Following steps are used to form the transition table:

1. Determine all feedback loops.
2. Designate each feedback loop output with Y_i and its corresponding input with y_i for $i = 1, 2, \ldots k$,.
   where K = No. of feedback loop.
3. Derive the boolean function for all Y's.
4. Plot each Y function in a map using the y variables for the rows and the external input for the rows and the external input for the columns.
5. Combine all; the maps into one table showing the value of $Y = Y_1, Y_2, \ldots Y_k$ inside each square.

6. Encircle those values of Y in each square that are equal to the value of $y = y_1, y_2, \ldots y_k$ in same row.

Once the transition table is available, the behaviour of the circuit can be analyzed by observing the state transition as a function of changes in the input variables.

Examples of ASLC:

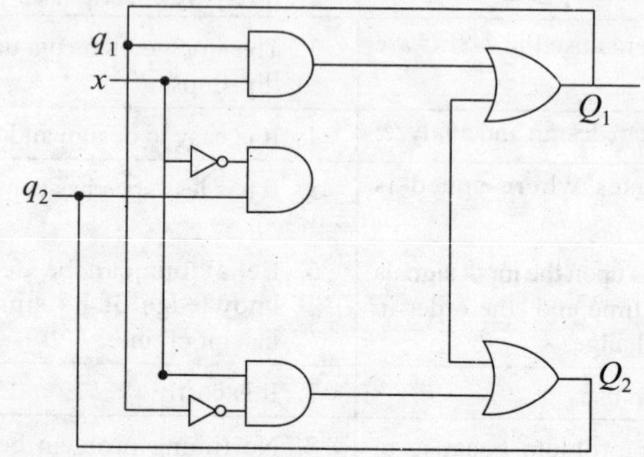

Here:

- One input is x.
- Two feedback loops
- Two excitation variables *i.e.*, Q_1 and Q_2
- Two secondary variables *i.e.*, q_1 and q_2.

Output is:

$$Q_1 = xq_1 + \bar{x}q_2$$

$$Q_2 = x\bar{q}_1 + \bar{x}q_2$$

The next step is to plot Q_1 and Q_2 functions in *k*-map.

$x \backslash q_1q_2$	00	01	11	10
0	0	1	1	0
1	0	0	1	1

$Q_1 = xq_1 + \bar{x}q_2$

$x \backslash q_1q_2$	00	01	11	10
0	0	1	1	0
1	1	1	0	0

$Q_1 = x\bar{q}_1 + \bar{x}q_2$

The next state is to form transition table. The transition table is obtained by combining the binary values in corresponding squares.

The transition table shows the value of $Q = Q_1\,Q_2$ inside each square. The first bit of Q is obtained from Q_1 and the second bit is obtained from the value of Q_2 in the same square position.

The state is stable only if the secondary variables must match the excitation variables (*i.e.*, the value of Q must match $q = q_1q_2$). There entries are encircled which means that the states are stable. An uncircled entry represents an unstable state.

Transition table is given as:

x \ q_1q_2	00	01	11	10
0	(00)	11	(11)	00
1	01	(01)	10	(10)

Stable States are:

$$q_1 q_2 x = \{000, 011, 110, 101\}$$

Unstable States are:

$$q_1q_2x = \{001, 010, 111, 100\}$$

If $q_1q_2x = 000$ and x makes a transition from $0 \rightarrow 1$, then $q_1q_2x = 011$. It means that when input makes a transition, the output state changes from one stable state to another which is the property of ASLC. This is shown:

Transition table

x \ q_1q_2	00	01	11	10
0	(00)	11	(11)	00
1	01	(01)	10	(10)

Transition from 00 to 01.

The transition table of ASLC is similar to state table used for synchronous circuit. Here we use secondary variables as present state and excitation as next state in order to state table.

Present state	Newt State	
	$x = 0$	$x = 1$
0 0	0 0	0 1
0 1	1 1	0 1
1 0	0 0	1 0
1 1	1 1	1 0

8.6 DESIGN PROCEDURE

The designing of Asynchronous sequential logic circuit is similar to the designing of asynchronous circuit that is counter. All the steps are similar to that.

8.7 REDUCTION OF STATE AND FLOW TABLE

State reduction procedure involves the method by which two or more states having the same output and same next states are combined together into single row.

Two states are equivalent if for each possible input, they give exactly the same output and same next state.

Example: In this state table having equivalent states

Present State	***Next state***		***Output***	
	x = 0	***x = 1***	***x = 0***	***x = 1***
P	*r*	*q*	0	1
q	*S*	*P*	0	1
r	*P*	*S*	0	0
S	*q*	*S*	0	0

Here, the present state of *P* and *q* have the same output 0 and 1. Their next states are *r* and *s* for $x = 0$ and *q* and *p* for $x = 1$.

If pair (*r, s*) are equivalent, then pair (*p, q*) are equivalent, because both have same output. It says that (*p, q*) imply (*r, s*) means if (*p, q*) are equivalent then (*r, s*) are also equivalent.

So state *P* and *q* are equivalent.

The present state of *r* and *s* have the same output.

So, (*q, p*) state are equivalent, so (*r, s*) imply (*q, p*) or (*r, s*) imply (*p, q*), then both pairs of states are equivalent.

Therefore the state table is reduced to two rows instead of four having two outputs *i.e.*, (0, 1) and (0, 0). The *p* and *q* are combined into first row and *r* and *s* are combined into second row.

This above reduction is simple as it has only 4 rows but when the state are increased then reduction is not feasible by this approach. So it can be reduced by making.

Implication Table, which is a chart that consists of squares, one for each possible pair of states, that provide spaces for listing nay possible implied states.

Example:

Present State	***Next State***		***Output***	
	x = 0	***x = 1***	***x = 0***	***x = 1***
p	*s*	*q*	0	0
q	*t*	*p*	0	0
r	*v*	*u*	0	1
s	*p*	*s*	1	0
t	*p*	*s*	1	:0
u	*r*	*q*	0	0
v	*p*	*t*	1	0

1st Step: To draw the implication table is drawn by taking states from q to v excluding the first state on the left side along the vertical.

And across the bottom horizontally the states are listed from p to u excluding the last state.

Here two states can be tested for equivalence *i.e.*, one from horizontal and one from vertical.

This is shown as:

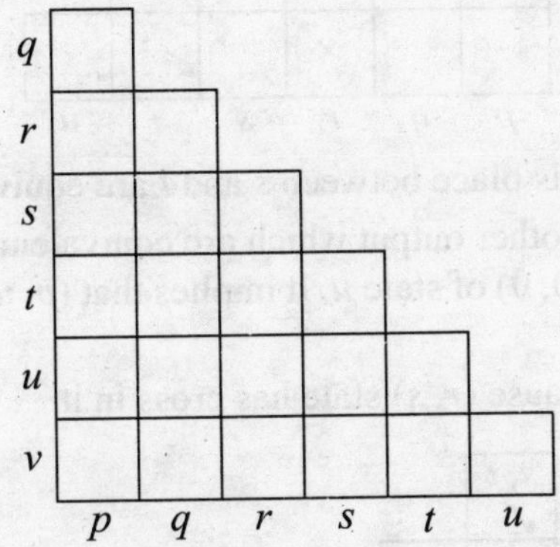

Here 21 squares are present.

2nd Step: Is to fill the squares. It can be filled by first checking the states that are not equivalent and are marked with cross (×), whereas there equivalence are marked with check mark (✓)

This is obtained by checking the output from state table. From table (0, 1) is not common to other so r is not equivalent. Put (×) to r.

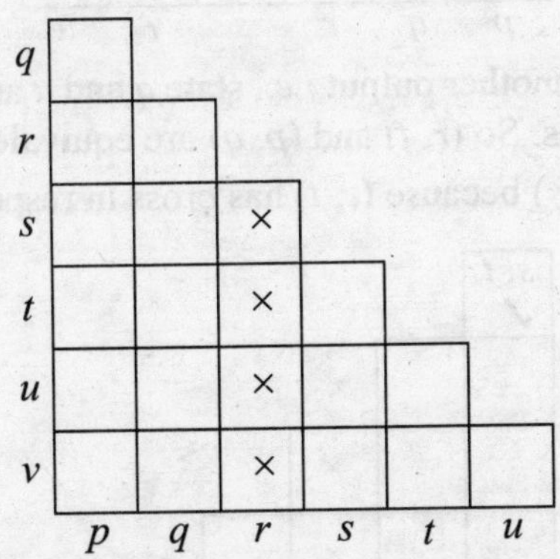

3rd Step: Is to check for equivalent state. From the state table, we see that pair (p, q) implies (s, t) is recorded in the square defined by column p and row q. It is shown as:

	p	q	r	s	t	u
q	s, t ✓					
r						
s			×			
t			×			
u			×			
v			×			

So, check mark (✓) is place between *s* and *t* are equivalent state.

4th Step: Is to check other output which are equivalent the output (0, 0) of state *p* is compared with output (0, 0) of state *u*. It implies that (*p*, *u*) state is equivalent with (*r*, *s*).

It is put with (×) because (*r*, *s*) state has cross in it.

	p	q	r	s	t	u
q	s, t ✓					
r						
s			×			
t			×			
u	r, s ×		×			
v			×			

5th Step: Check for another output *i.e.*, state *q* and *u* are equivalent because the ouput are same in both cases. So (*r*, *t*) and (*p*, *q*) are equivalent with (*q* and *u*).

Again it is marked (×) because (*r*, *t*) has cross in respective square.

	p	q	r	s	t	u
q	s, t ✓					
r						
s			×			
t			×			
u	r, s ×	r, t × q, u	×			
v			×			

6th Step: To check output (10). It shows that *s* and *t* states are equivalent. So it is marked with check mark in table shown.

q	s, t ✓					
r						
s			×			
t			×	✓		
u	r, s ×	r, t × q, u	×			
v			×			
	p	q	r	s	t	u

7th Step: The state s and v are equivalent because they have the same output (00) at $x = 0, x = 1$.

So (s, v) implies (s, t). They are equivalent and similarly t and v states are equivalent with (s,t) state.

q	s, t ✓					
r						
s			×			
t			×	✓		
u	r, s ×	r, t × q, u	×			
v			×	s, t ✓	s, t ✓	
	p	q	r	s	t	u

8th Step: To fill crosses in all other square boxes as shown below:

q	s, t ✓					
r	×	×				
s	×	×	×			
t	×	×	×	✓		
u	r, s ×	r, t × q, u	×	×	×	
v	×	×	×	s, t ✓	s, t ✓	×
	p	q	r	s	t	u

Implication Table

The equivalent states are represented as check marks. So, equivalent states are:

$$(p, q), (s, t), (s, v), (t, v)$$

we can combine pairs of states into larger groups of equivalent states. The last three states can be combined into three pair of equivalent state as (s, t, v).

The final partition of states consists of equivalent states from the implication table with the states not included in the equivalent state. Here *r* and *u* are not present in equivalent states. So, the group is given as:

$$(p, q), (r), (s, t, v), (u)$$

So table is reduced to 4 states, so, the reduced table is:

Present State	Next State		Output	
	x = 0	*x* = 1	*x* = 0	*x* = 1
(*p*, *q*)	(*s*, *t*)	(*q*, *p*)	0	0
(*r*)	*v*	*u*	0	1
(*s*, *t*, *v*)	*p*	(*s*, *t*)	1	0
u	*r*	*p*	0	0

We can reduce table further by replacing *q* with *p* and *t* and *v* by *s* and shown as:

Present State	Next State		Output	
	x = 0	*x* = 1	*x* = 0	*x* = 1
p	*s*	*p*	0	0
r	*v*	*u*	0	1
s	*p*	*s*	1	0
u	*r*	*q*	0	0

8.8 RACE FREE STATE ASSIGNMENT

To get race free state assignment, we assign a binary value to stable state in such a was that no race condition occur means only one variable can change at any given time when a state transition occurs.

After getting a reduced flow table, next step is to assign a binary value to each stable state. This process converts the flow table into transition table. To ensure that a transition table has no critical races, every possible state transition should be checked and verity that the binary state variable change one at a time.

This procedure is explained by taking examples as below:

3 Rows flow Table Example: Race problem does not occur in case of a single binary variable to a flow table with two rows. The reduced flow table is given as:

x_1x_2	00	01	11	10
p	(*p*)	*q*	*r*	*p*
q	*p*	(*q*)	(*q*)	*r*
r	*p*	(*r*)	(*r*)	(*r*)

Here, 3 rows are *p, q, r.*

In row *p*, the transition are from *p* to *q* in column 01 and *p* to *r* in column 11. This information is transferred into a transition diagram shown in figure 8.4.

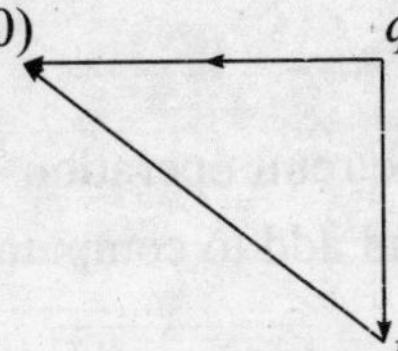

Fig. 8.4 Transition diagram

In this transition diagram, one problem of race occur when we goes from *p* to *r* then two states are changed, that is race condition. So, to avoid race, an extra row is assigned a state *s* and binary value 10. Modified Transition diagram is shown in figure 8.5.

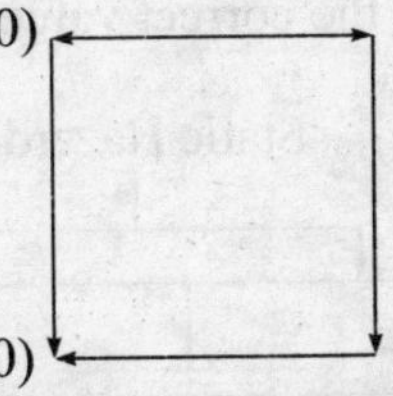

Fig. 8.5 Modified Transition Diagram

Now, 4th row is added in the reduced flow table and this new flow table is known as modified flow table

x_1x_2	00	01	11	10
p	(*p*)	*q*	*r*	*p*
q	*p*	(*q*)	(*q*)	*r*
r	*p*	(*r*)	(*r*)	(*r*)
s	*p*	–	*r*	–

Modified flow Table

This flow table is basically to avoid race condition.

8.9 HAZARDS

It is an undesirable effect caused by either a deficiency in system or external influences.

OR

It is a circuit which may produce a glitch, that is fast and small unwanted spike in output signals of digital circuit.

Glitch may occurs because the delay path through the circuit gives different propagation delay.

Cause of Hazard:

- Due to change in input and mainly occurs in asynchronous circuit.
- Due to change in input, same from of delay are caused by logic elements (NOT, AND, OR, gate).

Effect of Hazard:

- They can cause error in circuit operation.
- Also consume energy and add to computation time.

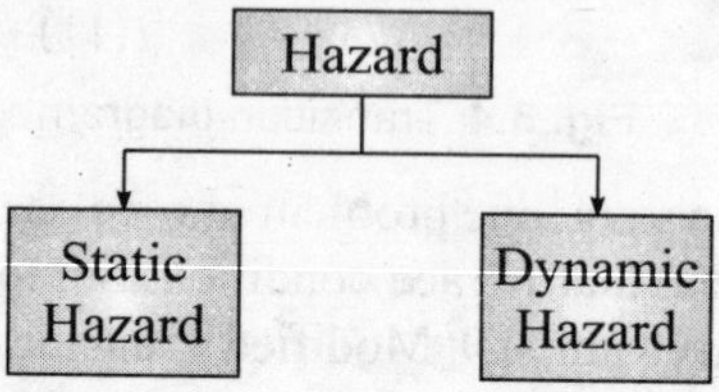

(*a*) **Static Hazard:** In this when one input variable changes the output changes momentarily before stabilizing to the correct value.

Static Hazard

Static-0 Hazard	***Static-1 Hazard***
In this output is currently '0' and after the inputs change, the output change to '1' before settling on '0'	In this the output is currently '1' and after the inputs change, the output changes to '0' before settling on '1'.
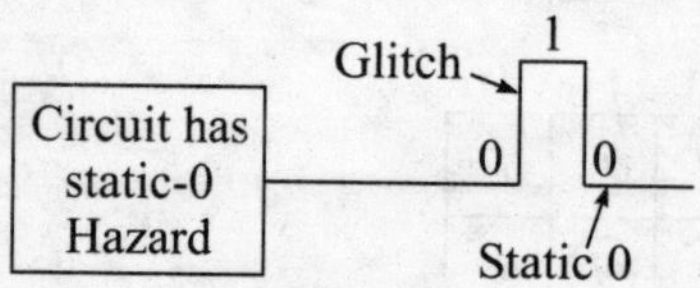	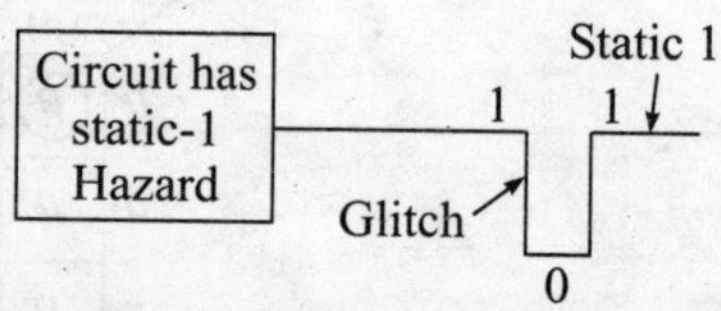

In SOP (sum of product) form there will be no static '0' hazards and in POS (Product of sum) there is no static–'1' hazads.

In order to eliminate static hazards, a redundant logic is added in the given output expression.

Example of static Hazard:

$$y = AB + \overline{A}C$$

we will draw it using *k*-map and error will occur when the input changes from 11 to 01 *i.e.*, when '*B*' changes state.

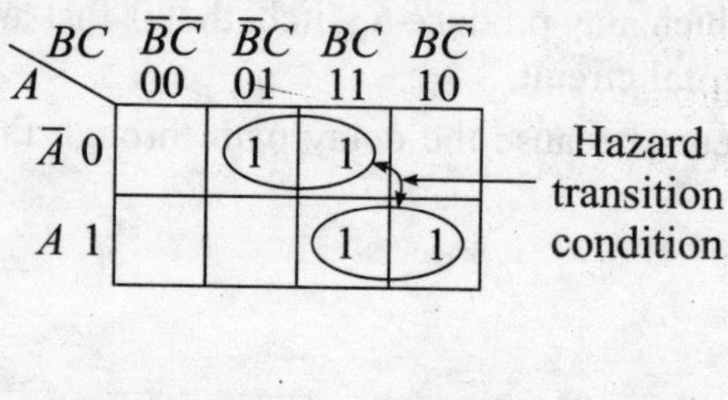

In order to remove this hazard condition a redundant loop is added which is shown below *i.e.*,

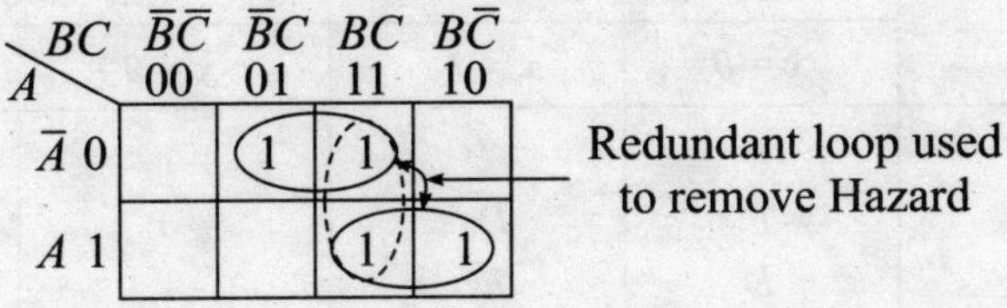

Required expression:

$$y = AB + \bar{A}C + BC \leftarrow \text{Redundant expression}$$

(*b*) **Dynamic Hazard:** In this output changes more than once as a result of a single input change. It occurs in larger logic circuits where there are different routes to output (from the input). If each route has different delay, then the output values changes from required output.

Example: a logic circuit is required to change output state from 0 to 1, but instead changes from 0 to 1 then 0 and finally resets at the correct value 1. This is dynamic hazard. Shown in figure 8.4

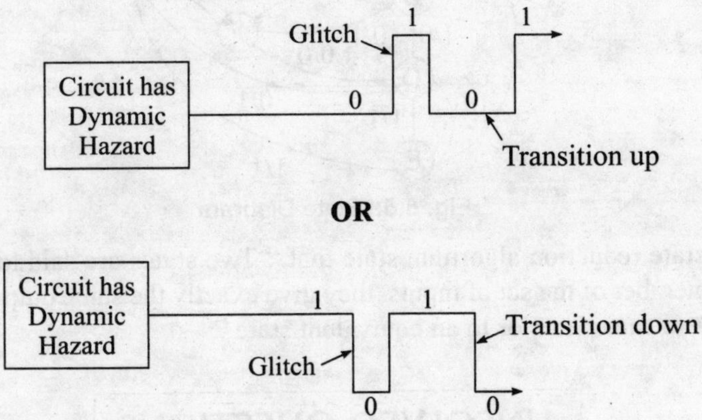

8.10 STATE REDUCTION AND ASSIGNMENT

Analysis of sequential circuit starts with circuit diagram and ends in state diagram. The design of sequential circuit starts from a set of specification and ends with logic diagram. State reduction helps in reducing the flip-flops and if flip-flop reduced then cost of circuit is also reduced.

State Reduction: In this certain algorithm are used which will help in reducing the number of states in state table, while keeping the external input-output requirements unchanged. In state reduction method, the equivalent states from a state table/diagram is eliminated.

Let us consider the state table of a sequential circuit:

Present State	*Next State*		*Output*	
	x = 0	***x = 1***	***x = 0***	***x = 1***
A	*B*	*C*	1	0
B	*F*	*D*	0	0
C	*D*	*E*	1	1
D	*F*	*E*	0	1
E	*A*	*D*	0	0
F	*B*	*C*	1	0

Fig. 1: State Table

Here, 6 states are present from *A* to *F*. The state diagram corresponding to the following state table is given as:

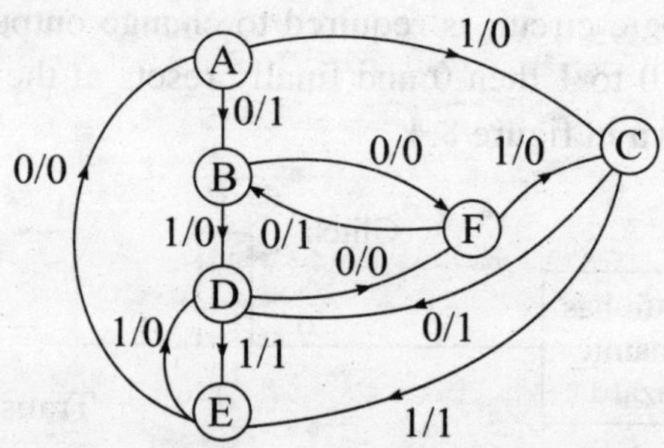

Fig. 8.5: State Diagram

The state reduction algorithm state that: "Two states are said to be equivalent if, for each member of the set of inputs, they give exactly the same output and send the circuit either to same state or to an equivalent state."

UNSOLVED QUESTIONS

1. Define RTL and its operation with the help of an example.
2. What is ASM? Explain it.
3. State and explain ASLC with block diagram.
4. Compare asynchronous and synchronous sequential circuit.
5. State the term "Reduction of state and flow table."
6. State the term "Race Free state Assignment."
7. What do you means by Hazard and explain it with types?
8. State state Reduction and assignment

EXPERIMENTS

Experiment No. I

AIM: To verify the operation of AND, OR, NOT, NAND, NOR, Ex-OR, Ex-NOR using TTL Kit.

APPARATUS: Connecting wires, connecting probe, Breadboard, TTL Trainer Kit, IC (7408 = NAND, 7432 = OR, 7404 = NOT, 7400 = NAND, 7402 = NOR, 7486 = Ex-OR) of gate, Regulated power supply etc.

THEORY:

(*i*) **AND Gate:** AND operation is defined as the output is 1 if and only if all the inputs are 1.

$$Y = A.B.C. \dots N$$

Logic Symbol:

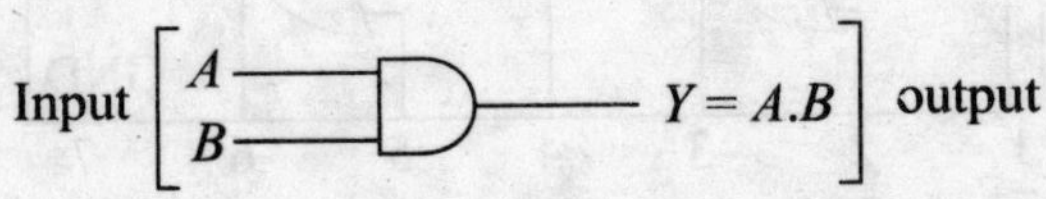

Truth Table:

Input		*Output*
A	***B***	$y = A \cdot B$
0	0	0
0	1	0
1	0	0
1	1	1

Circuit Diagram:

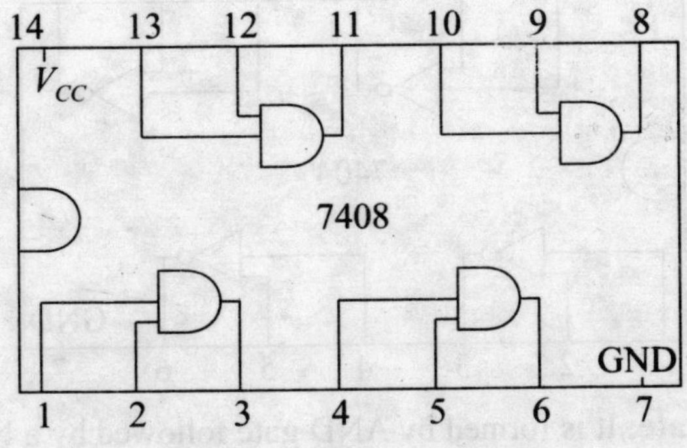

(*ii*) **OR Gate:** OR operation is defined as the output is 1 if and only if one or more inputs are 1.

$$Y = A + B + C + \ldots + N$$

Logic Symbol:

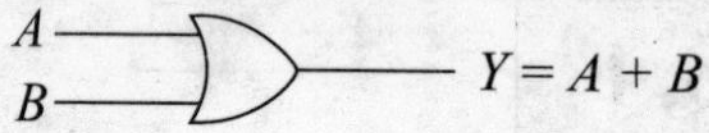

Truth Table:

Input		*Output*
A	***B***	***Y = A + B***
0	0	0
0	1	1
1	0	1
1	1	1

Circuit Diagram:

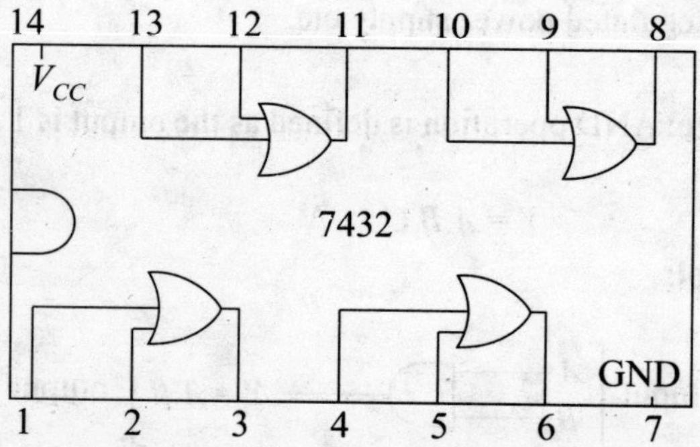

(*iii*) **NOT Gate:** It is defined as output is 1 if input is 0 and output is 0 if input is 1. It is also known as inverter.

$$Y = \overline{A}$$

Logic Symbol:

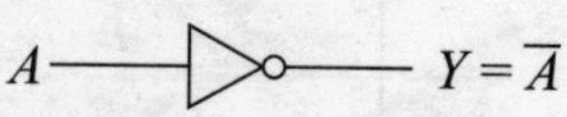

Truth Table:

Input	*Output*
A	***Y = $\overline{A}$***
0	1
1	0

Circuit Diagram:

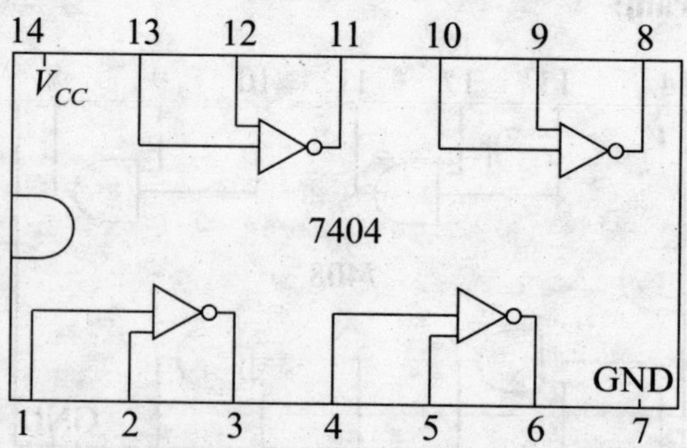

(*iv*) **NAND Gate:** It is formed by AND gate followed by a NOT gate.

$$Y = \overline{A \cdot B}$$

Logic Symbol:

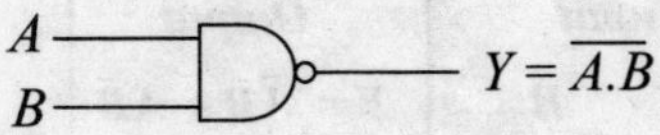

Truth Table:

Input		*Output*
A	*B*	$Y = \overline{A \cdot B}$
0	0	1
0	1	1
1	0	1
1	1	0

Circuit Diagram:

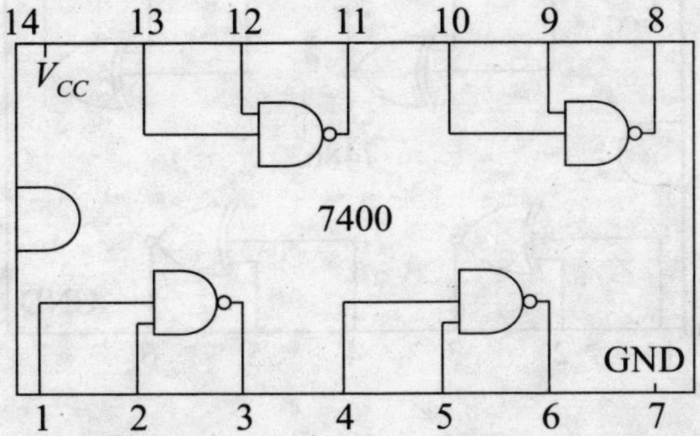

(*v*) **NOR Gate:** It is formed by OR gate followed by a NOT gate.

$$Y = \overline{A + B}$$

Logic Symbol:

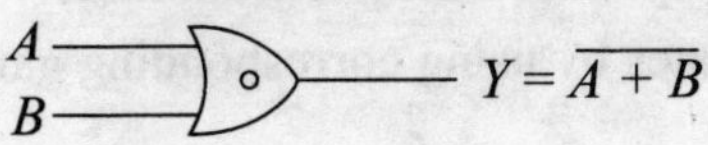

Truth Table:

Input		*Output*
A	*B*	$Y = \overline{A + B}$
0	0	1
0	1	0
1	0	0
1	1	0

Circuit Diagram:

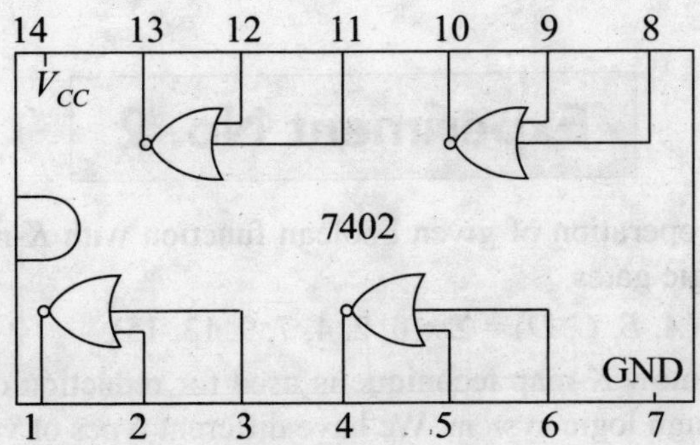

(*vi*) **Ex-OR Gate:** It is exponential-OR gate.

$$Y = \overline{A}B + A\overline{B}$$

Logic Symbol:

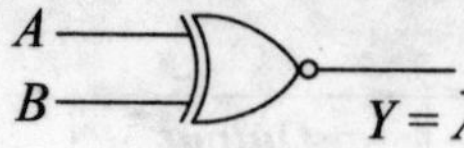

$Y = \overline{A}B + A\overline{B}$

Truth Table:

Input		Output
A	B	$Y = \overline{A}B + A\overline{B}$
0	0	0
0	1	1
1	0	1
1	1	0

Circuit Diagram:

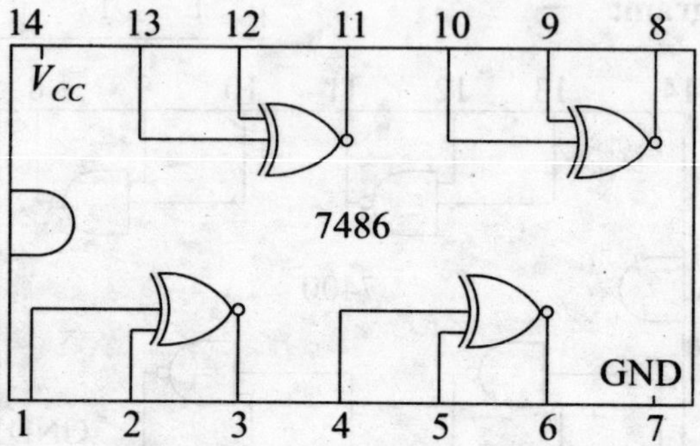

Procedure:

(*i*) Take the corresponding I.C. for which gate is to be studied.

(*ii*) Connect 14th pin to +5 V of the power supply.

(*iii*) Connect 7th pin to ground of power supply mean 0.

(*iv*) Put the input connections and check the output according to truth table.

(*v*) Now repeat all the steps to study all the gates by using corresponding gate IC's.

Precautions:

(*i*) Connection should be neat and clean.

(*ii*) V_{CC} should not exceed +5 V.

(*iii*) IC should keep carefully.

Result: We have studied verified all the gates successfully.

Experiment No. 2

AIM: To solve the operation of given boolean function with *K*-map and realize the expression using logic gates.

$$f(A, B, C, D) = \Sigma m(0, 2, 4, 7, 9, 13, 15).$$

THEORY: As we know *K*-map technique is used for reduction of boolean function and also used to design logic system. We have different types of variable in *K*-map:

(*a*) 2-variable

(*b*) 3-variable

(*c*) 4-variable

(*d*) 5-variable

(*e*) 6-variable.

But, to solve the above expression we will use the 4-variable K-map. These are minterms.

$$f(A, B, C, D) = \Sigma m(0, 2, 4, 7, 9, 13, 15)$$

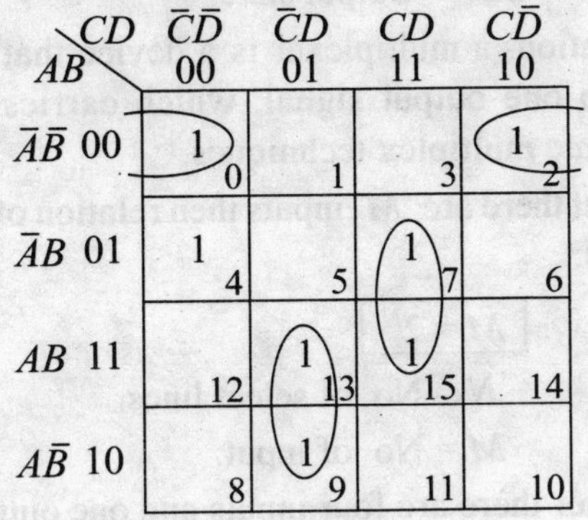

$$F = \bar{A}\bar{B} + \bar{C}D + CD$$

$$= \bar{A}\bar{B} + D(C + \bar{C})$$

$$\boxed{F = \bar{A}\bar{B} + D} \qquad [\because \quad C + \bar{C} = 1]$$

Realization of function using Logic Gates:

$$F = \bar{A}\bar{B} + D$$

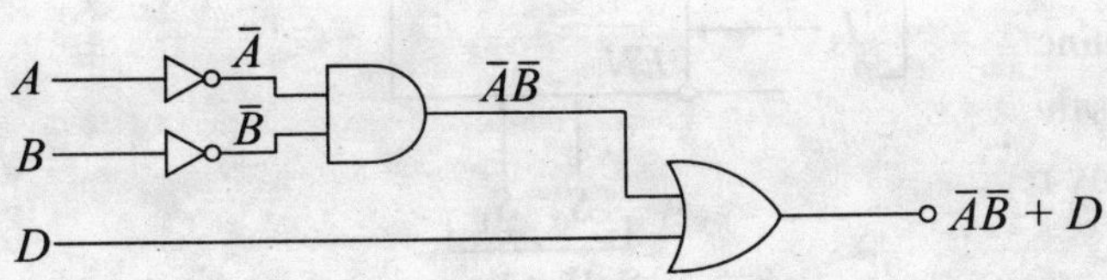

This is realization of function of K-map, using logic gates (AND, OR, NOT).

Result: We have solved the operation of given boolean function with K-map and realized the expression using logic gates.

Experiment No. 3

AIM: To verify the operation of 4:1 Multiplexer and realize its expression using logic gates.

APPARATUS: Connecting wires, D.C. Supply (5 V), logic probe, breadboard, IC 74153, Trainer kit.

THEORY: Multiplexer is a special combinational circuit, that is one of the most widely used standard circuit in digital design. It selects binary information from one of the many input channels and transmits to a single output line. So, due to this multiplexer is called. Data selector. The selection of particular input is controlled by select lines.

It is also considered as a multiple input single output switch or in other words Multiplexer means "many to one."

Many = Input combinations

One = Output state.

In telecommunication, a multiplexer is a device that combines several input information signals into one output signal, which carries several communication channel, by means of some multiplex techniques.

Let us suppose that there are '*M*' inputs then relation of multiplexer with respect to select lines is given as:

$$\boxed{M = 2^N}$$

N = No. of select lines.

M = No. of input.

In 4 : 1 multiplexer there are four inputs and one output, with two select lines input line will be selected.

Diagram:

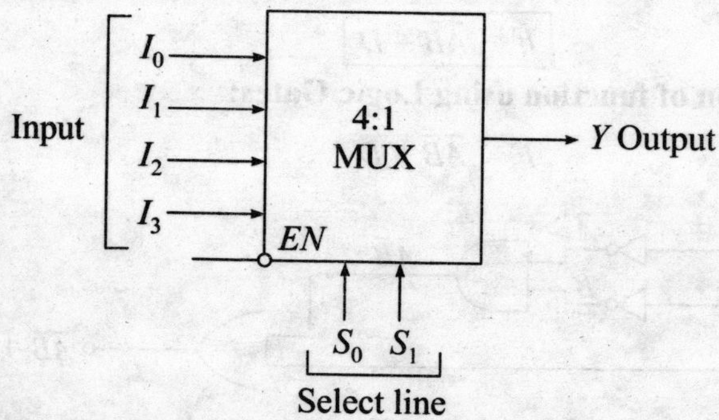

An enable (*EN*) input is used basically for cascading the other circuit and it is active low.

Truth Table:

Select line		Data Input	Output
S_0	S_1		
0	0	I_0	$\overline{S}_0\overline{S}_1I_0$
0	1	I_1	$\overline{S}_0S_1I_1$
1	0	I_2	$S_0\overline{S}_1I_2$
1	1	I_3	$S_0S_1I_3$

$$\text{Output } (Y) = \overline{S}_0\overline{S}_1I_0 + \overline{S}_0S_1I_1 + S_0\overline{S}_1I_2 + S_0S_1I_3$$

This is output of 4 : 1 multiplexer, now we will realize the output using logic gates.

Realization:

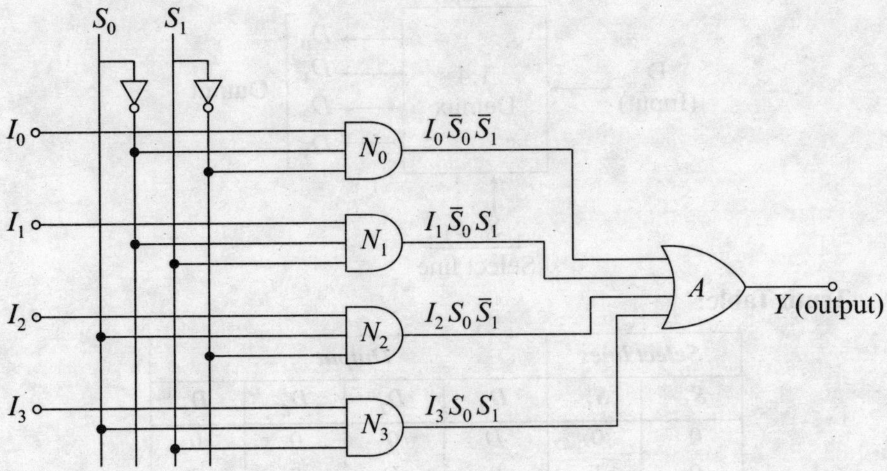

Procedure:

(*i*) Take the breadboard and insert 74153 IC in it or take the trainer kit.

(*ii*) Apply +5 V to TTL kit with IC 74153.

(*iii*) With select lines (S_0, S_1), select one of the inputs and check the output (I_0 to I_3) and verify it using truth table.

(*iv*) After that realize the network using logic gate.

Precautions:

(*i*) Insert and remove IC carefully or ON the switch of kit after right connection

(*ii*) Power supply should not exceed 5 V.

Result: We have verified the operation of 4 : 1 MUX and realized the expression using logic gates.

Experiment No. 4

AIM: To verify the operation of 1 : 4 demultiplexer and realize its expression using logic gates.

APPARATUS: Connecting wires, DC supply (5V), logic probe, breadboard, IC 74154, Trainer Kit.

THEORY: The demultiplexer is a combinational circuit which means "one to many". It states that its is having one input line and many outputs are present. It is a logic circuit which pass information through a single input line and transmits the same information over one of the possible 2^m output lines. The selection of specific output lines is controlled by bit combinations of selection lines.

The select input code determines to which output the data input will be transmitted.

Diagram:

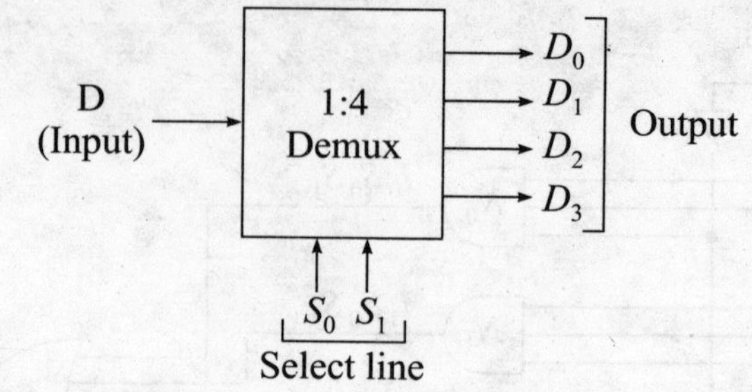

Truth Table:

Select lines		*Output*			
S_0	S_1	D_0	D_1	D_2	D_3
0	0	*D*	0	0	0
0	1	0	*D*	0	0
1	0	0	0	*D*	0
1	1	0	0	0	*D*

$$D_0 = \overline{S}_0\overline{S}_1 D$$
$$D_1 = \overline{S}_0 S_1 D$$
$$D_2 = S_0\overline{S}_1 D$$
$$D_3 = S_0 S_1 D$$

Realization:

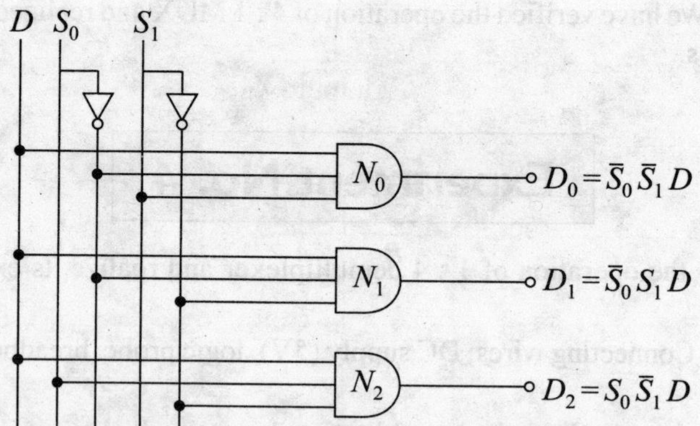

Procedure:

(*i*) Make the connections an breadboard or trainer Kit using IC or wires.

(*ii*) With the select lines (S_0, S_1); verify the output line from $D_0 - D_3$ as in truth table.

(*iii*) In case, if result is not verified then repeat the steps from (*i*) to (*ii*).

Precaution:

(*i*) Insert and remove IC carefully or switch ON the kit after right connection.

(*ii*) Power supply should not exceed 5 V.

Result: We have verified the operation of 1 : 4 De Mux and realized the expression using logic gates.

Experiment No. 5

AIM: To study magnitude comparator and design the circuit and verify it with the help of Kit.

APPARATUS: TTL Kit, breadboard, IC 7485, Power supply, connecting wires, connecting probe.

THEORY: Magnitude comparator is one of the useful combinational logic networks and has wide application. It compares two binary numbers and determines one number is greater than, less than or equal to the other numbers. It is a multiple output combinational logic circuit. If two binary numbers are considered as *A* and *B*, the magnitude compactor gives 3 output:

(*i*) $A > B$ (*ii*) $A < B$ (*iii*) $A = B$.

Truth Table and Diagram:

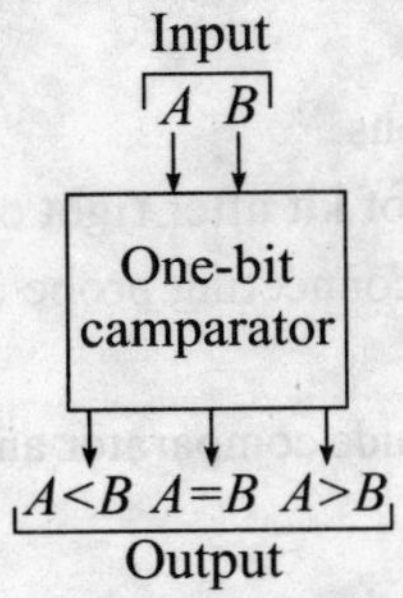

Truth Table:

Sr.	*Inputs*		*Outputs*		
No.	*A*	*B*	$A < B$	$A = B$	$A > B$
0	0	0	0	(1)	0
1	0	1	(1)	0	0
2	1	0	0	0	(1)
3	1	1	0	(1)	0

***K*-map:**

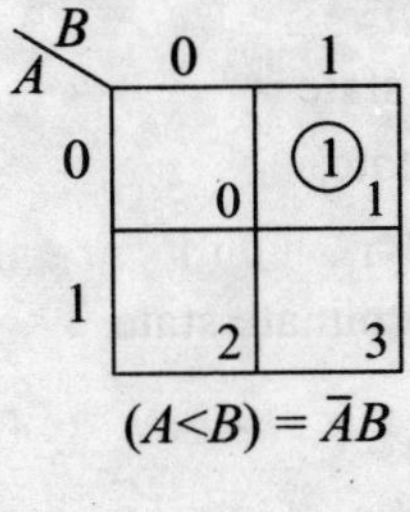

$(A<B) = \bar{A}B$

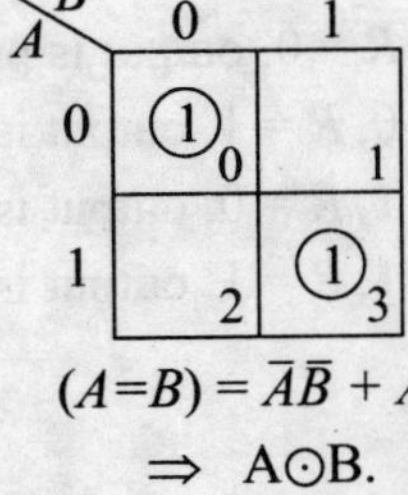

$(A=B) = \bar{A}\bar{B} + AB$

$\Rightarrow A \odot B.$

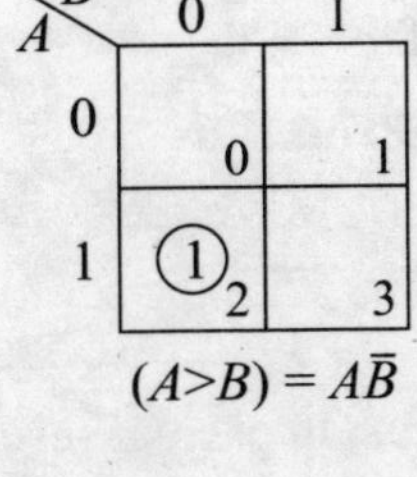

$(A>B) = A\bar{B}$

Realization:

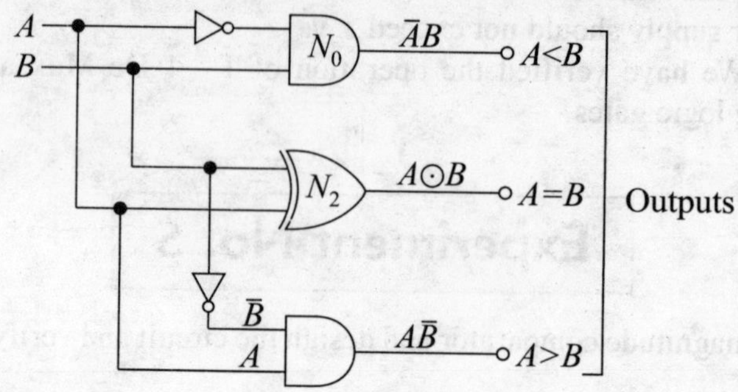

Procedure:

(*i*) Insert the IC 7485 in breadboard.

(*ii*) Connect 2nd pin to ground and 16th pin to power supply.

(*iii*) Connect first pin to *A* and 2nd pin to *B*.

(*iv*) Using connecting problem check the output and verify the truth table.

(*v*) The stage handling the LSB must have $A = B$ input connected to logic 1 level and ($A > B$), ($A < B$) are connected to logic 0 or logic 1 level.

(*vi*) Check the output and verify the table.

Precautions:

(*i*) Check all the connections.

(*ii*) Switch ON the bottom of kit after right connections.

(*iii*) Check the output with connecting probe carefully.

Result:

We have studied the magnitude comparator and designed the circuit and verified it with the help of trainer kit.

Experiment No. 6

AIM: To verify the operation of *S-R* and *J-K* flip-flop using trainer kit.

APPARATUS: Breadboard, connecting wires, connecting probe, power supply, IC (7476, 7474), trainer kit.

THEORY: (*i*) ***S-R* Flip-Flop:** In order to set or reset the memory cell in synchronization with train of pulses, we used clocked set-reset flip-flop. In the circuit if the clock pulse is present clk = 1, its operation is as follows,

if $S = R = 0$, output is previous state.

$S = 0, R = 1$, output is reset state.

$S = 1, R = 0$, output is set state.

$S = 1, R = 1$, output is undeterminate state.

Diagram:

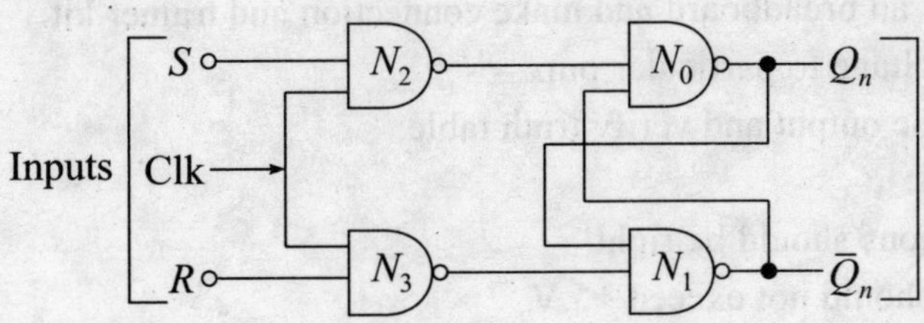

Truth Table:

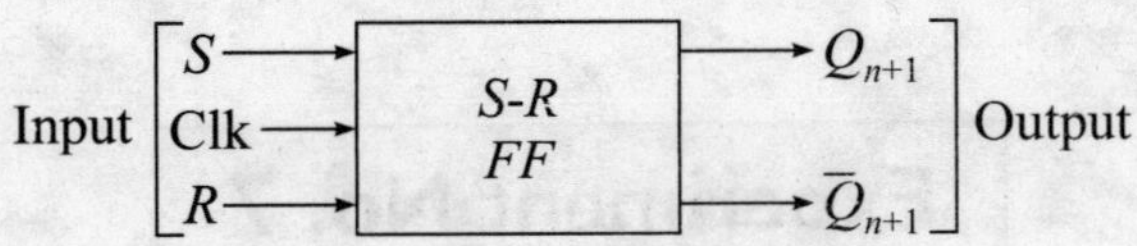

S	*R*	***Output*** (Q_{n+1})
0	0	Q_n
0	1	0 (Reset)
1	0	1 (Set)
1	1	? (not determine)

(*ii*) ***J-K*** **Flip-Flop:** Uncertainity in the state of *S-R* flip-flop when $S = R = 1$, can be eliminated by converting it in *J-K* flip-flop respectively to obtain *S-R* input *i.e.*:

$$S = J\bar{Q},\ R = k\,Q$$

Diagram:

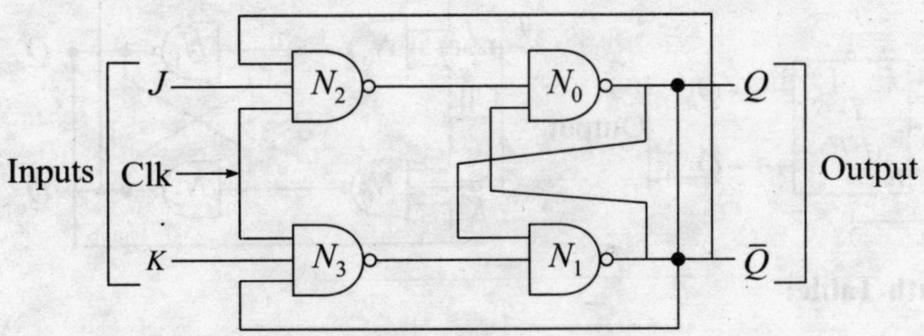

Block Diagram and Truth Table:

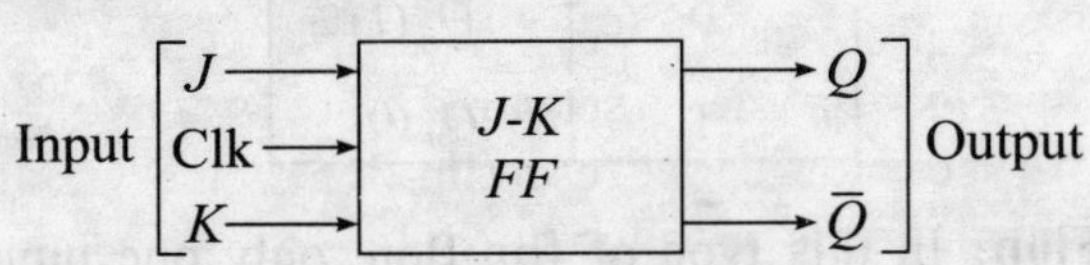

J	*K*	***Output*** (Q_{n+1})
0	0	Q_n
0	1	0 (reset)
1	0	1 (set)
1	1	$\bar{Q}_n$ (Toggle)

Procedure:

(*i*) Insert IC an breadboard and make connection and trainer kit.

(*ii*) Apply voltage to particular pin.

(*iii*) Obtain the output and verify truth table.

Precautions:

(*i*) Connections should be right.

(*ii*) Voltage should not exceed +5 V.

(*iii*) Insert IC carefully.

Result: We have verified the operation of *S-R* and *J- K* flip-flop.

Experiment No. 7

AIM: To verify the operation of Toggle (*T*) and Delay (*D*) flip-flop using trainer kit.
APPARATUS: Breadboard, connecting wires, connecting probe, IC (7476), Power supply.
THEORY: (*i*) **T (Toggle) Flip-Flop:** A *T*- flip flop can be constructed when both input of *J-K* flip-flop is combined ($J = K$) and named toggle (*T*).

Case I: When $T = 0$, then $J = K = 0$ and output $Q_{n+1} = Q_n$ and $\bar{Q}_{n+1} = \bar{Q}_n$.

Case II: When $T = 1$, then $J = K = 1$ and hence output $Q_{n+1} = \bar{Q}_n$ and $\bar{Q}_{n+1} = Q_n$.

Block Diagram and Circuit Diagram:

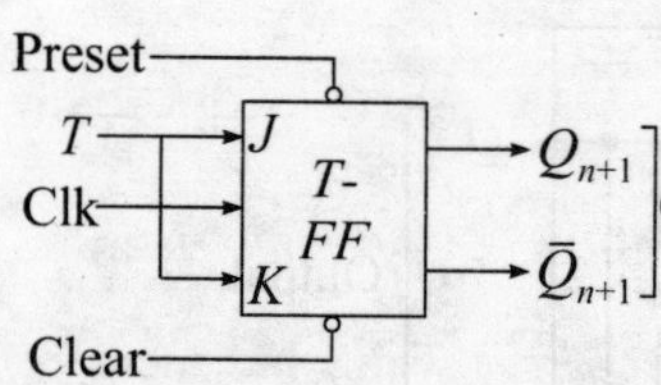

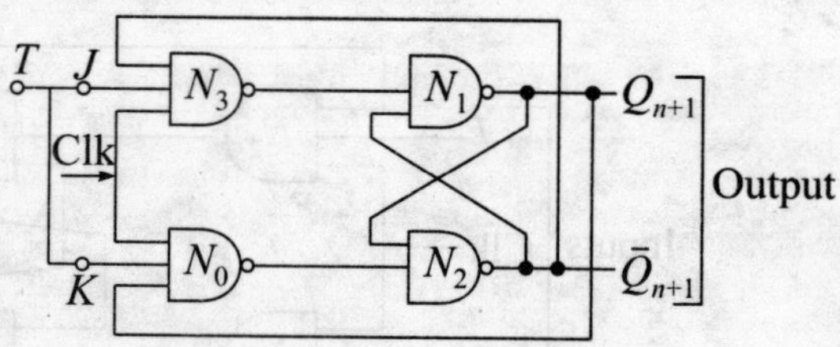

Truth Table:

T	Q_{n+1}
0	Q_n (1)
1	$\bar{Q}_n$ (0)

(*ii*) ***D* Flip-Flop:** In this type of flip-flop, only one input is referred the delay input, its truth table is given from which it is clear that the output quantity at the end of first pulse equals the input (*D*) before the clock pulse. This is equivalent to saying that input data from input to output is delayed. However it is named as *D* Flip-Flop. It is used either as delay device or as a latch store one bit of binary information.

Block Diagram and Circuit Diagram:

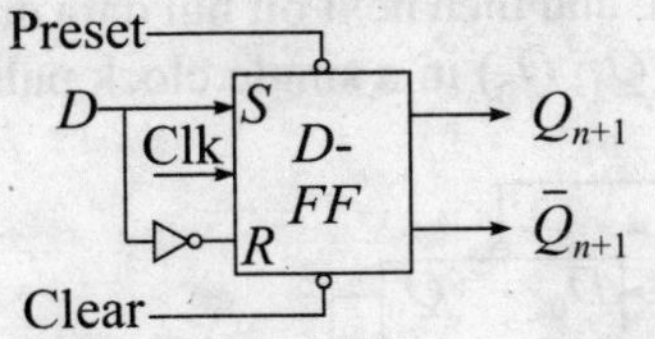

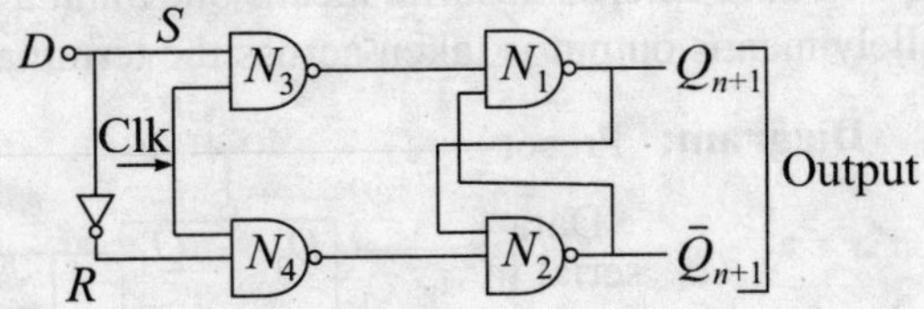

Truth Table:

D	Q_{n+1}
0	0
1	1

Procedure:

(*i*) Insert IC on the breadboard and make connections according to pin layout.

(*ii*) Apply voltage to appropriate pins.

(*iii*) Obtain the output and verify it with truth table.

Precautions:

(*i*) Do not exceed voltage greater than +5 V.

(*ii*) Insert IC carefully.

(*iii*) Connection should be tight.

(*iv*) Switch off power when not in use.

Result: We have verified the operation of *T* and *D* flip-flop.

Experiment No. 8

AIM: To study shift registers:

(*i*) SISO (*ii*) SIPO

(*iii*) PISO (*iv*) PIPO.

THEORY:

Shift Register: It is a type of sequential circuit (means which has memory) that is formed by combination of flip-flops and is capable of shifting data from left to tight or right to left.

(*i*) **SISO:** (Serial In Serial Out). In this data is entered serially *i.e.,* data is entered one bit at a time and then next bit and data is out serially also *i.e.*, one bit at a time and then next bit at next clock pulse.

Generally on clock pulse is consumed when one bit of data is stored.

Diagram:

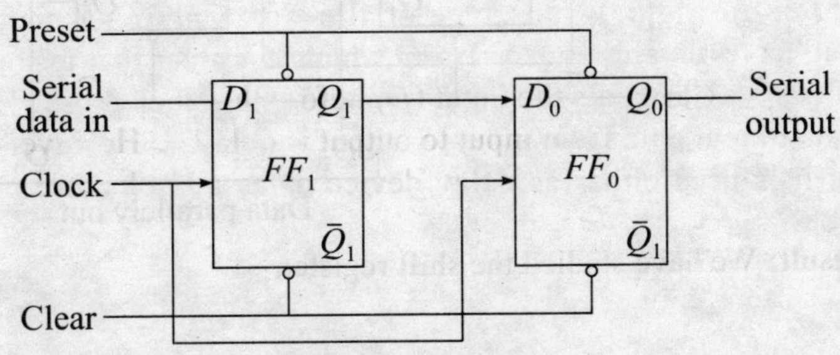

(*ii*) **SIPO:** (Serial In Parallel Out). In this data is entered in same style as in SISO *i.e.*, data entered in serial means one bit at a time and then next bit but data out is parallely means output is taken across the terminals (Q_1, Q_0) in a single clock pulse.

Diagram:

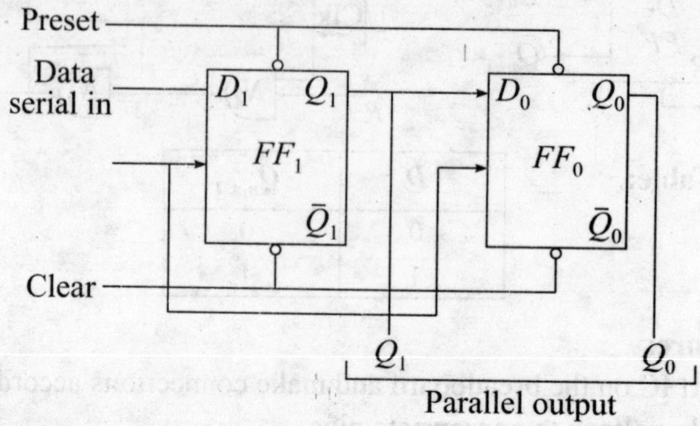

(*iii*) **PISO:** (Parallel in serial out). It work as reverse of SIPO. In this data is entered parallel across D_1, D_0 but data is out serially using single clock pulse. Data is out at the last value.

Diagram:

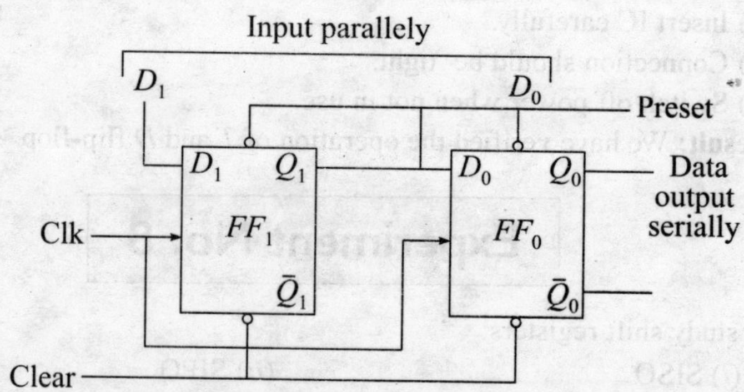

(*iv*) **PIPO:** (Parallel In Parallel out). In this we feed input parallely to all flip-flop and also take data out parallely from all flip-flop. In PIPO data entered simultaneously and data is out also simultaneously.

Diagram:

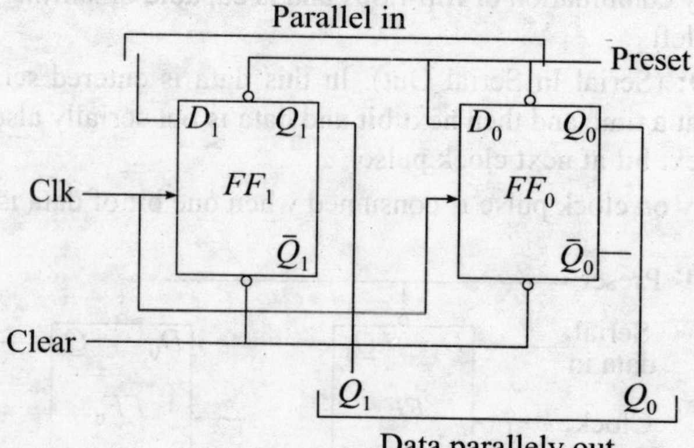

Result: We have studied the shift register.

Experiment No. 9

AIM: To study and verify the operation of 4-bit ring counter.

APPARATUS: Breadboard, connecting wires, power supply (5 V), IC 7496 (Shift Register), trainer kit, connecting probe. Current limiting resistor to LED,

THEORY: It is an synchronous counter in which last output (*i.e.*, LSB) is feeded back to the input os first flip-flop as shown in diagram. As it is synchronous one, the clock pulse of all flip-flop is same.

The output of FF-3 is set to one or arrival of I[st] clock pulse with the help of preset input and on reset all flip-flop are cleared to '0' with clear input = '0'. As the positive edge of clock is arrived flip-flop with shifts its output to the flip-flop 2 and similarly flip-flop 2 gives its output to the flip-flop 1 and so on.

The working can be easily understand by the help of state diagram and waveform as shown below:

State Diagram:

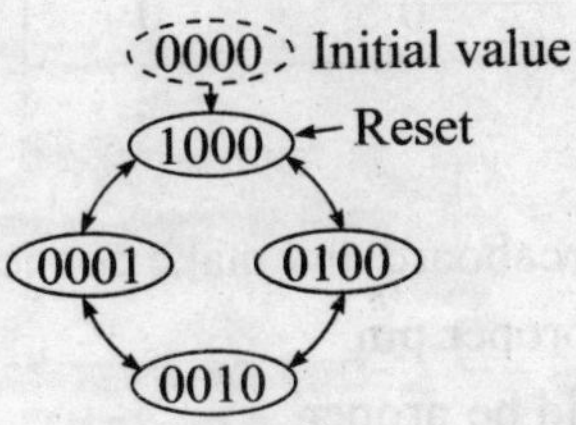

Circuit Diagram:

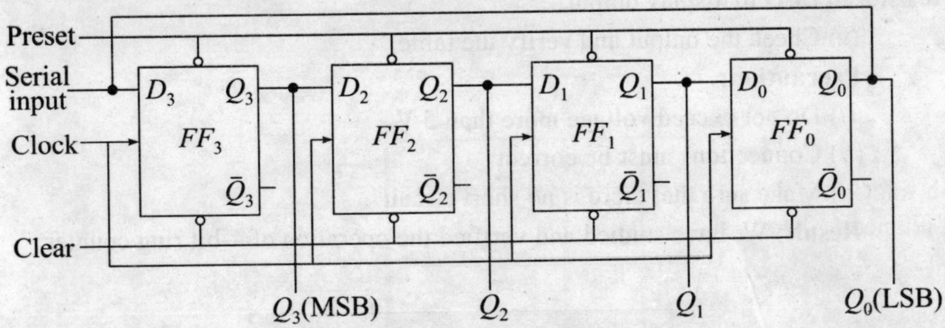

Truth Table:

Sr. No. Clock pulse	Q_3 *(MSB)*	Q_2	Q_1	Q_0 *(LSB)*	*Counts*
On Reset	0	0	0	0	0
1st clock pulse	1	0	0	0	1
2nd clock pulse	0	1	0	0	2
3rd clock pulse	0	0	1	0	3
4th clock pulse	0	0	0	1	4
5th clock pulse	0	0	0	0	Recycle

This is the truth table of ring counter in which after 4th count Recycle state will start.

Wavefrom:

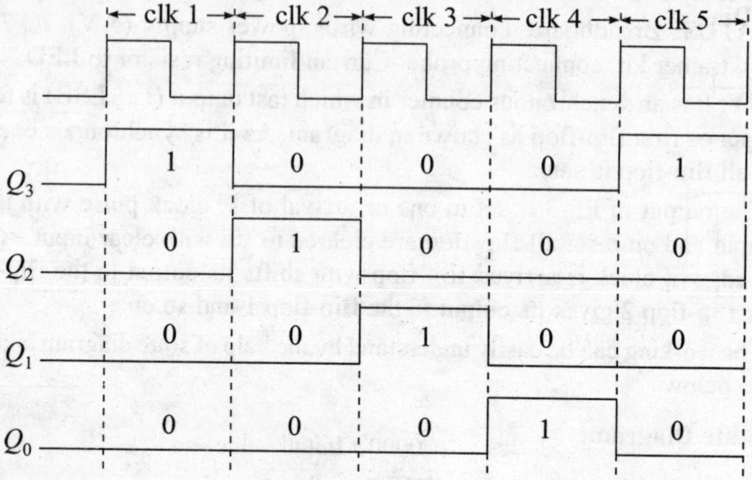

Procedure:

(*i*) Insert IC on the breaboard and make connection in given circuit.

(*ii*) Apply voltage at proper pin.

(*iii*) Connections should be proper.

(*iv*) All the outputs Q_0, Q_1, Q_2, Q_3 should be connected through current limiting resistor to LED to display output.

(*v*) Check the output and verify the table.

Precaution:

(*i*) Do not exceed voltage more than 5 V.

(*ii*) Connections must be correct.

(*iii*) Make sure that there is no short circuit.

Result: We have studied and verified the operation of 4-bit ring counter.

Experiment No. 10

AIM: To study and verify the operation of 4-bit Johnson counter.

APPARATUS: Breadboard, connecting wires, connecting probe, trainer kit, power supply, IC 7496, current limiting resistance for LED.

THEORY: This counter is similar to ring counter but only one difference is there that in it we feedback $\overline{Q}_0$ instead of Q_0 to serial input (D_3).

Initially all the flip-flop are in reset state *i.e.*, '0000'. After each clock pulse, the output of Q_3 is shifted to Q_2, the output of Q_2 is shifted to Q_1, Q_1 to Q_0 and the output of $\bar{Q}_0$ is shifted to Q_3 and so on. This sequence is repeated after every 8 clock pulse. It is shown in figure and in which last output ($\bar{Q}_0$) is feedback to serial input (D_3).

Diagram:

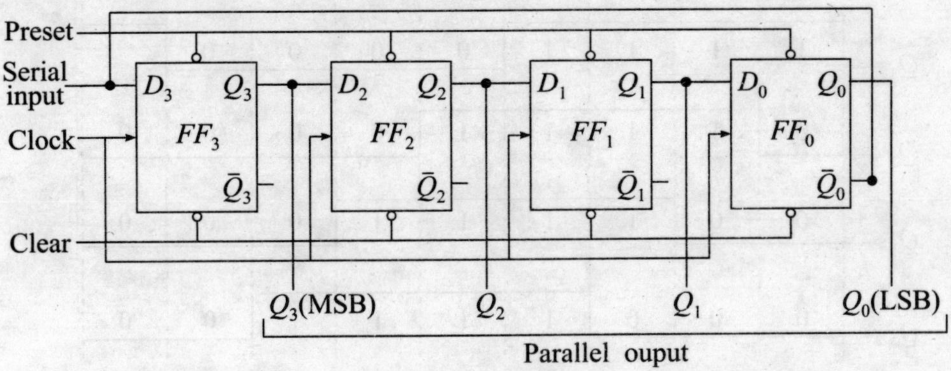

Truth Table:

Clock	Q_3	Q_2	Q_1	Q_0	*Counts*
Initially	0	0	0	0	0
1st clock	1	0	0	0	1
2nd clock	1	1	0	0	2
3rd clock	1	1	1	0	3
4th clock	1	1	1	1	4
5th clock	0	1	1	1	5
6th clock	0	0	1	1	6
7th clock	0	0	0	1	7
8th clock	0	0	0	0	0 → Recycle

State Diagram:

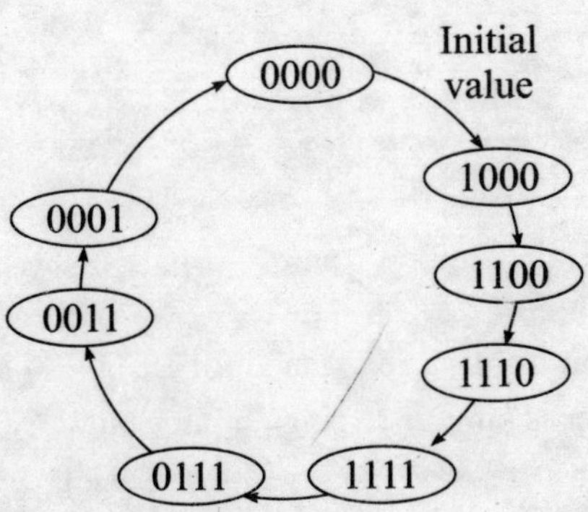

Waveform:

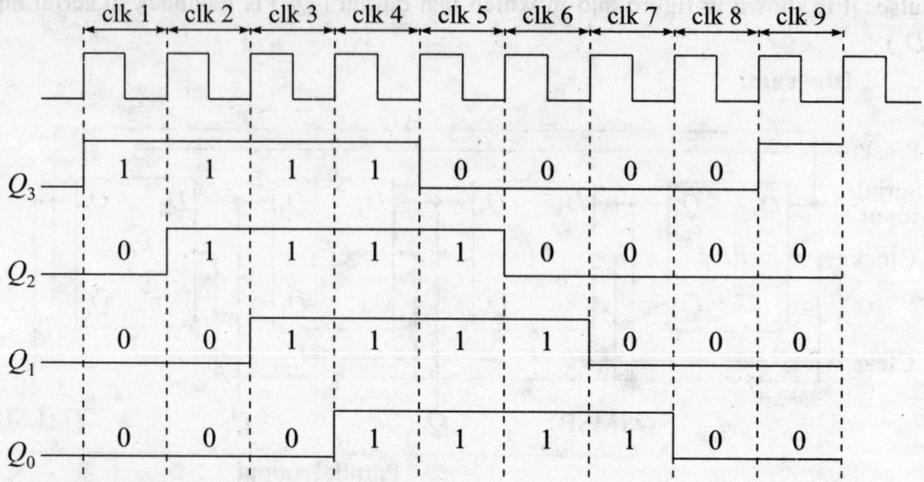

Procedure:

(*i*) Insert the IC and beradboard.
(*ii*) Make connection in given circuit.
(*iii*) First reset all flip-flops.
(*iv*) Apply +5 V to the circuit.
(*v*) Connect the LED in all the Q outputs through current limiting resistor.
(*vi*) Observe the glowing of LED.
(*vii*) Take the output and verify the table.

Precautions:

(*i*) Connection should be tight and correct.
(*ii*) Handle IC with care and voltage should not exceed 5 V.
(*iii*) Make sure that there is no short circuit.

Result: We have studied and verified the operation of Johnson counter.

UNIVERSITY PAPERS

University Papers

MAY-2005

1. (*a*) Explain universal and basic gates.
   (*b*) If $A = -60$, $B = -28$
   (*i*) Represent *A* and *B* in 8-bit 2's complement form
   (*ii*) Find $A + B$.
   (*iii*) Find $B - A$.
   (*c*) Divide $(110.11)_2$ by $(101)_2$
2. (*a*) Describe error detection and correction codes with their significance.
   (*b*) Why do we use gray code in *K*-maps?
3. (*a*) The seven bit Hamming code as received is 0010001. Assuming that even parity has been used, check it is correct or not. If not, find the correct code.
   (*b*) Simplify the Boolean expression using Q-M method:
   1. $Y = \pi M\,(0, 1, 3, 5, 6, 7, 10, 14, 15)$
   2. $Y = \pi M\,(1, 2, 3, 5, 6, 7, 9, 10, 13, 14)$

Implement the with circuit also.

4. (*a*) Explain the interfacing between following:
   (*i*) TTL-CMOS
   (*ii*) CMOS-TTL
   (*iii*) TTL interfaced to Buzzers and Relays.
   (*b*) What is tri-state logic? Explain its significance.
5. (*a*) Draw the circuit for finding 9's complement of any number.
   (*b*) What is Look Ahead Carry?
   (*c*) Discuss the difference between:
   (*i*) Encoder/Multiplexer. (*ii*) Decoder/Demultiplexer.
6. (*a*) What is difference between Latch and Flip-Flop?
   (*b*) What is Race Around condition?
   (*c*) Draw a 4-bit bidirectional shift register and explain the working in all four cases:
   (*i*) SISO (*ii*) SIPO
   (*iii*) PISO (*iv*) PIPO.
7. (*a*) What are the performance characteristics of *D*/*A* converter?
   (*b*) Explain both the technique for D/A conversion.
   (*c*) Why do we prefer SAR over dual slope for ADC?

8. Write short note on:

(*a*) PAL (*b*) Macrocell

(*c*) FPGA (*d*) NAND gate using CMOS Logic.

DEC., 2005

1. (*a*) Represent decimal number 6127 in BCD, Excess-3 and 2421 code.

(*b*) What does floating point number 0100101000010101 represent?

(*c*) Explain gray code. How do you convert it into binary number?

2. Minimize the following digital function:

$f(A, B, C, D) = \Sigma m\,(1, 3, 7, 11, 15) + d\,(0, 2, 5)$ using *k*-map and Q-M method. Draw the required circuit using NAND/NOR gates.

3. (*a*) Explain circuit of a tri-state TTL Logic.

(*b*) Write a note on the interfacing of TTL and CMOS IC's.

(*c*) What is major difference between TTL bipolar and MOSIC?

(*d*) What is CMOS? What are its merits and demerits?

4. (*a*) Explain the following:

(*i*) Multiplexer and Demultiplexer.

(*ii*) Decoder and Encoder.

(*b*) Design *BCD* arithmetic circuits.

5. (*a*) What is Edge Triggering?

(*b*) What is the difference between ripple counter, Ring counter and Johnson counter?

(*c*) Explain Master-slave JK Flip-Flop.

6. (*a*) Design a counter that will go through states 1, 3, 5, 6, 1, ... using J-K Flip-Flops.

(*b*) Explain CPLDs.

7. (*a*) A 5 bit DAC produces $V_0 = 1.4$ V for a digital input of 0011. Find its %age resolution, step size, V_0 for digital input of 0111 and full scale analog voltage.

(*b*) Explain R-2 R_L Ladder DAC.

(*c*) Explain successive Approximation type ADC.

8. (*a*) Explain, ROM, PLA and FPGA.

(*b*) design Excess-3 to BCD code counter using PLA and PAL.

MAY-2006

1. (*a*) Represent decimal number 6100 in *BCD*, Excess-3 and 2421 code.

(*b*) How do you represent –ve number?

(*c*) Explain Gray code. How do you convert it into binary and vice-versa?

2. Minimize the following digital function:

$$f(A, B, C, D) = \pi M\,(4, 5, 6, 7, 8, 12).\ d\,(1, 2, 3, 9, 11)$$

3. (*a*) Explain circuit of a TTL Not gate.

(*b*) Compare the different Logic families.

(*c*) Write a note on interfacing of CMOS and TTL families.

4. (*a*) Design a 4-bit full adder circuit. Modify it into subtractor.
(*b*) Design a 4-bit magnitude comparator.

5. (*a*) Explain Race Around condition.
(*b*) Describe universal shift Register.
(*c*) Explain functioning of synchronous Ring converter.

6. (*a*) Design a counter that will go through states 2, 7, 5, 6, 2, ... Using JK flip-flop.
(*b*) Explain PLD.

7. (*a*) Give specification of ADC.
(*b*) Explain Ladder network method of DAC.
(*c*) Describe Dual-slope ADC.

8. (*a*) Explain ROM, PAL, FPGA.
(*b*) Design *a* BCD to 7-segment decoder using PLA and PAL

December-2006

1. (*a*) Find largest binary number that can be expressed with 16 bits. Write its equivalent in decimal and Hexadecimal.
(*b*) $(8620)_D = ()_{BCD} = ()_{Excess\text{-}3}$
(*c*) Convert Gray code 1011 into binary and 1001011 into Gray code.
(*d*) Convert binary word 1011 into 7-bit even parity Hamming code.

2. Minimize the following using *k*-map and verify it using Q-M method:
$$f(A, B, C, D) = \Sigma m\ (0, 1, 2, 3, 7, 8, 9, 10, 11) + d\ (6, 14, 15)$$
Implement it using NAND gate.

3. (*a*) What is CMOS? What are its merits and Demerits?
(*b*) Explain Tristate TTL gate.
(*c*) Compare RTL and TTL.

4. Explain the working of following:
(*a*) BCD Adder.
(*b*) Multiplexer and Demultiplexer.

5. (*a*) What is difference between asynchronous and synchronous sequential circuits? Give example.
(*b*) What is difference between shift Register Ring counter and Johnson counter? Explain with help of circuit diagram and output waveform.

6. (*a*) Design a counter that will go through 0, 2, 4, 6, 0 ... using J-K flip-flop. Give necessary state diagram.
(*b*) Explain ROM.

7. (*a*) 12 bit DAC has full scale analog voltage of 5*v*. Determine step size, % resolution and analog output voltage to 1001000111.
(*b*) Explain *R*-2*R* Ladder DAC.
(*c*) Describe parallel A/D converter.

8. Explain ROM, PLA and CPLD.

MAY-2007

1. (*a*) Convert:

(*i*) $(34.4375)_D = ()_{Binary}$

(*ii*) $(8690)_D = ()_{Excess\text{-}3}$

(*b*) Convert binary code 1010 into 7 bit even parity Hamming code.

(*c*) Find 8 bit 2's complement form of – 49 and – 107.

(*d*) What is EBCDIC?

2. Minimize using Q-M method.

$f(A, B, C, D) = \Sigma m\ (1, 2, 6, 7, 9, 13, 14, 15, 17, 22, 23, 25, 29, 30, 31)$ verify it using *k*-map and implement it using NAND gate.

3. (*a*) Explain the circuit of a Tristate TTL gate.

(*b*) Describe major difference between combinational and sequential logic circuits.

(*c*) Write a note on interfacing of TTL and CMOS gate.

4. Explain following with circuit diagram:

(*a*) Decoder and Encoder.

(*b*) 4-bit comparator.

5. (*a*) What is major difference between bipolar and MOS devices?

(*b*) What is race around condition? How do you remove it.

(*c*) Explain Master-slave J-K Flip-Flop.

(*d*) What is edge triggering?

6. (*a*) Design a decade ripple counter.

(*b*) Explain MOD-8 synchronous counter.

7. (*a*) Describe dual slope *A*/*D* converter.

(*b*) Determine the analog output voltage of 6-bit *R*-2*R* ladder network DAC with V_{ref} as 5*v* when the digital input is 011100.

(*c*) Explain binary weighted Resistor method for DAC.

8. Write short note on:

(*a*) PLA

(*b*) CPLD

(*c*) EX-OR and EX-NOR gate.

QUESTION BANK

Question Bank

Long-Short Questions

1. Convert the following decimal numbers to binary, octal and hexadecimal numbers:
   (*a*) 36 (*b*) 79
   (*c*) 147 (*d*) 588
   (*e*) 746 (*f*) 1086.
2. Convert the following octal numbers to binary, decimal and Hexadecial format:
   (*a*) 36.25 (*b*) 27.125
   (*c*) 376.4
3. Obtain the following operations using NOR gates:
   (*a*) NOT (*b*) AND
   (*c*) OR.
4. Prove using Demorgan's Theorem:
   $$AB + CD = \overline{\overline{AB}\,.\,\overline{CD}}$$
5. What is digital signal?
6. Give the BCD, Excess-3 and Gray code of following numbers:
   (*a*) 236 (*b*) 841
   (*c*) 902 (*d*) 57.23
   (*e*) 34
7. Explain with example the difference between the following:
   (*a*) Weighted and non-weighted code.
   (*b*) Sequential and non sequential code.
   (*c*) Reflective and non reflective code.
8. What are error detection and correction codes?
9. Explain the parity method of detecting error?
10. Find a 5 bit binary number which is equal to its own 2's complement.
11. Perform the following additions:
    (*a*) $(107)_8 + (73)_8$ (*b*) $(2c3)_{16} + (DA9)_{16}$
    (*c*) $(164)_8 + (756)_8$ (*d*) $(A48)_{16} + (B97)_{16}$
12. Add the following binary numbers:
    (*a*) 11001 + 1001 (*b*) 101100 + 110011
    (*c*) 10011 + 111 (*d*) 10010 + 1101
13. Solve the binary using 1's complement system.
    (*a*) $(11 - 7)_{10}$ (*b*) $(86 - 39)_{10}$
    (*c*) $(17 - 30)_{10}$ (*d*) $(35 - 57)_{10}$

14. Solve in binary using 2's complement system.

(*a*) $(62-29)_{10}$ (*b*) $(57-37)_{10}$

(*c*) $(43-60)_{10}$ (*d*) $(21-55)_{10}$

15. Solve using 9's complment:

(*a*) $(135-89)_{10}$ (*b*) $(2179-364)_{10}$

(*c*) $(75-863)_{10}$

16. Solve following using 10's complement system:

(*a*) $(39-87)_{10}$ (*b*) $(265-1023)_{10}$

(*c*) $(367-245)_{10}$

17. Evaluate using BCD arithmetic.

(*a*) 139 + 973 (*b*) 48 + 964.

18. Solve using 2's complment arithmetic:

(*a*) 1101.001 – 100.11 (*b*) 1000.101 – 111.01

19. Simplify using Boolean Algebra:

(*a*) $B(A+B)(B+C)$ (*b*) $A[B+\overline{C}(AB+A\overline{C})]$

(*c*) $(A+\overline{B}\,\overline{C})(A\overline{B}+ABC)$ (*d*) $ABC+\overline{A}B+AB\overline{C}$

(*e*) $\overline{(A+\overline{B})}\;\overline{(\overline{A}+\overline{B})}$ (*f*) $A+\overline{A}B\overline{C}+\overline{(B+C)}$

20. Realize the following functions without simplification, using Basic logic gates

(*a*) $AB+\overline{A}C$ (*b*) $(A+\overline{B})(\overline{A}+B)$

(*c*) $\overline{A}\,\overline{B}+A$ (*d*) $\overline{A}\,(B\overline{C}A)$

(*e*) $\overline{A(BC+\overline{\overline{A}B})}$

21. Find the complement of following functions:

(*a*) $y=A\overline{B}+\overline{A}B$ (*b*) $y=AB+\overline{A}\,\overline{B}$

(*c*) $y=(A+\overline{B})(\overline{A}+B)$.

22. Give truth table for XOR gate and hence derive the function and realize the circuit diagram in SOP and POS form.

23. Define and explain with example the following:

(*a*) Canonical expression.

(*b*) Maxterm and Minterm.

24. Convert the following SOP expression in canonical form and hence give the equivalent POS form:

(*a*) $y(ABC)=A+\overline{B}\,\overline{C}$ (*b*) $y(ABC)=\overline{A}B+\overline{B}C$

25. Convert the following POS expression in canonical form and hence give the equivalent SOP form.

(*a*) $y(ABC)=(\overline{A}+B)(B+\overline{C})$ (*b*) $y(ABC)=(A+C)(B+\overline{C})$

26. Simplify using *K*-map:
(*a*) $y = \Sigma m\ (1, 3, 4, 9, 11, 12, 13)$
(*b*) $y = \Sigma m\ (3, 4, 5, 7, 9, 13, 14, 15)$
(*c*) $y = \Sigma m\ (0, 1, 4, 5, 6, 7, 12, 14)$
(*d*) $y = \pi M\ (3, 4, 6, 7, 11, 12, 13, 14, 15)$
(*e*) $y = (A, B, C, D) = \Sigma m\ (0, 2, 3, 6, 7, 8, 9, 10, 13)$
27. Simplify using *QM* method.
$y\,(A\,B\,C\,D) = A\,B + A\,\bar{B}\,C\,\bar{D} + A\,\bar{B}\,\bar{C}\,D + \bar{A}\,B\,C\,D + \bar{A}\,B\,C\,\bar{D}$
28. Simplify the following using *Q-M* method:
(*a*) $y = \Sigma m\ (0, 4, 6, 7, 13, 16, 20, 22, 23, 29)$
(*b*) $y = \Sigma m\ (1, 3, 5, 7, 8, 9, 17, 19, 21, 23, 24, 25)$
29. Give the Truth Table for a full adder, using 2 Half adder.
30. Draw and explain the circuit diagram for a 1 digit BCD adder.
31. Give the circuit for full subtractor and derive it using *k*-map.
32. Design a combinational circuit which generates output 1 if a 4 bit number contains even number of 1 *s* otherwise output is 0.
33. Design the following logic circuit using 4 × 1 and 8 × 1 multiplexer.
$F\,(A\,B\,C\,D) = \text{sum}\ (0, 1, 3, 4, 6, 8, 12, 13, 15)$
34. Implement the following functions using a decoder
(*a*) $m\ (2, 3, 5, 7) + m\ (1, 3, 6)$
(*b*) $m\ (1, 3, 5, 7, 8, 9) + M\ (0, 1, 2, 3, 4, 5, 8, 9, 13, 14)$
35. Difference between Ripple and parallel counter.
36. Design a counter with binary sequence 0,1, 3, 7, 6, 4 and repeat. Use *T* Flip-Flop.
37. Design a Decade asynchronous counter.
38. Explain the difference between the following:
(*a*) Level Triggering and Edge Triggering.
(*b*) Latch and Flip-Flop
(*c*) SR and JK Flip-Flop.
(*d*) Sequential and combinational circuits.
39. Design a 3 bit Gray counter using '*D*' Flip-Flop.
40. Explain the working of JK Flip-Flop and compare it with master flip-flops?
41. Explain the working of 4 bit up-down ripple counter.
42. Design the following binary ripple counter:
(*a*) MOD 3 (*b*) MOD 9
(*c*) MOD 5
43. Explain the working of 4 bit up-down ripple counter.
44. Design the circuit for following conversions:
(*a*) BCD to Gray (*b*) Gray to BCD.
(*c*) Binary to *X*53 (*d*) *X*53 to BCD.
(*e*) 4 bit Binary to BCD (*f*) 2421 to 8421 code.

45. Write the advantage of TTL Logic Families over other Logic Families.

46. Write short note on:

(*a*) TTL Logic Families (*b*) Wired AND

(*c*) Wired Logic.

47. Explain 3 state Logic.

48. With the help of neat circuit diagram explain the working of:

(*a*) CMOS Inverter. (*b*) 2 input TTL NAND gate.

(*c*) ECL basis OR-NOR gate. (*d*) 2 input CMOS NAND gate.

(*e*) 2 input CMOS NOR gate.

49. Explain w.r.t. Digital ICs:

(*a*) Figure of Merit

(*b*) Wired Logic

(*c*) Noise Immunity.

50. Define Resolution, settling time and Linearity error of DAC.

51. Explain the working of successive approximation ADC with the help of a neat block diagram.

52. Compare counter Ramp and Successive approximation type ADC.

53. Explain the working of Dual slope ADC with the help of a neat block diagram.

54. Explain PLA, PAL, ROM.

55. What is CPLD? Explain it.

56. Define Macrocell?

MODEL TEST PAPERS

Model Test Paper 1

Note: *Attempt any five questions out of Eight questions.*

1. (*a*) Represent decimal number 7123 in *BCD, XS*-3 and 2421 code.
(*b*) How do you represent positive number?
(*c*) Explain Gray code. How do you convert it into binary and vice-versa?

2. Minimize the following digital function:
$F(A, B, C, D, E) = \Sigma m(0, 2, 5, 7, 9, 20, 23) + d(1, 14, 17, 27, 30)$
Realize it using NAND gate only.

3. (*a*) Design full subtractor using two Half Subtractor.
(*b*) Explain Priority Encoder with example.

4. (*a*) Design a 4-bit magnitude comparator.
(*b*) What is decoder? Explain any one type of decoder.

5. (*a*) Describe universal shift register with clean circuit diagram.
(*b*) Explain functioning of synchronous ring counter.

6. (*a*) Design a UP/DOWN Counter that will go through states 2, 7, 5, 6, 2, ... Using SR flip-flop.
(*b*) Compare the different logical families.

7. (*a*) Give specification of DAC.
(*b*) Explain Dual-Slope ADC.
(*c*) Explain circuit of a TTL NOT gate.

8. (*a*) Explain CPLD and draw the circuit diagram of Macrocell.
(*b*) Design a Full adder using ROM, PLA.

Model Test Paper 2

Note: *Attempt any five questions out of Eight questions.*

1. (*a*) (23-17) using 2's complement.
   (*b*) Define EBCDIC code.
   (*c*) State and explain Demorgan's Theorem.
2. (*a*) Convert binary word 1101 into 7 bit odd parity Hamming code.
   (*b*) Circuit diagram of 9's complement.
3. Minimize the following using *k*-map and verify it using Q-M method:
   $f(A, B, C, D) = \pi M(0, 1, 3, 7, 9, 10) + d(6, 14, 15)$ and implement it using NAND gate.
4. (*a*) What do you mean by Demultiplexer Tree? Explain it with an example.
   (*b*) Explain working of BCD adder.
5. (*a*) What is difference between asynchronous sequential and synchronous sequential circuits.
   (*b*) What is difference between Latch and flip-flop.
6. (*a*) Convert SR to JK flip-flop.
   (*b*) Explain FPGA with types.
7. (*a*) 12 bit DAC has full scale analog voltage of 5 V. Determine step size, percentage resolution and analog output voltage to input of 1100100011.
   (*b*) Explain R-2-R ladder DAC.
8. (*a*) Explain tristate TTL gate.
   (*b*) Compare DTC and DCTL logic families.

Index

A

Adder, 78
Algorithmic State Machines, 227
Alphanumeric Code, 31
Analog Signal, 3
Analog to Digital Converter, 195
Analysis Procedure, 231
AND Gate, 5
Arithmetic Logic Unit (ALU), 116
ASCII Code, 31
Asynchronous Sequential Logic, 229

B

Basic Concept of Counter, 144
Basic Gate, 5
BCD Adder, 114
BCD Addition, 37
BCD Arithmetic Element, 114
BCD Subtractor, 115
BCD to Decimal Decoder, 108
BCD to Excess-3 Encoder, 100
BCD to Gray Code Decoder, 109
Bidirectional Shift Register, 140
Binary Arithmetic, 23
Binary Codes, 27
Binary to Octal, 107
Bipolar and MOS Devices, 163
Bipolar Devices, 163
Bipolar Logic Family, 165
Block Diagram of CPLD, 220
Boolean Algebra, 14

C

Cascading of Demultiplexer, 90
Cascading of Multiplexer, 84
Checksum Method, 34
CMOS Driving TTL, 180
CMOS Logic Family, 179
Combinational Circuits, 77, 122
Complex Programmable Logic Device, 219
Counter, 143

D

D-Type Flip-Flop, 128
Data Distributor, 88
Data Selector, 82
De Morgan's Theorem, 16
Decimal to BCD Encoder, 96
Decimal to BCD Priority Encoder, 104
Decoder, 106
Demultiplexer, 88, 91
Design at RTL, 225
Design of Modulus Counter, 149
Design Procedure, 233
Design Using Gates, 51
Designing of Sequential Circuits, 149
Digital Electronics, 4
Digital Logic Families, 159
Digital Signal, 3
Digital to Analog Converter, 189
Diode Transistor Logic, 169
Direct Coupled Transistor Logic, 170
Dual Lope A/D Converter, 201

E

EBCDIC Code, 32
Edge Triggering, 124
Edge-Triggered Flip-Flop, 129
Emitter Coupled Logic, 176
Encoder, 95
Error Correction Method, 35
Error Detection Code, 33
EX-NOR Gate, 12
Excess-3 to BCD Encoder, 102
Excitation Table of Flip-Flops, 131
Exclusive Gate, 10
Exclusive–OR Gate, 10

F

Field Programmable Gate Array, 217
Fixed Point Arithmetic, 23
Flip-Flop, 123
Fractional Part Conversion, 21
Full Adder, 78
Full Subtractor, 81
Fundamental of Digital Techniques, 3

G

Gray to Binary Encoder, 97

H

Half Adder, 78
Half Subtractor, 81
Hamming Code, 35
Hazards, 239
High Threshold Logic, 171

I

Integrator Injection Logic, 176

J

J-K Flip-Flop, 126
Johnson Counter, 145

K

Karnaugh-MAP (K-Map), 54

L

Latch, 122
Level Triggering, 124
Logic Design, 51
Logic Gates, 4

M

Magnitude Comparator, 91
Master-Slave Flip-Flop, 130
Maxterm, 64
Minterm, 64
MOS Devices, 165
MoS family, 177
Multiplexer, 82, 91

N

NAND Gate, 7
New IEEE Symbols, 41
NMOS Logic Family, 177
Non-Positional Number System, 17
NOR Gate, 8
NOT Gate, 6
Number System, 17
Numeric Code, 27

O

Octal to Binary Encoder, 95
Octal to Binary Priority Encoder, 106
OR Gate, 5

P

Pairing of 1's/0's, 57
Parallel Binary Adder, 79
Parallel Comparator, 197
Parity Method, 33
PIPO (Buffer Register), 139
PISO, 139
PMOS Logic Family, 178
Positional Number System, 17
Positive Logic and Negative Logic, 14
Preset and Clear Signal, 127
Priority Encoder, 103
Product of Sum Form (POS), 53
Programmable Array Logic, 213
Programmable Logic Array, 210
Programmable Logic Devices, 207

Q

Quantization, 195
Quine-Mc Clusky Method, 65

R

Race Around Condition, 129
Race Free State Assignment, 238
Read Only Memory, 208
REduction of State and Flow Table, 233
Register, 135
Register Transfer Level, 225
Representation of Boolean Function, 52
Resistor-Transistor Logic, 168
Ring Counter, 144

S

S-R Flip-Flop, 125
Sample and Hold, 188
Sample and Hold Circuit, 189
Sequence Generator, 149
Sequential Circuits, 121, 122
Sequential Logic Design, 225
Shift Register, 136
Signed Arithmetic, 24
Simplification of Boolean Function, 52
SIPO, 138
SISO, 136
Specification of D/A Converter, 194
Steps in Q-M Method, 65
Subtractor, 80
Sum of Product Form (SOP), 52

T

T-Type Flip-Flop, 128
Tabular Method, 65
Transistor-transistor Logic, 172
TTL Driving CMOs, 181

U

Universal Gate, 7
Universal Shift Register, 142
Unsigned Arithmetic, 23

W

Weighted Resistor D/A Converter, 190

Win Prizes

ATTENTION READERS

Win Prizes

We request for your frank assessment regarding various aspects of the book as given below :

Digital Electronics

BY : Preeti Sehrawat

Please Fill up the Sheet in Neat and Clean Writing

(i) What topics of your syllabus, important for your examination point of view, are not covered in this book ?

....

....

....

....

(ii) What are the chapters wherein the treatment of the subject matter is not systematic or organised or updated ?

....

....

....

....

(iii) Have you come across misprints/mistakes/factual inaccuracies in the book ? Please specify the chapters and the page numbers.

....

....

....

....

(iv) What other books on the same subject have you found/heard better than the present book ? Please specify in terms of price and quality.

....

....

....

....

(v) Further suggestions and comments for the improvement of this book.

....

....

....

....

(PS : If need be please attach a separate sheet to write your views)

OTHER DETAILS

(i) Who recommended you this book ? Please tick (✓) in the box near the option relevant to you.

☐ Teacher ☐ Friends ☐ Bookseller

....

....

....

....

(ii) Name of the recommending teacher, his designation and address:

....

....

....

....

(iii) Name and address of the bookseller you purchased the book from :

....

....

....

....

(iv) Name and address of your institution (Please mention the University or Board as the case may be)

....

....

....

....

(v) Your name and complete address :

....

....

....

....

(vi) Course you are enrolled in :

....

....

....

....

The best assessment will be awarded each month : The award will be in the form of our publications as decided by the Editorial Board.

Please mail the filled up coupon at your earliest to :

S.K. KATARIA & SONS®
4760-61/23, Ansari Road, Daryaganj, New Delhi-110002
☎ +91-11-23243489, 23269324; Telefax: +91-11-23243489
e-mail: katariabooks@yahoo.com; katariabook@gmail.com
Website : www.skkatariaandsons.com